Micky mouse

Welcome to Disneyland!

Solid State
Pulse Circuits

SECOND EDITION

Solid State Pulse Circuits

SECOND EDITION

David A. Bell
Lambton College of
Applied Arts and Technology
Sarnia, Ontario, Canada

Reston Publishing Company, Inc.
A Prentice-Hall Company
Reston, Virginia

Library of Congress Cataloging in Publication Data

Bell, David A
  Solid state pulse circuits.

  Includes index.
  1. Pulse circuits. 2. Digital electronics. 3. Semiconductors. I. Title.
TK7868.P8B44   1981   621.3815'34   80-24532
ISBN 0-8359-7057-4

10 9 8 7 6 5

Printed in the United States of America

DEDICATED TO
my mother and father

Contents

Preface

This book attempts to explain the operation, design, and analysis of all the basic semiconductor pulse circuits. The design approach is a simple step-by-step procedure in which the designer knows exactly why each component value is selected. Many design examples are included in the text, device data sheets in Appendix 1 are referred to when appropriate, and standard value components are selected. The mathematics employed does not go beyond algebraic equations and logarithms.

As well as discrete component circuits, the design procedure for using IC operational amplifiers in the various pulse circuits is covered. Digital integrated circuit families are also studied. However, this is a text on *pulse circuits*, not a book on computer logic.

The text progresses through: *pulse waveforms*; *RC circuits*; *diode switching*; *transistor switching*; *transistor and IC inverter circuits*. Then it treats: *Schmitt trigger circuits*; *voltage comparators*; *ramp generators*; *timers*; *monostable, astable, and bistable multivibrators*; *logic gates*; *sampling gates*. After the individual circuits are fully explained, they are used as building blocks to describe: *digital counting, digital frequency meters, digital voltmeters, pulse modulation*, and *time division multiplexing*. The various *seven-segment numerical display* devices are covered, as are *IC flip-flops* and *counting circuits*.

It is believed that this book shows that pulse circuits are easy to understand, and that their design is fairly simple.

I wish to express my sincere appreciation to all those who made suggestions for improvement to the first edition. Special thanks go to Richard Furbacher of Niagara College of Applied Arts and Technology in Welland, Ontario; and to Jack L. Waintraub of Middlesex County College in Edison, New Jersey.

David A. Bell

CHAPTER 1

Waveforms

INTRODUCTION

The term **pulse waveform** normally is applied only to approximately rectangular waveshapes. However, many different types of waveforms are involved in the study of pulse circuits. Waveforms are defined in terms of amplitude and time interval measurements. Each of the various waveforms can be shown to contain many higher frequency sinusoidal components, known as **harmonics**. A study of the harmonics shows a definite relationship between the bandwidth of a circuit and the distortion produced in a square wave output from the circuit.

1-1 TYPES OF WAVEFORM

1-1.1 Repetitive Waveforms and Transients

When one quantity varies in relation to another quantity, the relationship can be represented by plotting a graph. Thus, for a semiconductor diode, I_F plotted against V_F gives the forward characteristics of the device [Figure 1-1(a)]. Similarly, graphs may be plotted to show how certain quantities vary with respect to time. A plot of dc voltage or current *versus* time normally produces a straight line graph, as in Figure 1-1(b). An alternating voltage, as its name implies, increases and decreases with respect to time.

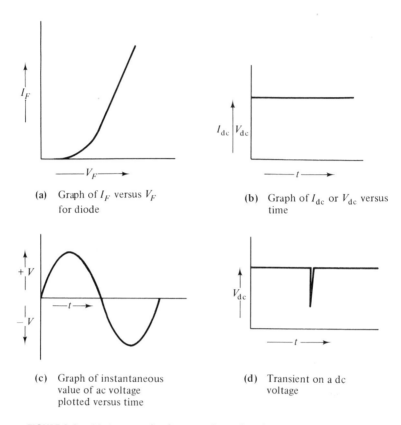

(a) Graph of I_F versus V_F
    for diode

(b) Graph of I_{dc} or V_{dc} versus
    time

(c) Graph of instantaneous
    value of ac voltage
    plotted versus time

(d) Transient on a dc
    voltage

FIGURE 1-1.   Various graphs showing relationships between two-quantities.

When the instantaneous voltage levels v are plotted against time t, the graph that results is called the waveform of the voltage. In Figure 1-1(c) the instantaneous values of a sinusoidal alternating voltage are plotted to a base of time. It is seen that the voltage increases positively to a peak value, decreases through zero to a negative peak value, then returns to zero; then the cycle recommences. The sine wave is a repeating cycle of voltage (or current) with a sinusoidal relationship to time. All waveforms which are composed of identical cycles that keep repeating are termed *repetitive waveforms* or *periodic waveforms*. It is necessary to study only one cycle of such a waveform to gain an understanding of the behavior of the voltage or current involved. When successive cycles of an alternating voltage are *not* identical, the waveform is described as *aperiodic*.

Sometimes a direct voltage suddenly decreases (or increases) for a brief instant and then returns to its normal level [Figure 1-1(d)]. This may happen, for example, when a load is suddenly switched onto a power supply. Such brief *nonrepetitive* waveforms are termed *transients*.

1-1.2   Display Methods

Since electrical waveforms (repetitive and nonrepetitive) usually occur during a period of milliseconds or microseconds, manual measurements cannot be taken of such instantaneous levels for plotting on a graph. Instead, instruments are employed to do the actual plotting of voltage or current to a base of time. One such instrument is the *strip chart recorder* (Figure 1-2) in which a pen or another marking device traces the waveform on a moving strip of paper. The vertical movement of the pen is directly proportional to the instantaneous value of the applied voltage and, since the paper moves at a constant speed, the horizontal trace is directly proportional to time. Similar instruments replace the moving arm and pen with a beam of light reflected from a moving mirror onto photographically treated paper.

Although the chart recorder provides a permanent record of the waveform under study, its major drawback is that it is essentially a low-frequency instrument. The *cathode-ray oscilloscope*, which can display much higher frequencies, is widely used in the study of electrical waveforms. In the oscilloscope, an electron beam striking a fluorescent screen produces a tiny spot of light. The light spot becomes a line when the electron beam is deflected vertically by the voltage to be displayed and horizontally in proportion to time (Figure 1-3). The light spot starts at the left-hand side of the screen and moves to the right. At the end of one (or more) cycles of

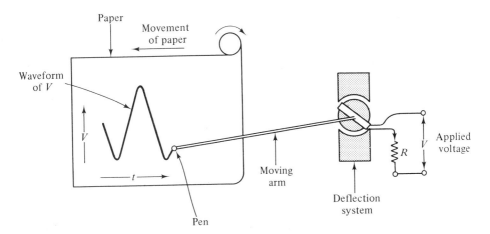

FIGURE 1-2.   Waveform display by strip chart recorder.

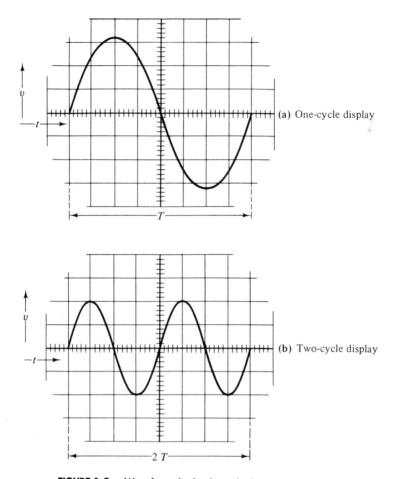

FIGURE 1-3. Waveform display by cathode ray oscilloscope.

the waveform, the light spot returns almost instantaneously to the left-hand side of the screen. Thus, a repetitive waveform is traced on the screen again and again. When a permanent record of the waveform is required, a camera is used to photograph the display on the oscilloscope. A camera can be employed also to obtain photographs of any transient waveform that might be displayed briefly on the oscilloscope screen.

1-1.3 Miscellaneous Waveforms

Sinusoidal. The most common electrical waveform is the sine wave, shown in Figure 1-3. Half-wave rectification removes the negative (or positive) half-cycles of a sine wave [Figure 1-4(a)], while full-wave

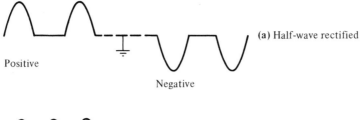

(a) Half-wave rectified

Positive

Negative

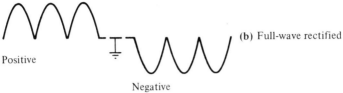

(b) Full-wave rectified

Positive

Negative

FIGURE 1-4. Rectified sine waves.

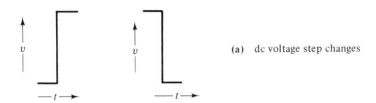

(a) dc voltage step changes

Positive-going step   Negative-going step

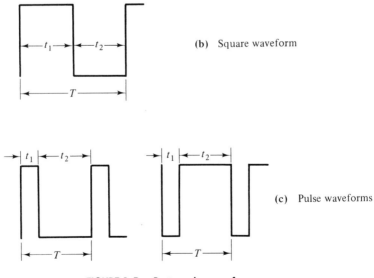

(b) Square waveform

(c) Pulse waveforms

FIGURE 1-5. Rectangular waveforms.

rectification produces a train of unidirectional half-sine waves, as shown in Figure 1-4(b).

Rectangular.   When a dc voltage suddenly changes from one level to another, the change is referred to as a *step change*. The change might be positive or negative, as shown in Figure 1-5(a). A *rectangular waveform* consists simply of successive cycles of positive step changes followed by negative step changes. Where the time duration t_1 for the upper dc level is equal to the time duration t_2 for the lower level, the waveform is termed a *square wave* [Figure 1-5(b)]. When t_1 and t_2 are unequal, as illustrated in Figure 1-5(c), the wave is usually referred to as a *pulse waveform*.

Ramp.   A voltage that increases or decreases at a constant rate with respect to time, has a graph that is a positive or negative *ramp* [Figure 1-6 (a)]. A repetitive cycle of positive ramp followed by negative ramp is known as a *triangular waveform* [Figure 1-6(b)]. When one ramp is much

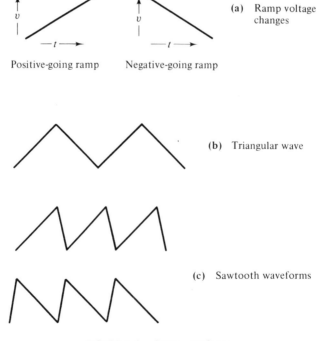

(a)   Ramp voltage changes

Positive-going ramp   Negative-going ramp

(b)   Triangular wave

(c)   Sawtooth waveforms

FIGURE 1-6.   Ramp waveforms.

Positive-going exponential   Negative-going exponential

(a) Exponential
    voltage changes

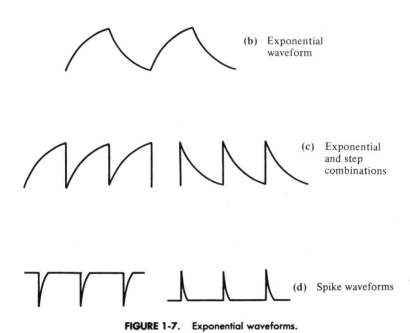

(b) Exponential
    waveform

(c) Exponential
    and step
    combinations

(d) Spike waveforms

FIGURE 1-7.   Exponential waveforms.

steeper than the other, as illustrated in Figure 1-6(c), the waveform is usually termed a *sawtooth waveform*.

Exponential.   In this case the voltage varies with respect to time according to the equation $V = \epsilon^{kt}$ or $V = \epsilon^{-kt}$, where t is time, k is a constant, and ϵ is the *exponential constant* ($\epsilon = 2.718$). The resultant graphs of voltage *versus* time are of the form shown in Figure 1-7(a). Repetitive cycles of positive and negative exponentials produce an *exponential waveform*, [Figure 1-7(b)]. An exponential change followed by a step change gives the waveforms shown in Figure 1-7(c). Introduction of a gap results in the *spike waveforms* of Figure 1-7(d). Obviously a great variety of waveforms can be produced by combining two or more of the various voltage changes discussed above.

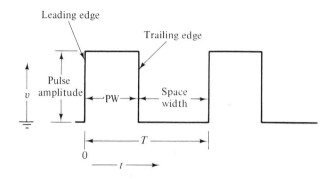

FIGURE 1-8.   Ideal pulse waveform.

1-2 CHARACTERISTICS OF PULSE WAVEFORMS

Consider the ideal pulse waveform shown in Figure 1-8. In this particular case the pulses are positive with respect to ground. The pulse amplitude is simply the voltage level of the top of the pulse measured from ground. The first edge of the pulse (at $t=0$) is referred to as the *leading edge*, and the second edge is termed the *trailing edge* or *lagging edge*.

The *time period* T is the time measured from the leading edge of one pulse to the leading edge of the next pulse. If $T=1$ sec, then the *pulse repetition frequency* (PRF) is 1 cycle/sec, or 1 pulse per sec (pps), or $PRF = 1/T$ pps. Instead of pulse repetition frequency, the term *pulse repetition rate* (PRR) is sometimes used.

The time measured from the leading edge to the trailing edge of one pulse is known as the *pulse width* (PW), the *pulse duration* (PD), or sometimes as the *mark length*. The time between pulses is simply referred to as the *space width*. The proportion of the time period occupied by the pulse is defined as the *duty cycle*, or as the *mark-to-space* (M/S) *ratio*:

$$\text{Duty cycle} = (PW/T) \times 100\% \qquad (1\text{-}1)$$

and

$$\text{M/S ratio} = PW/(\text{space width}) \qquad (1\text{-}2)$$

The duty cycle usually is expressed as a percentage, while the mark-to-space ratio is expressed simply as a ratio.

EXAMPLE 1-1 _____

For the pulse waveform displayed in Figure 1-9, determine the pulse amplitude, PRF, PW, duty cycle, and M/S ratio. The vertical scale is 1 V per division, and the horizontal scale is 0.1 ms per division.

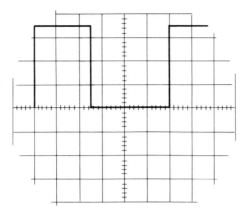

FIGURE 1-9.   Pulse waveform on oscilloscope.

solution

$$\text{Pulse amplitude} = (3.5 \text{ divisions}) \times (1\text{V}/\text{division})$$
$$= 3.5 \text{ V}$$
$$T = (6 \text{ divisions}) \times (0.1 \text{ ms}/\text{division})$$
$$= 0.6 \text{ ms}$$
$$\text{PRF} = 1/T = 1/0.6 \text{ ms} = 1666 \text{ pps}$$
$$\text{PW} = (2.5 \text{ divisions}) \times (0.1 \text{ ms}/\text{division})$$
$$= 0.25 \text{ ms}$$
$$\text{Space width} = 3.5 \times 0.1 \text{ ms} = 0.35 \text{ ms}$$
$$\text{Duty cycle} = \frac{\text{PW}}{T} \times 100\%$$
$$= \frac{0.25 \text{ ms}}{0.6 \text{ ms}} \times 100\% = 41.6\%$$
$$\text{M}/\text{S ratio} = \frac{\text{PW}}{\text{Space width}} = \frac{0.25 \text{ ms}}{0.35 \text{ ms}}$$
$$= 0.71$$

The pulse displayed in Figure 1-9 appears to have a perfectly flat top and perfectly vertical sides. When pulses are examined very carefully, however, it is found that the top is never perfectly flat. The amplitude of the lagging edge normally is less than that of the leading edge. In many cases the slope at the top of the pulse may be so small that it cannot be easily measured. In other cases, as in Figure 1-10, the slope may be very obvious. The pulse voltage does not go from zero to its maximum level instantaneously, and from maximum to zero instantaneously. In fact, there

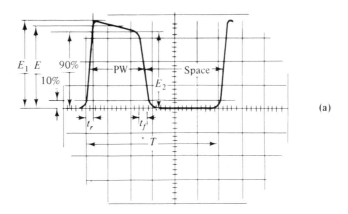

(a)

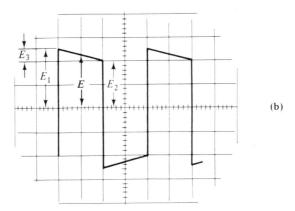

(b)

FIGURE 1-10.   Waveforms with rise and fall times and tilt.

is a definite *rise time* t_r and *fall time* t_f at the leading and lagging edges of the pulse. This is illustrated in Figure 1-10(a).

If the pulse width (PW) is measured near the top of the pulse in Figure 1-10(a), it would be quite different from the PW as measured close to the bottom of the pulse. Therefore, the PW is defined as the *average* pulse width, and is normally measured at half the average amplitude, [see Figure 1-10(a).]. The space width (SW) is measured at the same amplitude as the pulse width. The sum of pulse width and space width is always equal to the time period (T):

$$PW + SW = T$$

In Figure 1-10(a), E_1 is the maximum pulse amplitude, E_2 is the minimum amplitude, and E is the average pulse amplitude:

$$E = \frac{E_1 + E_2}{2}$$

The rise time is defined as the time required for the voltage to go from 10% to 90% of the average amplitude. Similarly, the fall time is the time required for the pulse to fall from 90% to 10% of the average amplitude. The *slope* or *tilt* at the top of the waveform is defined in terms of the average amplitude:

$$\text{Tilt} = \frac{E_3}{E} \times 100\% \qquad (1\text{-}3)$$

$$= \frac{E_1 - E_2}{E} \times 100\%$$

In Figure 1-10(b) the wave is symmetrical above and below ground level. Although there is obvious tilt, the leading and trailing edges are equal in amplitude. Thus if E_1 was measured as the whole leading edge and E_2 as the whole trailing edge, Equation (1-3) would give zero tilt. Instead E_1 and E_2 are each measured with respect to ground, as shown. Then,

$$\text{Tilt} = \frac{E_1 - E_2}{2E} \times 100\%$$

EXAMPLE 1-2

For the waveform displayed in Figure 1-10(a), determine pulse amplitude, tilt, t_r, t_f, PW, PRF, mark-to-space ratio, and duty cycle. For the square waveform in figure 1-10(b), determine tilt. The vertical scale is 100 mV/division, and the horizontal scale is 100 μs/division in each case.

solution (a)

$$\text{Pulse amplitude, } E = \frac{E_1 + E_2}{2} = \frac{380 \text{ mV} + 330 \text{ mV}}{2} = 355 \text{ mV}$$

$$\text{Tilt} = \frac{E_1 - E_2}{E} \times 100\%$$

$$= \frac{380 \text{ mV} - 330 \text{mV}}{355 \text{ mV}} \times 100\% = 14.1\%$$

$$t_r = (0.3 \text{ divisions}) \times (100 \text{ } \mu\text{s/division}) = 30 \text{ } \mu\text{s}$$

$$t_f = (0.4 \text{ divisions}) \times (100 \text{ } \mu\text{s/division}) = 40 \text{ } \mu\text{s}$$

$$T = (6.1 \text{ divisions}) \times (100 \ \mu s/\text{division}) = 610 \ \mu s$$

$$\text{PRF} = 1/T = 1/610 \ \mu s = 1639 \text{ pps}$$

$$\text{PW} = (2.2 \text{ divisions}) \times (100 \ \mu s/\text{division}) = 220 \ \mu s$$

$$\text{Space width} = (3.9 \text{ divisions}) \times (100 \ \mu s/\text{division}) = 390 \ \mu s$$

$$\text{M/S ratio} = \frac{220 \ \mu s}{390 \ \mu s} = 0.564$$

$$\text{duty cycle} = \frac{220 \ \mu s}{610 \ \mu s} \times 100\% = 36.1\%$$

solution (b)

$$E_1 = (2.5 \text{ divisions}) \times (100 \text{ mV}/\text{division}) = 250 \text{ mV}$$

$$E_2 = (2 \text{ divisions}) \times (100 \text{ mV}/\text{division}) = 200 \text{ mV}$$

$$\text{Average voltage} = E = \frac{E_1 + E_2}{2} = \frac{250 \text{ mV} + 200 \text{ mV}}{2}$$

$$= 225 \text{ mV}$$

$$\text{Tilt} = \frac{250 \text{ mV} - 200 \text{ mV}}{2 \times 225 \text{ mV}} \times 100\% = 11.1\%$$

The square waveform shown in Figure 1-11(a) is symmetrical above and below ground level. The positive and negative peaks are of equal amplitudes and equal widths (*i.e.*, $t_1 = t_2$). This means that the average value of the waveform is zero. If this waveform were applied to a dc voltmeter, the instrument would indicate zero. The average value of the waveform is found simply by summing the positive and negative areas enclosed by one cycle and dividing by the time period.

$$\text{Average voltage} = V_{av} = \frac{(V_+ \times t_1) + (V_- \times t_2)}{T} \qquad (1\text{-}4)$$

For Figure 1-11(a):

$$\text{Average voltage} = \frac{(6 \text{ V} \times 1 \text{ ms}) + (-6 \text{ V} \times 1 \text{ ms})}{2 \text{ ms}} = 0 \text{ V}$$

The waveform of Figure 1-11(b), has no negative portion, and the average value is

$$V_{av} = \frac{(12 \text{ V} \times 1 \text{ ms}) - (0)}{2 \text{ ms}} = 6 \text{ V}$$

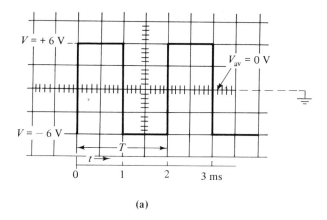

(a)

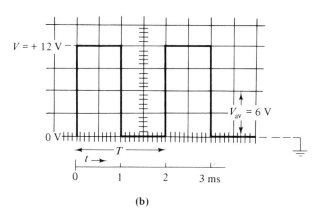

(b)

FIGURE 1-11.   Waveforms with the same peak-to-peak amplitude but different
average values.

EXAMPLE 1-3 _____

Determine the average values of the pulse waveforms shown in Figures
1-12(a), (b), and (c).

solution (a)

$$V_{av} = \frac{(V_1 \times t_1) + (V_2 \times t_2)}{T}$$

$$= \frac{(12 \text{ V} \times 1 \text{ ms}) + (0 \times 3 \text{ ms})}{4 \text{ ms}}$$

$$= 3 \text{ V}$$

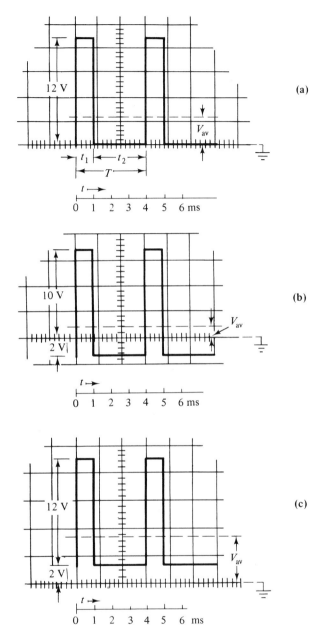

FIGURE 1-12. Pulse waveforms with different average values.

solution (b)

$$V_{av} = \frac{(10\ V \times 1\ ms) + (-2\ V \times 3\ ms)}{4\ ms}$$

$$= 1\ V$$

solution (c)

$$V_{av} = \frac{(14\ V \times 1\ ms) + (2\ V \times 3\ ms)}{4\ ms}$$

$$= 5\ V$$

1-3 HARMONIC CONTENT OF WAVEFORMS

In Figure 1-13(a) two signal generators are shown connected in series. Generator A is producing a sinusoidal output waveform as shown. The output of generator B is also a sine wave, but its frequency is three times the frequency from signal generator A. The amplitude of the output from B is also less than that from A. The waveform produced by the two generators in series is the larger amplitude (and lower frequency) signal, with the smaller amplitude signal superimposed. It is seen that the combination approximately resembles a square wave with its peaks dented. Figure 1-13(b) shows a third generator connected in series with A and B. The output from generator C is smaller in amplitude than that from B, and the frequency of this third signal is five times the frequency of the output of generator A. The waveform produced by the three generators in series now more closely resembles a square wave. It is important to note that the resultant waveforms shown in Figures 1-13(a) and (b) are produced only when *the generators are synchronized*; i.e., all component waves must commence exactly at the same instant.

The building up of the approximate square waveform is easily seen by referring to Figure 1-14 where the instantaneous amplitudes of waveforms A and B are added. At time t_1, for example, the amplitude of waveform A is 6.5 V, and that of B is approximately 2.5 V. Therefore, the amplitude of the resultant waveform at time t_1 is 9 V (point 1). At time t_2, the amplitude of B is zero, so the resultant amplitude is 8.5 V, that is, the amplitude of A at t_2 (point 2). At t_3, the amplitudes of B, -3 V, and of A, 10 V, are added together to produce an amplitude of 7 V. When this process is continued, it is seen how the final waveform is constructed.

(a) Fundamental and
third harmonic

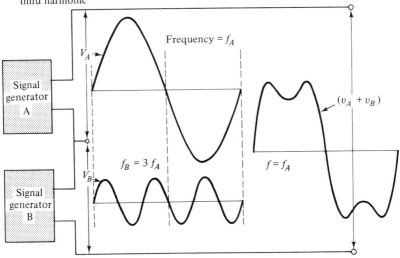

(b) Fundamental, third
and fifth harmonic

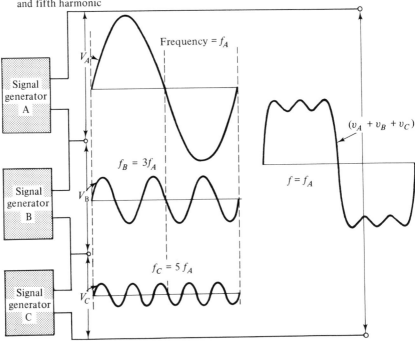

FIGURE 1-13. Combination of fundamental and harmonics to form approximate
square wave. (Note that the signal generators must be synchronized.)

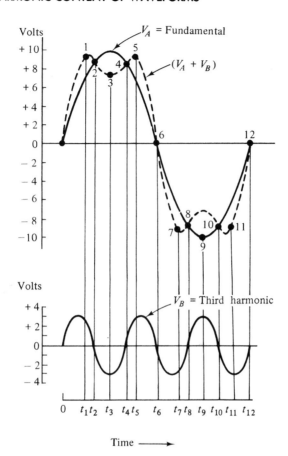

FIGURE 1-14. Addition of the instantaneous levels of a fundamental and third harmonic.

The process of building up a particular waveform by combining several sine waves of different frequencies and amplitudes is referred to as *frequency synthesis*. If the process were continued and appropriate higher frequency waveforms were added, the resultant each time would more closely resemble a square wave.

The converse of frequency synthesis is *harmonic analysis*. In this process a waveform is analyzed to discover the sine wave frequencies it contains. By harmonic analysis, it can be shown that periodic non-sinusoidal waveforms are composed of combinations of pure sine waves. Some waveforms can also have dc components. One major component, a large amplitude sine wave of the same frequency as the periodic wave under consideration, is termed the *fundamental*. The other components of a periodic waveform are sine waves with frequencies which are exact multiples of the frequency of the fundamental. These waves, referred to as *harmonics*, are numbered according to the ratio between their frequencies and that of the fundamental. For example, a harmonic with a frequency exactly double that of the fundamental is called the *second harmonic*. The frequency of the *third harmonic*, obviously, is three times the fundamental frequency.

By the mathematical operation known as *Fourier analysis* waveforms can be analyzed to determine their harmonic content. The amplitude of each harmonic and its phase relationship to the fundamental can be found. Also, the amplitude of any dc component can be calculated. A perfect square wave which is symmetrical above and below ground can be shown, by Fourier analysis, to have a fundamental component and odd-numbered harmonics, but no even-numbered harmonics and no dc component. A pulse waveform is found to contain both odd- and even-numbered harmonics and (usually) a dc component. Sawtooth waveforms, triangular waveforms, and rectified sine waves are made up of more complicated combinations of odd- and even-numbered harmonics. In all cases, the harmonic content actually goes to infinity, but the amplitudes of the harmonics decrease as their frequencies increase. Thus the higher frequency components are the least important.

Information which can be derived by harmonic analysis becomes very important when considering the circuitry through which various waveforms are processed. Suppose a square wave with a frequency of 1 kHz is applied to an amplifier with an upper frequency limit of 15 kHz. In this case, the amplifier will not reproduce waveforms with frequencies greater than 15 kHz. Thus the amplifier will not pass harmonics of 1 kHz greater than the fifteenth. If the square wave applied to the same amplifier had a frequency of 5 kHz, only the first, second, and third harmonics would be passed. If a 5 kHz square wave were to be amplified, and if all harmonics up to the thirty-third were to be reproduced, then the amplifier must have an upper frequency limit greater than 33×5 kHz. Fifteen to twenty harmonics are usually required to reproduce a waveform approximately in its original shape. For accurate reproduction, however, many more harmonics may be required.

If a square wave is applied to circuitry that does not pass all the necessary frequency components, the resultant output is a distorted square wave. The type of distortion depends upon whether the circuitry has poor low-frequency response or poor high-frequency response. In Figure 1-15(a) the long rise and fall times of the square wave show that the high-frequency harmonics are attenuated and thus the circuit has poor high-frequency response. Figure 1-15(b) shows the output from a circuit which has good high-frequency response but poor low-frequency response. The *tilt* on the

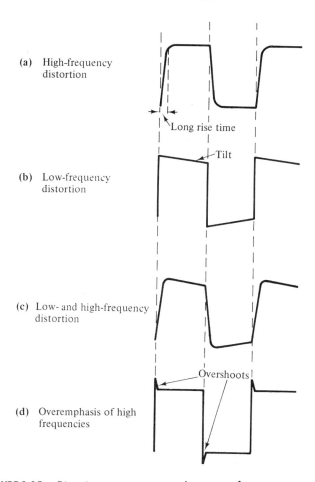

(a)   High-frequency
      distortion

Long rise time

Tilt

(b)   Low-frequency
      distortion

(c)   Low- and high-frequency
      distortion

Overshoots

(d)   Overemphasis of high
      frequencies

FIGURE 1-15.   Distortion on square waves due to poor frequency response of
circuitry.

top and bottom of the square wave results because the low-frequency
components were not passed by the circuit. The waveform in 1-15(c) shows
both long rise times and tilt. This result is obtained when the involved
circuitry has neither a low enough nor a high enough frequency response
for the applied square wave. When circuits overemphasize some of the
high-frequency harmonics, *overshoots* are produced, as shown in Figure
1-15(d).

When a square wave is applied to an amplifier the rise time of the
output waveform is limited by the time taken for the highest harmonic
frequency to go from zero to its peak value (Figure 1-16). Of course, the

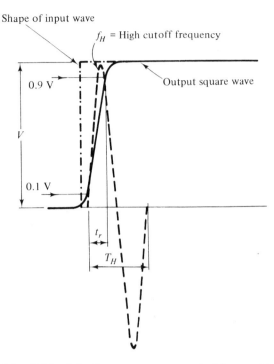

FIGURE 1-16.   Origin of rise time on square wave output from an amplifier.

highest harmonic frequency that can be reproduced is the upper cutoff frequency f_H. It is found that:

$$t_r = 0.35 \times (\text{Time period of } f_H)$$

$$t_r = \frac{0.35}{f_H} \tag{1-5}$$

By Equation (1-5) the output rise time can be predicted when the upper cutoff frequency of the circuitry is known. Equation (1-5) also affords a fast means of pulse testing to determine the cutoff frequency of any circuit or device.

EXAMPLE 1-4 _____

The output waveform from an amplifier under pulse test has a rise time of 1 μs. Determine the upper 3 dB frequency of the amplifier.

solution

From Equation (1-5)

$$f_H = \frac{0.35}{t_r} = \frac{0.35}{1 \, \mu s}$$
$$= 350 \text{ kHz}$$

It is possible, to decide the upper cutoff frequency for a circuit that must pass pulse waveforms with an *acceptable* amount of high frequency distortion. The question that arises, of course, is just what is an acceptable amount of distortion. Refer to Fig. 1-15(a). The rise time shown (and the fall time) is approximately one tenth of the pulse width. If the rise and fall times were much greater than PW/10, the pulse would be severely distorted. Therefore, in many cases $t_r \simeq$ PW/10 might be used as a guide for acceptable high frequency distortion. In some other circumstance there may be a requirement for less distortion.

EXAMPLE 1-5 _____

A pulse waveform has a PRF of 1.5 kHz and a duty cycle of 3%. (a) Determine the frequency of the highest harmonic required for accurate reconstruction of the waveform. (b) If the 1.5 kHz pulse is to be amplified by equipment with a high frequency limit of 1 MHz, calculate the minimum pulse width and duty cycle that can be reproduced accurately.

solution (a)
For a duty cycle of 3%:

$$PW = 0.03 \times 1/f$$
$$= 0.03 \times 1/1.5 \text{ kHz}$$
$$= 20 \, \mu s$$

For a $t_r = 10\%$ of PW:

$$t_r = 0.1 \times 20 \, \mu s$$
$$= 2 \, \mu s$$

From Eq. (1-5),

$$f_H = 0.35/t_r$$
$$= 0.35/2 \, \mu s$$
$$= 175 \text{ kHz}$$

solution (b)
Eq. (1-5),

$$t_r = 0.35/f_H$$
$$= 0.35/1 \text{ MHz}$$
$$= 0.35 \ \mu s$$

For $t_r = 10\%$ of PW:

$$\text{Minimum PW} = 10 \times t_r$$
$$= 10 \times 0.35 \ \mu s$$
$$= 3.5 \ \mu s$$
$$\text{Duty cycle} = \frac{\text{PW}}{\text{T}} \times 100\%$$
$$= \text{PW} \times f \times 100\%$$
$$= 3.5 \ \mu s \times 1.5 \text{ kHz} \times 100\%$$
$$\simeq 0.5\%$$

If a square wave is applied as input to an amplifier with a lower cutoff frequency of $f_L = 0$, then the top of the output square wave is perfectly flat. When f_L is greater than zero, however, tilt is present on the output waveform. As illustrated in Figure 1-17, the tilt is proportional to the ratio of the square wave time period T to the time period T_L of the lower cutoff frequency.

It is found that:

$$\text{Fractional tilt} = \pi \times \frac{T}{T_L}$$

or

$$\text{Fractional tilt} = \pi \frac{f_L}{f} \tag{1-6}$$

EXAMPLE 1-6 _____

An amplifier with a low cutoff frequency of 10 Hz is to be employed for amplification of square waves. For the tilt on the output waveform to not exceed 2%, calculate the lowest input frequency that can be amplified.

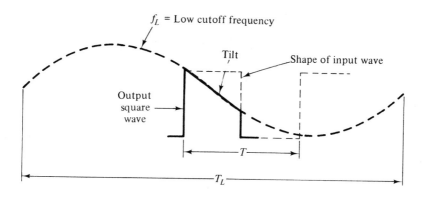

FIGURE 1-17.   Origin of tilt on square wave output from an amplifier.

solution

From Equation (1-6):

$$f = \frac{\pi f_L}{\text{Fractional tilt}}$$
$$= \frac{\pi \times 10 \text{ Hz}}{0.02}$$
$$= 1.57 \text{ kHz}$$

EXAMPLE 1-7

Determine the bandwidth required to amplify a 1-kHz square wave, if the rise time of the output is not to exceed 200 ns and 3% tilt is acceptable.

solution

From Equation (1-5):

$$f_H = \frac{0.35}{t_r} = \frac{0.35}{200 \text{ ns}} = 1.75 \text{ MHz}$$

From Equation (1-6):

$$f_L = \frac{f \times \text{Fractional tilt}}{\pi} = \frac{1 \text{ kHz} \times 0.03}{\pi}$$
$$= 9.5 \text{ Hz}$$

EXAMPLE 1-8 _____

Determine the upper and lower 3 dB frequencies of the circuitry which produced the output waveform shown in Figure 1-10(a).

solution

From Example 1-2:

$$t_r = 30 \ \mu s \qquad PRF = 1639 \ pps \qquad Tilt = 14.1\%$$

From Equation (1-5):

$$f_H = \frac{0.35}{t_r} = \frac{0.35}{30 \ \mu s} = 11.7 \ kHz$$

From Equation (1-6):

$$f_L = \frac{f \times (\text{Fractional tilt})}{\pi} = \frac{1639 \times 0.141}{\pi} = 73.6 \ Hz$$

REVIEW QUESTIONS AND PROBLEMS

1-1 Define: repetitive waveforms, periodic waveforms, aperiodic waveforms, transients.

1-2 Draw sketches to show the shapes of the following waveforms: square, pulse, triangular, sawtooth, exponential.

1-3 For a pulse waveform, define: leading edge, lagging edge, trailing edge, T, PRF, PRR, PW, PD, M/S ratio, duty cycle.

1-4 For the pulse waveform illustrated in Figure 1-18, determine: pulse amplitude, PRF, PW, duty cycle, and M/S ratio. The vertical scale is 0.1 V per division, and the horizontal scale is 1 ms per division.

1-5 (a) Define rise time, fall time, and tilt. (b) Determine the percentage tilt on the square wave shown in Figure 1-19.

1-6 For the waveform displayed in Figure 1-20, determine: pulse amplitude, tilt, t_r, t_f, PW, PRF, M/S ratio, and duty cycle. The vertical scale is 1 V/division, and the horizontal scale is 10 μs/division.

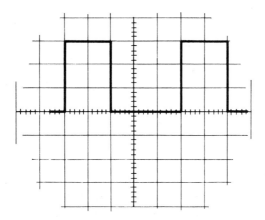

FIGURE 1-18. Problem 1-4.

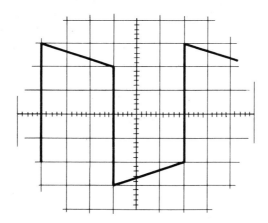

FIGURE 1-19. Problem 1-5.

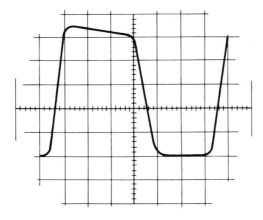

FIGURE 1-20. Problem 1-6.

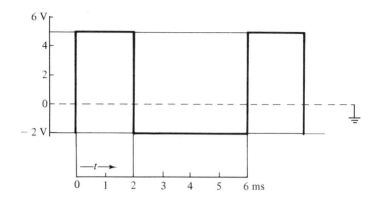

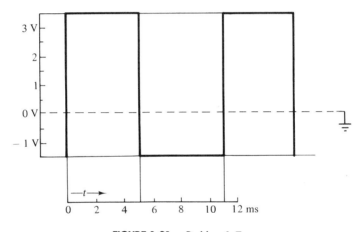

FIGURE 1-21.   Problem 1-7.

1-7   If the pulse waveforms shown in Figure 1-21 were applied to a dc voltmeter, determine the voltages that would be indicated in each case.

1-8   (a) Define the following terms: fundamental, harmonic, frequency synthesis, harmonic analysis. (b) What harmonics will be passed by an amplifier which has an upper cutoff frequency of 1 MHz: (i) when a 10 kHz square wave is applied to it? and (ii) when the input is a 150 kHz square wave?

1-9   (a) A 12 kHz pulse waveform is amplified by a circuit having a high-frequency limit of 1 MHz. Determine the minimum pulse width that can be reproduced accurately. (b) If the duty cycle of the 12 kHz pulse waveform becomes 0.5%, determine the approximate upper cutoff frequency of a circuit that will reproduce the waveform accurately.

1-10 Sketch a square wave that is amplified by equipment which: (a) has poor low-frequency response; (b) has poor high-frequency response; (c) overemphasizes high frequencies; (d) has a combination of poor low-frequency and poor high-frequency responses.

1-11 A 1 kHz square wave output from an amplifier has $t_r = 350$ ns and tilt $= 5\%$. Determine the upper and lower 3 dB frequencies of the amplifier.

1-12 Calculate the rise time and tilt that may be expected on the square wave output of an amplifier with a bandwidth extending from 10 Hz to 500 kHz. The applied square wave has a frequency of 5 kHz.

1-13 Determine the bandwidth of the circuitry that produced the output waveform shown in Figure 1-20 (Problem 1-6).

1-14 Construct the waveform that results when the fundamental and harmonics shown in Figure 1-22 are added together.

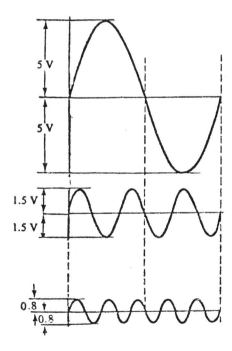

FIGURE 1-22. Problem 1-14.

CHAPTER 2

Capacitive Resistive (CR) Circuits

INTRODUCTION

When a capacitor is charged from a dc voltage source via a resistor, the instantaneous level of capacitor voltage may be calculated at any given time. There is a definite relationship between the **time constant** of a CR circuit, and the times required for the capacitor to charge to approximately 63% and 99% of the input voltage. Also, an important relationship exists between the time constant of a circuit and the rise time of the output voltage from the circuit. Depending upon the arrangement of the CR circuit, it may be employed as an **integrator or a differentiator**. In each case, the circuit time constant must be related to the time period of the input waveform.

2-1 CR CIRCUIT OPERATION

Consider the circuit and graph shown in Figure 2-1. If the charge on capacitor C is zero at the instant that switch S is closed, then the voltage

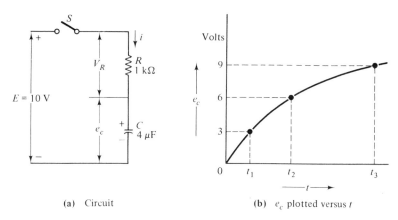

(a) Circuit   (b) e_c plotted versus t

FIGURE 2-1. Circuit for charging capacitor via resistor, and graph of capacitor voltage variation with respect to time.

across R at $t=0$ is

$$V_R = E - e_c$$

where E is the supply voltage and e_c is the capacitor voltage. The current through R at $t=0$ is

$$i_c = \frac{V_R}{R}$$

$$= \frac{E - e_c}{R} \qquad (2\text{-}1)$$

$$= \frac{10\ \text{V} - 0}{1\ \text{k}\Omega}$$

$$= 10\ \text{mA}$$

This current causes capacitor C to charge with the polarity shown, so that at some time t_1 the capacitor voltage e_c might be 3 V [see Figure 2-1(b)]. This alters V_R:

$$V_R = E - e_c$$
$$= 10\ \text{V} - 3\ \text{V} = 7\ \text{V}$$

Now,

$$i_c = (10\ \text{V} - 3\ \text{V})/1\ \text{k}\Omega = 7\ \text{mA}$$

Because C accumulates some charge, e_c is increased and the voltage across R is reduced; thus the charging current through R is reduced. Since the current is reduced, C is being charged at a slower rate than before. After some longer time period, e_c increases to 6 V (t_2 on the graph). Now,

$$V_R = 10\text{ V} - 6\text{ V} = 4\text{ V}$$

and

$$i_c = \frac{4\text{ V}}{1\text{ k}\Omega} = 4\text{ mA}$$

The charging current has now been reduced more. Consequently, even a longer time period is required to charge C by another 3 V.

The capacitor does not receive its charge at a constant rate. Instead, e_c is continuously increasing, so the voltage across R is continuously decreasing, and the charging current is decreasing. This means that C is charged at a rapid rate initially, and then the rate decreases as the capacitor voltage grows.

It can be shown that the capacitor voltage follows an exponential law:

$$e_c = E - (E - E_o)\epsilon^{-t/CR} \qquad (2\text{-}2)$$

where

e_c = capacitor voltage at instant t

E = charging voltage

E_o = initial charge on the capacitor

ϵ = exponential constant = 2.718

t = time from commencement of charge

C = capacitance being charged

R = charging resistance

When there is no initial charge on the capacitor,

$$E_O = 0$$

$$e_c = E - [E - 0]\epsilon^{-t/CR}$$

or

$$e_c = E(1 - \epsilon^{-t/CR}) \qquad (2\text{-}3)$$

and, since

$$i_c = \frac{E - e_c}{R}$$

$$i_c = \frac{E - E(1 - \epsilon^{-t/CR})}{R}$$

then

$$i_c = \frac{E}{R} \epsilon^{-t/CR}$$

and

$$i_c = I\epsilon^{-t/CR} \qquad (2\text{-}4)$$

where $I = E/R$ is the initial level of charging current when $t = 0$.

EXAMPLE 2-1

Calculate the levels of e_c, the capacitor voltage across C in the circuit of Figure 2-1(a), at 2 ms intervals from the instant when switch S is closed. Plot a graph of e_c versus time.

solution

Since $E_O = 0$, Equation (2-3) may be used to calculate e_c.
At $t = 0$,

$$e_c = E(1 - \epsilon^0) = 0 \text{ V} \qquad \text{point 1}$$
$$\text{(in Fig. 2-2)}$$

At $t = 2$ ms,

$$e_c = 10 \text{ V}(1 - \epsilon^{-2\,\text{ms}/(4\,\mu\text{F}\times 1\,\text{k}\Omega)}) = 3.93 \text{ V} \qquad \text{point 2}$$

At $t = 4$ ms,

$$e_c = 10 \text{ V}(1 - \epsilon^{-4\,\text{ms}/(4\,\mu\text{F}\times 1\,\text{k}\Omega)}) = 6.32 \text{ V} \qquad \text{point 3}$$

At $t = 6$ ms,

$$e_c = 7.77 \text{ V} \qquad \text{point 4}$$

At $t = 8$ ms,

$$e_c = 8.65 \text{ V} \qquad \text{point 5}$$

At $t = 10$ ms,

$$e_c = 9.18 \text{ V} \qquad \text{point 6}$$

At $t = 12$ ms,

$$e_c = 9.5 \text{ V} \qquad \text{point 7}$$

At $t = 14$ ms,

$$e_c = 9.7 \text{ V} \qquad\qquad \text{point 8}$$

At $t = 16$ ms,

$$e_c = 9.82 \text{ V} \qquad\qquad \text{point 9}$$

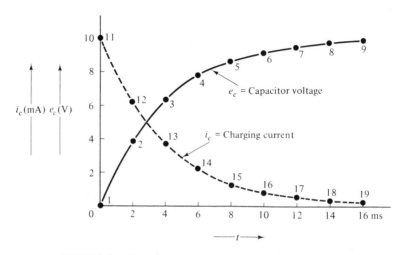

FIGURE 2-2.   Capacitor current and voltage plotted versus time.

EXAMPLE 2-2

Determine the instantaneous levels of charging current, in the circuit of Figure 2-1(a) at 2 ms time intervals from the instant that switch S is closed. Plot a graph showing i_c versus time.

solution

By Equation (2-1);

$$i_c = \frac{E - e_c}{R}$$

At $t = 0$,

$$i_c = \frac{10 \text{ V} - 0}{1 \text{ k}\Omega} = 10 \text{ ma}, \qquad\qquad \text{point 11}$$

$t = 2$ ms,
$$\qquad\qquad\qquad\qquad\qquad\qquad\qquad (\text{in Fig. 2-2})$$

$$e_c = 3.93 \text{ V (from Example 2-1)}$$

$$i_c = \frac{10 \text{ V} - 3.93 \text{ V}}{1 \text{ k}\Omega} = 6.07 \text{ mA} \qquad\qquad \text{point 12}$$

$t = 4$ ms,

$$i_c = \frac{10\ V - 6.32\ V}{1\ k\Omega} = 3.68\ ma \qquad\qquad \text{point 13}$$

$t = 6$ ms,

$$i_c = \frac{10\ V - 7.77\ V}{1\ k\Omega} = 2.23\ mA \qquad\qquad \text{point 14}$$

$t = 8$ ms,

$$i_c = \frac{10\ V - 8.65\ V}{1\ k\Omega} = 1.35\ mA \qquad\qquad \text{point 15}$$

$t = 10$ ms,

$$i_c = \frac{10\ V - 9.18\ V}{1\ k\Omega} = 0.82\ mA \qquad\qquad \text{point 16}$$

$t = 12$ ms,

$$i_c = \frac{10\ V - 9.5\ V}{1\ k\Omega} = 0.5\ mA \qquad\qquad \text{point 17}$$

$t = 14$ ms,

$$i_c = \frac{10\ V - 9.7\ V}{1\ k\Omega} = 0.3\ mA \qquad\qquad \text{point 18}$$

$t = 16$ ms,

$$i_c = \frac{10\ V - 9.82\ V}{1\ k\Omega} = 0.18\ mA \qquad\qquad \text{point 19}$$

Refer again to Figure 2-1(a). Suppose the capacitor becomes completely charged to the level of the supply voltage. Now assume that the input voltage is reduced to zero, while the switch S is still closed. The result is that the capacitor discharges through resistor R_1. Equation (2-2) can be used to calculate the capacitor voltage at any time during discharge. The initial charge on the capacitor is E (i.e., the level of E before it went to zero). But during discharge E becomes zero. Therefore, Equation (2-2) can be simplified:

$$e_c = 0 - (0 - E)\epsilon^{-t/CR}$$

or $\qquad\qquad e_c = E\epsilon^{-t/CR} \qquad\qquad\qquad\qquad (2\text{-}5)$

When a capacitor discharge curve is plotted [using Equation (2-5)] it is found to be similar in shape to the charging current graph in Figure 2-2.

In Figure 2-3 *normalized* charge and discharge curves are presented. These can be employed to graphically solve many problems. The normalized curves are plotted for the case of: $E = 1$ V, $C = 1$ F, and $R = 1$ Ω. For these values, the capacitor voltage can be determined at any given time t after commencement of charge or discharge.

When the supply voltage is not 1 V, the capacitor voltage at any given time can be found simply by multiplying the voltage from the graph by the value of E. For example, when $E = 5$ V and $t = 0.7$ s on the charge curve:

$$e_c = 0.5 \text{ V} \times 5 \text{ V} = 2.5 \text{ V}$$

Similarly, when C and R are other than 1 F and 1 Ω, the time at any instant is multiplied by $C \times R$. As an example of this take the case of $C = 1$ μF and $R = 1$ kΩ when $e_c = 0.5$ V.

The time for e_c to reach 0.5 V is

$$t = 0.7 \text{ s} \times 1 \text{ } \mu\text{F} \times 1 \text{ k}\Omega$$
$$= 0.7 \text{ ms}$$

EXAMPLE 2-3

Using the normalized charge and discharge curves in Figure 2-3, determine:
(a) e_c at 1.5 ms starting from $e_c = 0$ V, when R $= 1$ kΩ, $C = 1$ μF, and $E = 10$ V.
(b) e_c at 6 ms from full charge when $R = 20$ kΩ, $C = 0.1$ μF, and $E = 12$ V.

solution (a)
     Each second on the time scale becomes

$$1 \text{ s} \times 1 \text{ } \mu\text{F} \times 1 \text{ k}\Omega = 1 \text{ ms}$$

At $t = 1.5$ ms (point a on the charge curve):

$$e_c = 10 \text{ V} \times 0.78$$
$$= 7.8 \text{ V}$$

solution (b)
     Each second on the time scale becomes

$$1 \text{ s} \times 0.1 \text{ } \mu\text{F} \times 20 \text{ k}\Omega = 2 \text{ ms}$$

At $t = 6$ ms (point b on the discharge curve):

$$e_c = 12 \text{ V} \times 0.05$$
$$= 0.6 \text{ V}$$

E (volts)

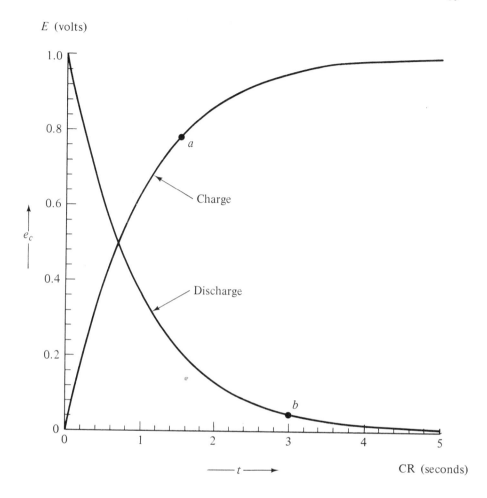

FIGURE 2-3.   Normalized charge and discharge curves for a *CR* circuit.

2-2   CR CIRCUIT EQUATIONS

Refer to the graphs in Figure 2-4, where charging current and capacitor voltage are plotted *versus* time for the circuit of Fig. 2-1(a). It is seen that when $t = 4$ ms, e_c is 6.32 V. In this case 4 ms is equal to the product of capacitance and resistance. That is,

$$t = C \times R = 4\ \mu\text{F} \times 1\ \text{k}\Omega = 4 \times 10^{-3}$$

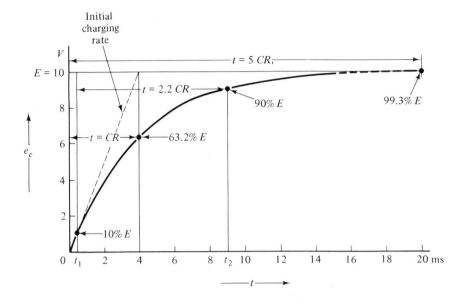

(a)   CR and t relationships to e_c

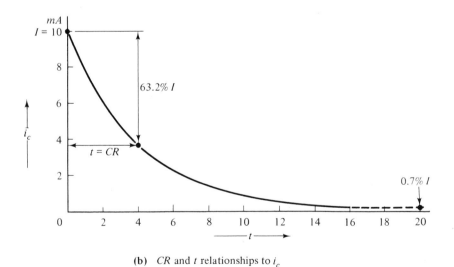

(b)   CR and t relationships to i_c

FIGURE 2-4.   CR and t relationships to e_c and i_c.

Therefore, when $t = CR$, e_c is 6.32 V or 63.2% of E.
    Consider Equation (2-3) once again:

$$e_c = E(1 - \epsilon^{-t/CR})$$

When $t = CR$,

$$
\begin{aligned}
e_c &= E(1 - \epsilon^{-CR/CR}) \\
&= E(1 - \epsilon^{-1}) \\
&= E \times 0.632
\end{aligned}
\tag{2-6}
$$

    Thus, when $t = CR$, $e_c = 63.2\%$ of E, no matter what the value of E, C, or R.
    The product CR is termed the *time constant* of a circuit. As will be seen, the time constant is a very important quantity. It can be used to classify a circuit, and it can be related to the rise time of an output pulse. The Greek letter τ is frequently employed as the symbol for the time constant. For a resistive capacitive circuit, $\tau = CR$.
    Now consider the equation for instantaneous charging current [Equation (2-4)]:

$$i_c = I\epsilon^{-t/CR}$$

and again let $t = CR$.

$$
\begin{aligned}
i_c &= I\epsilon^{-CR/CR} = I\epsilon^{-1} \\
&= I \times 0.368
\end{aligned}
\tag{2-7}
$$

or, when $t = CR$,

$$i_c = I(1 - 0.632)$$

    Thus, after time $t = CR$, the charging current is reduced by 63.2% of its initial value [see Figure 2-4(b)].
    If the charging current were to remain constant at its initial level, the quantity of charge contained in the capacitor would be

$$Q = I \times t \text{ coulombs}$$

Also,

$$Q = C \times V \text{ coulombs}$$

where C is the capacitance (in farads), and V is the capacitor voltage (in volts).

Therefore,

$$It = CV$$

or

$$V = \frac{It}{C} \qquad (2\text{-}8)$$

It is important to note that Equation (2-8) applies only to circuits in which the charging current is held at a constant level. It does not apply to the circuit of Figure 2-1(a). However, this equation can be employed to learn a little more about the time constant CR.

The initial level of charging current in an ordinary resistive capacitive circuit [as in Figure 2-1(a)] is $I = E/R$. If this were to remain constant then the time for the capacitor to become completely charged could be calculated from Equation (2-8).

$$t = \frac{CV}{I}$$

for $V = E$, and $I = E/R$,

$$t = \frac{C \times E}{E/R} = CR$$

As illustrated by the broken line Figure 2-4(a), with the initial charging current constant, the capacitor would be completely charged in a time period of $t = CR$. Again, note that the time constant is involved.

Once again refer to Figure 2-4(a). It is seen that even after 16 ms, the capacitor is not completely charged to the level of the supply voltage. Theoretically, because the charging current continuously decreases, the capacitor cannot become completely charged to the supply voltage level. However, after a period of five time constants, that is, $t = 5 \times CR$, the capacitor is more than 99% charged and, for all practical purposes, can be regarded as completely charged. For the circuit of Figure 2-1(a), after $t = 5 \times CR$ the capacitor voltage is

$$e_c = 10 \text{ V}(1 - \epsilon^{-5CR/CR}) = 9.93 \text{ V}$$

Similarly, it can be shown that i_c reduces to less than 1% of its initial level (I) after a time period of 5 CR [see Figure 2-4(b)].

It was pointed out in Chapter 1 that the rise time of a pulse output from a circuit is determined as the time it takes for the output to go from 10% to 90% of maximum output level. The CR circuit shown in Figure

2-1(a) has a step input of 10 V applied to it when switch S is closed. It has been shown that the output eventually approaches the maximum level of the step input. The rise time of the output can then be calculated as

$$(t \text{ at } e_c = 90\% \ E) - (t \text{ at } e_c = 10\% \ E)$$

An expression for t at a given level of e_c can be derived from Equation (2-2):

$$e_c = E - (E - E_O) \epsilon^{-t/CR}$$

$$(E - E_O) \epsilon^{-t/CR} = E - e_c$$

$$\epsilon^{-t/CR} = \left(\frac{E - e_c}{E - E_O} \right)$$

$$\epsilon^{t/CR} = \left(\frac{E - E_O}{E - e_c} \right)$$

$$\frac{t}{CR} = \ln \left(\frac{E - E_O}{E - e_c} \right)$$

$$t = CR \ln \left(\frac{E - E_O}{E - e_c} \right) \tag{2-9}$$

For $e_c = 90\%$ of E [t_2 on Figure 2-4(a)]:

$$t_2 = CR \ln \left(\frac{E - E_O}{E - 0.9 \ E} \right)$$

$$= 2.3 \ CR \tag{2.10}$$

For $e_c = 10\%$ of E [t_1 on Figure 2-4(a)]:

$$t_1 = CR \ln \left(\frac{E - 0}{E - 0.1 \ E} \right)$$

$$= 0.1 \ CR \tag{2-11}$$

Therefore, the rise time of the output is

$$t_r = (t_2 - t_1) = CR(2.3 - 0.1)$$

$$t_r = 2.2 \ CR \tag{2-12}$$

Equation (2-12) can be applied to any resistive capacitive circuit when the time constant CR is known. Sometimes a time constant is calculated for an amplifier or for a single transistor. Then Equation (2-12) can be used to determine the rise time of an output pulse. It can also be shown that the upper cutoff frequency (f_H) of a circuit is related to CR by the equation:

$$f_H = \frac{1}{2\pi CR} \tag{2-13}$$

To summarize the relationships between CR and e_c and time: when $t = CR$,

$$e_c = 0.632\,E$$
$$i_c = I(1 - 0.632)$$

when $t = 5\ CR$,

$$e_c = 0.993\,E$$
$$i_c = I(1 - 0.993)$$

and,

$$t_r = 2.2\ CR$$
$$f_H = \frac{1}{2\pi CR}$$

EXAMPLE 2-4

A 1 μF capacitor is charged from a 6 V source through a 10 kΩ resistor. If the capacitor has an initial charge of -3 V, calculate its voltage after 8 ms.

solution

By Equation (2-2),

$$e_c = E - (E - E_O)\epsilon^{-t/CR}$$

At $t = 8$ ms,

$$e_c = 6\ \text{V} - [6\ \text{V} - (-3\ \text{V})]\epsilon^{-8\ \text{ms}/(1\ \mu\text{F} \times 10\ \text{k}\Omega)}$$

$$= 6\ \text{V} - [9\ \text{V}]\epsilon^{-0.8} = 1.96\ \text{V}$$

EXAMPLE 2-5 _____

A 5 V step is switched *on* to a 39 kΩ resistor in series with a 500 pF capacitor. Calculate the rise time of the capacitor voltage, the time for the capacitor to charge to 63.2% of its maximum voltage, and the time for the capacitor to become competely charged.

solution

Rise time:

$$t_r = 2.2 \ CR$$
$$= 2.2 \times 500 \ \text{pF} \times 39 \ \text{k}\Omega$$
$$= 42.9 \ \mu\text{s}$$

$e_c = 0.632 \ E$, at $t = CR$:

$$t = 500 \ \text{pF} \times 39 \ \text{k}\Omega$$
$$= 19.5 \ \mu\text{s}$$

Capacitor is 99.3% charged at $t = 5 \ CR$:

$$t = 97.5 \ \mu\text{s}$$

2-3   CR CIRCUIT RESPONSE TO SQUARE WAVES

A resistive capacitive circuit with an input square wave is shown in Figure 2-5(a). The capacitor voltage first increases from zero to a level e_1 at time t_1 [see Figure 2-5(b)]. Between t_1 and t_2 the applied voltage is zero, so the capacitor discharges to e_2 volts. Then the capacitor charges to a new level e_3 at time t_3. To determine the level of e_c at any time greater than t_2 it is necessary first to calculate e_1 at time t_1. Then e_2 must be calculated by using e_1 as the initial voltage on the capacitor, and noting that the input voltage is zero from t_1 to t_2. Between t_2 and t_3, the initial voltage is e_2 volts, and the input voltage is again greater than zero.

EXAMPLE 2-6 _____

Calculate the capacitor voltage in the circuit of Figure 2-5(a) at 14 ms from $t = 0$.

solution

$$e_c = E - (E - E_O)\epsilon^{-t/CR}$$

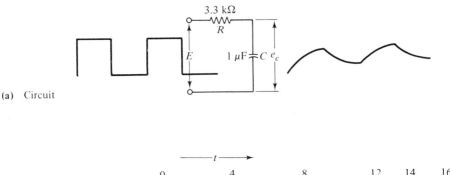

(a) Circuit

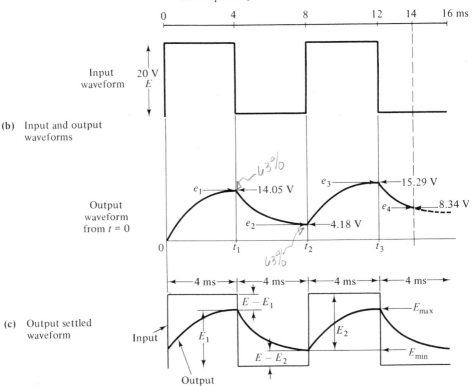

(b) Input and output waveforms

(c) Output settled waveform

FIGURE 2-5. *CR* circuit response to a square wave input.

At $t = 4$ ms,

$$e_c = 20 \text{ V} - (20 \text{ V} - 0)\epsilon^{-4 \text{ ms}/1 \,\mu\text{F} \times 3.3 \text{ k}\Omega}$$

$$= 14.05 \text{ V} \left[e_1 \text{ in Figure 2-5(b)} \right]$$

From $t = 4$ ms to $t = 8$ ms, $E = 0$ V and $E_O = 14.05$ V.
At $t = 8$ ms,

$$e_c = 0 - (0 - 14.05 \text{ V})\epsilon^{-4 \text{ ms}/1 \,\mu\text{F} \times 3.3 \text{ k}\Omega}$$

$$= 4.18 \text{ V} \left[e_2 \text{ in Figure 2-5(b)} \right]$$

From $t = 8$ ms to $t = 12$ ms, $E = 20$ V and $E_O = 4.18$ V.
At $t = 12$ ms,

$$e_c = 20 \text{ V} - (20 \text{ V} - 4.18 \text{ V})\epsilon^{-4 \text{ ms}/1 \,\mu\text{F} \times 3.3 \text{ k}\Omega}$$

$$= 15.29 \text{ V} \left[e_3 \text{ in Figure 2-5(b)} \right]$$

From $t = 12$ ms to $t = 16$ ms, $E = 0$ and $E_O = 15.29$ V.
At $t = 14$ ms,

$$e_c = 0 - (0 - 15.29 \text{ V})\epsilon^{-2 \text{ ms}/1 \,\mu\text{F} \times 3.3 \text{ k}\Omega}$$

$$= 8.34 \text{ V} [e_4 \text{ in Figure 2-5(b)}]$$

After several intervals of charging, partially discharging, and recharging, the capacitor voltage will eventually arrive at a settled condition. When this occurs, the capacitor always charges to a maximum voltage level, E_{max}, and discharges to a minimum level, E_{min}, as shown in Figure 2-5(c). These final levels occur when the charging and discharging voltages are equal. Thus, in Figure 2-5(c) $E_1 = E_2$. Also $E_{max} = E_1$, and $E_{min} = (E - E_{max})$.

Calculating E_{min}, starting from $E_O = E_{max}$ and $E = 0$ V,

$$E_{min} = e_c = 0 - (0 - E_{max})\epsilon^{-t/CR}$$

$$= E_{max}\epsilon^{-t/CR}$$

and since $E_{min} = (E - E_{max})$

$$E - E_{max} = E_{max}\epsilon^{-t/CR}$$

$$E = E_{max}\epsilon^{-t/CR} + E_{max}$$

$$= E_{max}(\epsilon^{-t/CR} + 1)$$

and

$$E_{max} = \frac{E}{1 + \epsilon^{-t/CR}} \qquad (2\text{-}14)$$

EXAMPLE 2-7 _____

For the circuit of Figure 2-5(a) determine the maximum and minimum levels at which the capacitor voltage will settle.

solution

Refer to Equation (2-14):

$$E_{max} = \frac{E}{1 + \epsilon^{-t/CR}}$$

$$= \frac{20 \text{ V}}{1 + \epsilon^{-4 \text{ ms}/(1 \text{ } \mu F \times 3.3 \text{ k}\Omega)}}$$

$$= 15.41 \text{ V}$$

$$E_{min} = E - E_{max}$$

$$= 20 - 15.41 = 4.59 \text{ V}$$

The charging current for the circuit of Figure 2-5(a) may be most easily calculated at any instant as:

$$i_c = \frac{E - e_c}{R}$$

Note that during the time intervals when $E = 0$, i_c is a negative quantity. Because the capacitor is discharging, the current through R is reversed. The charging current is also discussed in Section 2-5.

2-4 INTEGRATING CIRCUITS

Figure 2-6 shows a CR circuit with a square wave input and the output voltage taken across the capacitor. The shape of the output (capacitor) voltage waveform is dependent upon the relationship between the time constant (CR) and the pulse with (PW).

Consider the case where CR is much smaller than PW [waveform (a) in Figure 2-6]. In Sec. 2-1, it was demonstrated that the capacitor is charged to 99.3% of the input voltage after time $t = 5$ CR:
Let

$$CR = \tfrac{1}{10} \text{ PW}$$

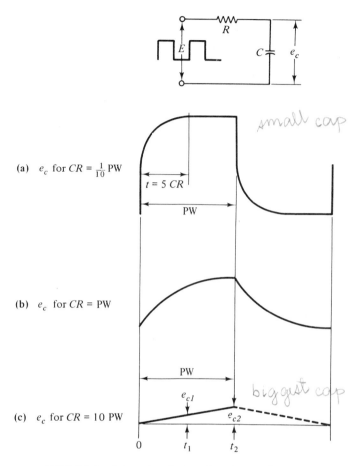

FIGURE 2-6.   Integrating circuit and output waveforms.

Then,

$$e_c = 99.3\% \text{ of } E \text{ at } t = 5\left(\tfrac{1}{10} \text{ PW}\right)$$

That is,

$$e_c \simeq E \text{ at } t = \tfrac{1}{2} \text{ PW}$$

In this case the output roughly approximates the square wave input. If CR is made smaller than $\tfrac{1}{10}$ PW, then the output even more closely resembles the square wave input.

For the waveform (b) in Figure 2-6, CR is equal to the pulse width. In Sec. 2-1 it was shown that the capacitor is charged to 63.2% of the input voltage after time $t = CR$. However, the settled waveform has an amplitude which is less than 63.2% of E. Under these conditions the waveform of capacitor voltage begins to approach a triangular shape.

When CR is made equal to ten times the pulse width, waveform of Figure 2-6(c) results. In this case the CR circuit, as arranged, is referred to as an *integrator*. To understand how the circuit integrates, it is necessary to calculate the output voltage levels in relation to time. At $CR = 10 \times PW$, use Equation (2-2) to obtain:

$$e_{c2} = E - (E - E_O)\epsilon^{-PW/10\,PW}$$
$$\simeq E - E\epsilon^{-1/10}, \text{ for } E_O = 0$$
$$\simeq E(1 - 0.9)$$
$$\simeq 0.1\,E$$

To calculate e_{c1} at $\frac{1}{2}$ PW:

$$e_{c1} = E - E\epsilon^{-1/2\,PW/10\,PW}$$
$$= E - E\epsilon^{-1/20}$$
$$\simeq E(1 - 0.95)$$
$$\simeq 0.05\,E$$

This result shows that after time t_1, $e_{c1} = 0.05\,E$ and after $t_2 = 2t_1$, $e_{c2} = 0.1\,E$. That is, $e_{c2} \simeq 2e_{c1}$ when $t_2 = 2t_1$ [see Figure 2-6(c)]. Thus the capacitor voltage is growing almost linearly. To further examine the output waveform when CR is 10 (or more) times the pulse width, consider Example 2-8.

EXAMPLE 2-8 _____

The circuit shown in Figure 2-7 has the following pulse inputs applied: (a) $E = 10$ V, PW $= 1$ ms; (b) $E = 10$ V, PW $= 2$ ms; (c) $E = 20$ V, PW $= 1$ ms. Calculate the level of e_c at the end of each pulse. The initial voltage on C is assumed to be zero.

solution

From Equation (2-2),

$$e_c = E - (E - E_O)\epsilon^{-t/CR}$$

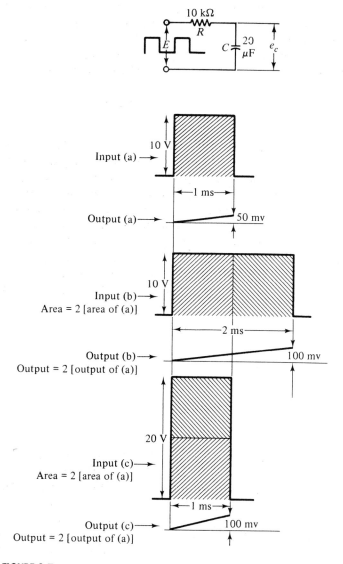

FIGURE 2-7. Integration of pulses having different widths and amplitudes.

For input (a):

$$e_{c(a)} = 10 \text{ V} - (10 \text{ V} - 0 \text{ V})\epsilon^{-1 \text{ ms}/(20 \mu\text{F} \times 10 \text{ k}\Omega)}$$
$$\simeq 50 \text{ mV}$$

For input (b):

$$e_{c(b)} = 10 \text{ V} - (10 \text{ V} - 0 \text{ V})\epsilon^{-2 \text{ ms}/(20 \mu\text{F} \times 10 \text{ k}\Omega)}$$
$$\simeq 100 \text{ mV}$$

For input (c):

$$e_{c(c)} = 20 \text{ V} - (20 \text{ V} - 0 \text{ V})\epsilon^{-1 \text{ ms}/(20 \mu\text{F} \times 10 \text{ k}\Omega)}$$
$$\simeq 100 \text{ mV}$$

Since the charging current remains substantially constant during the input PW, this problem can also be solved by using Equation (2-8):

$$V = \frac{It}{C}$$
$$= \frac{E}{R} \times \frac{t}{C}$$

For input (a):

$$e_{c(a)} = \frac{10 \text{ V} \times 1 \text{ ms}}{10 \text{ k}\Omega \times 20 \mu\text{F}} = 50 \text{ mV}$$

For input (b):

$$e_{c(b)} = \frac{10 \text{ V} \times 2 \text{ ms}}{10 \text{ k}\Omega \times 20 \mu\text{F}} = 100 \text{ mV}$$

For input (c):

$$e_{c(c)} = \frac{20 \text{ V} \times 1 \text{ ms}}{10 \text{ k}\Omega \times 20 \mu\text{F}} = 100 \text{ mV}$$

Example 2-8 shows that when the pulse width is doubled, the output voltage is doubled. Because the charging rate is (almost) linear, the output amplitude is proportional to the pulse width. Also, when the pulse amplitude is doubled, the output voltage is doubled. In this case, the charging rate is increased in proportion to the input voltage. Thus, the output voltage is proportional to the *pulse area*, that is, to the product of the pulse width (PW) and the pulse amplitude (PA).

$$e_c \propto PA \times PW$$

Figure 2-7 and example 2-8 show that the output voltage from an integrating circuit is proportional to the area of the pulse expressed as

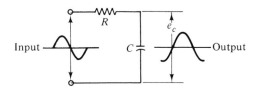

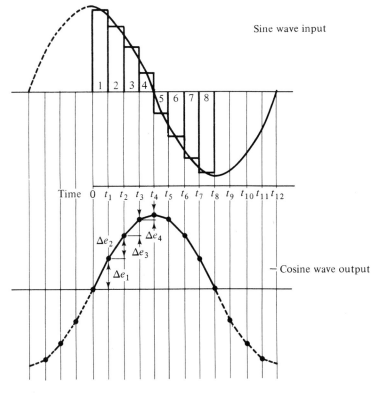

FIGURE 2-8.   Integration of a sine wave.

(volts $\times$ time). An *integrating circuit* is an *CR* circuit with the output taken across the capacitor, and $CR \geqslant (10 \times \text{PW})$. In other integrator circuits to be discussed in Chapter 7, the capacitor charging current is held constant by the use of additional components.

Figure 2-8 illustrates the integration of a sine wave. From time zero at the peak of the sine wave, the wave is divided into sections of equal widths. The height of each section corresponds approximately to the instantaneous sine wave amplitude. Thus the sine wave is represented by a series of pulses of varying amplitudes. The first pulse causes a linear increase in capacitor voltage from time 0 to t_1. This produces output voltage Δe_1. The second pulse, from t_1 to t_2, also produces a linear increase in the capacitor voltage. However, the pulse amplitude is now smaller, so the rate of increase in capacitor voltage is reduced. Thus Δe_2 is less than Δe_1. Similarly, the third and fourth pulses produce linear voltage increases, at decreasing rates. Since pulse 5 is negative, it causes the capacitor voltage to decrease by a small amount. Pulse 5 and pulse 4 are equal in amplitude, so the decrease in e_c due to pulse 5 is equal to the increase (Δe_4) produced by pulse 4. Also, negative pulses 6, 7, and 8 linearly decrease the output (capacitor) voltage. Extending the waveforms (broken lines), we see that integration of the sine wave input produces a negative cosine wave output.

2-5   DIFFERENTIATING CIRCUITS

When the output from a *CR* circuit is taken across *R*, the output voltage is the differential of the input. As in the case of the integrating circuit, the relationship between *CR* and the pulse width is important. Figure 2-9 shows the various output waveforms that can occur, depending upon PW and *CR*.

The voltage across *R* is the product of the charging current and the resistance; that is $e_R = i_c \times R$. When the time constant *CR* is 10 times the pulse width (or greater), the capacitor charges very little during the pulse time. The charging current falls only a small amount from its initial level [see waveform (a) in Figure 2-9]. Thus, e_R remains almost constant during the PW. During the space width the capacitor is discharged, and i_c is a negative quantity. The resistor voltage is now negative and, again, remains nearly constant for the discharge time.

If *CR* is made equal to the pulse width, the capacitor is charged to approximately 60% of the input voltage during the pulse time. Consequently the charging current falls by about 60% of its initial value, giving an output waveform with a very pronounced tilt [see the waveform in Figure 2-9(b)].

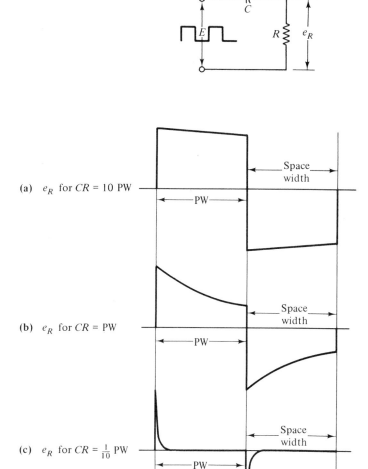

FIGURE 2-9.   Differentiating circuit and output waveform.

When CR is less than one-tenth of the pulse width, the capacitor is charged very rapidly. Only a brief pulse of current is necessary to charge and discharge the capacitor at the beginning and end of the pulse. The resultant waveform of resistor voltage is a series of positive and negative spikes at the pulse leading and lagging edges, respectively [see Figure 2-9(c)]. The *differential* of a quantity is a measure of the rate of change of the quantity. At the leading edge of the pulse, the input voltage is changing rapidly in a positive direction. At the lagging edge of the pulse, the input voltage is changing rapidly in a negative direction. During both the pulse width and the space width, the input voltage does not change at all. Thus, it is seen that the positive and negative spikes with intervening spaces [Fig. 2-9 (c)] do indeed represent a differentiated square wave.

When a ramp voltage is applied to the input of a differentiating circuit, the resultant output is a constant dc voltage level (Figure 2-10). While the input voltage continuously increases, the capacitor cannot become com-

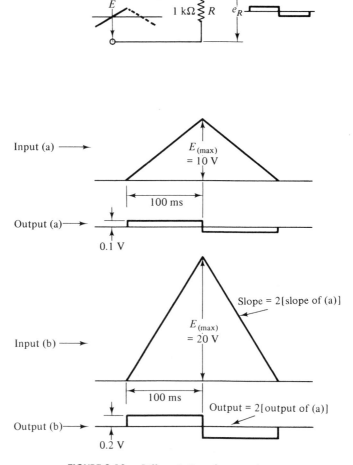

FIGURE 2-10.   Differentiation of ramp voltage.

pletely charged. Hence, the instantaneous capacitor voltage is always slightly less than the instantaneous input voltage. This small difference in E and e_c is developed across R, giving a constant level of charging current, and thus a constant level of e_R. While the ramp increases positively, the capacitor is charged with the polarity shown and i_c produces a positive level of e_R. When the ramp goes negative, i_c is reversed and, consequently, e_R is negative.

EXAMPLE 2-9

For inputs (a) and (b), calculate the level of the outputs from the differentiating circuit shown in Figure 2-10.

solution

At the end of input ramp $e_c \simeq E_{max}$. Since the charging current is constant, Equation (2-8), $V = It/C$, may be applied.
(a) For the 10 V ramp:

$$I = \frac{CE_{max}}{t} = \frac{1\,\mu F \times 10\,V}{100\,ms} = 0.1\,mA$$

and

$$e_R = I \times R = 0.1\,mA \times 1\,k\Omega = 0.1\,V$$

(b) For the 20 V ramp:

$$I = \frac{CE_{max}}{t} = \frac{1\,\mu F \times 20\,V}{100\,ms} = 0.2\,mA$$

and

$$e_R = I \times R = 0.2\,mA \times 1\,k\Omega = 0.2\,V$$

In Example 2-9, the rate of change of the 10 V ramp is 10 V/100 ms, that is, 0.1 V/ms. For the 20 V ramp, the rate of change is 0.2 V/ms. Thus, as shown in the example, the differentiated output doubles when the rate of change of input voltage is doubled.

From Example 2-9 an equation for the output from a differentiating circuit is

$$e_R = CR \times \frac{E_{max}}{t} \qquad (2\text{-}15)$$

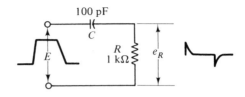

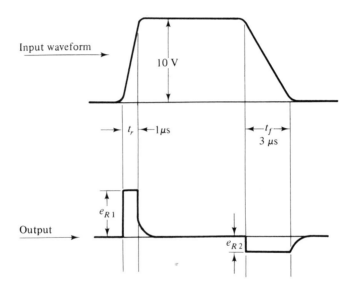

FIGURE 2-11.   Differentiation of pulse wave.

That is, $e_R = CR \times$ (rate of change of input). This equation may be applied in the case of pulse waveforms with known rise and fall times, as illustrated in Figure 2-11.

EXAMPLE 2-10

Calculate the amplitude of the differentiated output waveform for the circuit and input pulse shown in Figure 2-11.

solution

$$e_R = CR \times (\text{slope})$$

$$e_{R1} = CR \times \frac{E_{\max}}{t_r} = 100 \text{ pF} \times 1 \text{ k}\Omega \times 10 \text{ V}/\mu\text{s}$$

$$= 1 \text{ V}$$

$$e_{R2} = CR \times \frac{E_{\max}}{t_f} = 100 \text{ pF} \times 1 \text{ k}\Omega \times (-10 \text{ V}/3 \ \mu\text{s})$$

$$= -0.3 \text{ V}$$

Sometimes the input waveform to a differentiator has very fast rise and fall times, and it is intended to use the differentiator only as a means of generating positive and negative spikes. In this case the circuit time constant is simply made much smaller than the input pulse width (i.e., instead of smaller than t_r and t_f). The spike waveform that results usually has a peak-to-peak amplitude which is twice the peak-to-peak amplitude of the input wave.

Consider Figure 2-12. When the input is positive at $+E$ volts, the capacitor charges to E volts with the polarity shown. When the input goes to $-E$, the output spike amplitude is (briefly) the sum of the input voltage and the capacitor voltage, i.e., $-2 E$ volts. Similarly, while the input is negative, the capacitor charges with the opposite polarity and the (positive) output spike amplitude is again 2 E, the sum of the input and capacitor voltages.

The process by which a sine wave is differentiated is illustrated in Figure 2-13. At the peak of the sine wave ($t=0$), the rate of change of the

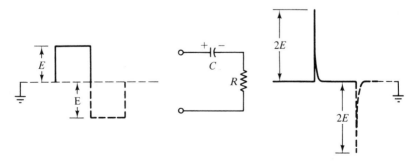

FIGURE 2-12. Differentiating circuit with an input which has very fast t_r and t_f.

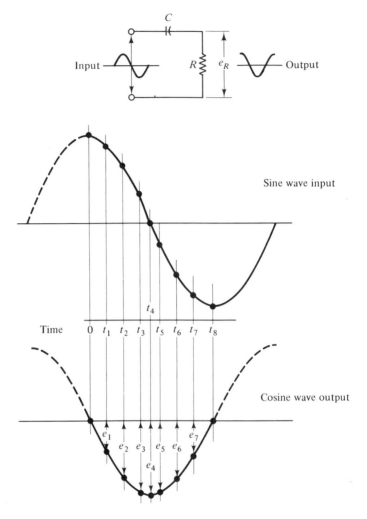

FIGURE 2-13.   Differentiation of a sine wave.

voltage is zero. Thus the differentiated output voltage is zero. At time t_1 the sine wave amplitude decreases, producing a negative rate of change. Consequently, the differentiated output voltage is $-e_1$. At t_2 and t_3 the negative rate of change increases and becomes maximum at t_4. The differentiated output, therefore, increases negatively through $-e_2$ and $-e_3$ to $-e_4$. Beyond t_4, the negative rate of change decreases to zero at t_8. Extending the waveform (broken lines) shows the differential of a sine wave to be a cosine wave.

2-6   LOADING EFFECTS ON DIFFERENTIATING AND INTEGRATING CIRCUITS

If a circuit connected to the output of a differentiator has an input resistance of (say) 1 kΩ, then that 1 kΩ might be used as R in the differentiator (see Figure 2-9). If the input resistance of the circuit is too small to be used as R in the differentiator, an *emitter follower* or a *voltage follower circuit* (see Section 7-7) must be used to increase the circuit input

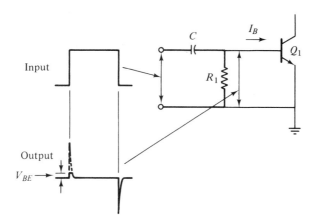

(a)   Positive spike clipped by transistor

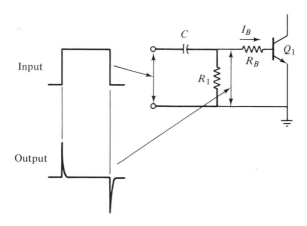

(b) $R_1 \ll R_B$, waveform not clipped

FIGURE 2-14.   Transistor loading effect on a differentiating circuit.

resistance. When the input resistance is very high compared to likely values for R, R is selected at any convenient (much lower) value. In this event, the connected circuit has no significant effect on the operation of the differentiator.

One loading problem that occurs frequently is the case of a transistor which is to be switched *on* by the output of a differentiating circuit. Figure 2-14(a) illustrates the situation. When the transistor is not connected, the differentiated waveform developed across R_1 is positive and negative spikes. With Q_1 connected, the voltage across R_1 cannot exceed the V_{BE} of the transistor. (This is typically 0.7 V for a silicon transistor, 0.2 V for a germanium device). Of course, this affects only the positive spikes when an *npn* transistor is used. During the negative spikes the device is *off* and the spike amplitude is unaltered. Whenever a *pnp* transistor is involved, the reverse is true: negative spikes clipped, positive spikes unaltered.

If it is intended that the transistor should be triggered *on* at the edge of the input pulse, then the spike-clipping may not be important. However, when the spikes are not to be clipped the circuit must be modified slightly. Figure 2-14(b) illustrates the necessary modification. Resistor R_B limits the transistor base current to any desired level, and allows the voltage across R_1 to climb to the required amplitude.

R_B is calculated as explained in Chapters 4 and 5, and the total input resistance offered by the transistor and R_B is

$$R_i = R_B + \frac{V_{BE}}{I_B}$$

The differentiating circuit resistance (during positive pulses when an *npn* transistor is used) now becomes

$$R = R_1 \| R_i$$

During negative-going spikes Q_2 is biased *off*, I_B ceases to flow, and R becomes equal to R_1. If R_1 is made much smaller than R_i, R is always approximately equal to R_1, and the transistor has a negligible effect on the performance of the differentiating circuit.

The same kind of loading effect may present a problem with integrating circuit. If the transistor is directly connected, then, as before, the output waveform is clipped [see Figure 2-15(a)]. Again, the solution is to employ a base resistor with the transistor; R_B in Figure 2-15(b). In this case the capacitor current I must be made very much larger than I_B if the transistor is to have a negligible effect on the performance of the integrating circuit.

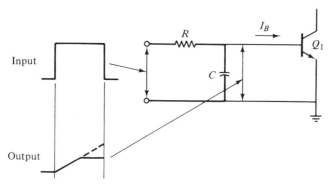

(a) Output wave clipped

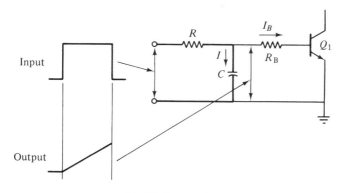

(b) $I \gg I_B$, waveform not clipped

FIGURE 2-15. Transistor loading effect on an integrating circuit.

Alternatively, an emitter follower or a voltage follower may be used to minimize the loading effect on the integrator.

REVIEW QUESTIONS AND PROBLEMS

2-1 A capacitor C is charged from a voltage source E, via a resistance R. The general formula for the capacitor voltage is

$$e_c = E - (E - E_O)\epsilon^{-t/CR}$$

(a) Derive an expression for the time required for the capacitor voltage to go from 10% to 90% of its maximum level. (b) Derive expressions for e_c when $t = CR$ and when $t = 5\ CR$.

2-2   The CR circuit shown in Figure 2-7 has a 20 V dc voltage input. Plot the graphs of e_c and e_R *versus* time.

2-3   A 10 μF capacitor is charged via a 10 kΩ resistance from a 5 V source. If the capacitor has an initial charge of -2 V, calculate its charge after 50 ms.

2-4   For Problem 2-3, determine the level of charging current after 35 ms.

2-5   For Problem 2-3, calculate the time required for the capacitor to charge to 4.5 V.

2-6   A 10 V step is switched on to a 22 kΩ resistor in series with a 300 pF capacitor. Calculate the rise time of the capacitor voltage, the time for the capacitor to charge to 63.2% of its maximum voltage, and the time for the capacitor to become completely charged.

2-7   If the square wave input to the circuit shown in Figure 2-5(a) has a frequency of 250 Hz, and an amplitude of 15 V, determine the capacitor voltage at $t = 7$ ms.

2-8   For Problem 2-7, determine the maximum and minimum levels of capacitor voltage when the waveform has settled. Sketch the waveform of e_c to show its relationship to the input square wave.

2-9   For Problem 2-7, determine the maximum and minimum levels of charging current when the waveform has settled. Sketch the waveform of i_c to show its relationship to the e_c waveform.

2-10 A 1,000 pF capacitor is charged from a 5 V source via a 4.7 kΩ resistor. Using the normalized charge and discharge graphs in Figure 2-3, determine: (a) the time required for the capacitor to charge to 3.5 V, (b) the capacitor voltage 15 μs after commencing to discharge from its fully charged level.

2-11 For the circuit described in Problem 2-6, use the normalized charge and discharge graphs in Figure 2-3 to determine (a) the capacitor voltage 10 μs after $e_c = 0$ V, (b) the time required for the capacitor to charge to $e_c = 9$ V.

2-12 For the circuit and input shown in Figure 2-16; (a) determine the level of e_c and i_c at $t = 2.5$ ms. (b) Sketch the settled waveform of e_c.

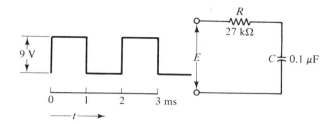

FIGURE 2-16. Problem 2-12.

2-13 Sketch an integrating circuit with a square wave input. Show the output waveforms for (a) $CR \simeq 10 \times PW$, (b) $CR \simeq \frac{1}{10}$ PW, (c) $CR \simeq PW$.

2-14 Explain why the output of an integrating circuit represents the integration of the input waveform.

2-15 Sketch the shape of the output waveform from an integrator when the input is a cosine wave.

2-16 A pulse having an amplitude of 5 V and a PW of 100 μs is applied to the CR circuit shown in Figure 2-16. Determine the amplitude of e_c at the end of the pulse. If the input PW goes to 150 μs and the amplitude goes to 7.5 V, calculate the new level of e_c at the end of the pulse time.

2-17 Sketch a differentiating circuit with a square wave input. Show the waveforms for (a) $CR \simeq \frac{1}{10}$ PW, (b) $CR \simeq 10 \times PW$, (c) $CR \simeq PW$.

2-18 Explain why the output of a differentiating circuit represents the differential of the input waveform.

2-19 Sketch the shape of the output waveform from a differentiator when the input is a cosine wave.

2-20 A 100 Hz triangular wave with a peak-to-peak amplitude of 9 V is applied to a differentiating circuit. $R = 1$ MΩ and $C = 100$ pF. Calculate the output amplitude and sketch the waveform of the output.

2-21 A pulse with a rise time of $t_r = 500$ ns, a fall time of $t_f = 1$ μs, PA $= 12$ V, and PW $= 10$ μs is applied to a differentiating circuit with $C = 200$ pF and $R = 470$ Ω. Determine the amplitude of the differentiated outputs, and sketch the output waveform.

2-22 Discuss the loading problems that occur with differentiating and integrating circuits, and explain how they may be overcome.

Diode Switching

INTRODUCTION

Because it passes a large current when forward-biased and an extremely small current when reverse-biased, a semiconductor diode can be employed as a switch. The speed with which a diode can be switched is determined by the *reverse recovery time* of the device. Diodes are widely applied to *clip* unwanted portions from a waveform, or to *clamp* the peak of a waveform to a desired dc level. Zener diodes may be used as reference voltage sources in clipping and clamping circuits.

3-1 THE DIODE AS A SWITCH

Typical characteristics for a low-current silicon diode are shown in Figure 3-1. It is seen that when the *forward bias voltage* V_F is approximately 0.7 V the *forward current* I_F is approximately 10 mA. Above the 10 mA level, I_F increases substantially with very small increases in V_F. The reverse characteristics show an approximately constant *reverse saturation current* I_S of 0.05 μA for *reverse voltages* ranging up to *reverse breakdown* at 75 V. (Some diodes can survive reverse biases much greater than 75 V.) Since the typical reverse current is on the order of 1/200,000 of the forward current,

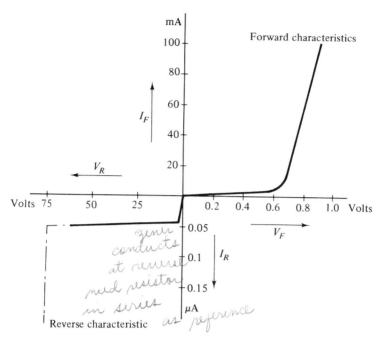

FIGURE 3-1.   Forward and reverse characteristics for a typical low current silicon diode.

the reverse current can be neglected for many purposes. The diode is then thought of as a *one-way device*. It simulates a switch, which is closed when the device is forward-biased and open when it is reverse-biased.

A switch is characterized by zero voltage drop when closed and zero current when open. This is also true of an ideal diode. Figure 3-2(a) illustrates the characteristics of a switch, and Figures 3-2(b) and (c) show similar approximate diode characteristics. The silicon device has a typical forward voltage drop of 0.7 V, while V_F for the germanium diode is approximately 0.3 V. For both silicon and germanium, the reverse current is normally very small.

In a great many applications diode characteristics can be ignored. The device is assumed to have a constant forward voltage drop V_F when forward-biased, and a constant (temperature dependent) *reverse leakage current* I_S when reverse-biased. To select a diode for a particular application, it is necessary to determine the forward current that must be passed, the power dissipation, the reverse voltage, and the maximum leakage current that can be tolerated. Another item that must be considered is the required operating frequency of the diode.

The effect of a sudden change from forward bias to reverse bias is illustrated in Figure 3-3(a). Instead of switching *off* sharply when the input

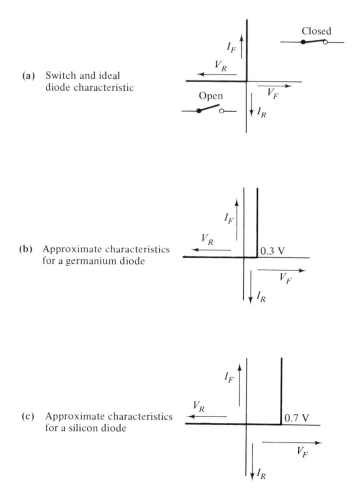

(a) Switch and ideal diode characteristic

(b) Approximate characteristics for a germanium diode

(c) Approximate characteristics for a silicon diode

FIGURE 3-2. Switch characteristics and approximate diode characteristics.

becomes negative, the diode initially conducts in reverse. The reverse current I_R is at first equal to I_F; then it gradually falls off to the reverse saturation level I_S. At the instant of reverse bias there are charge carriers crossing the junction depletion region, and these must be removed. This removal of charge carriers constitutes the reverse current I_R. The *reverse recovery time* t_{rr} is the time required for the reverse current to go to I_S. Typical values of t_{rr} range from 4 ns to 50 ns.

If the diode forward current is reduced to a very small value before the device is reverse-biased, then the reverse current will also be very small. Even when a large forward current is flowing, the reverse current can be kept small if the forward current is reduced slowly. This means that for

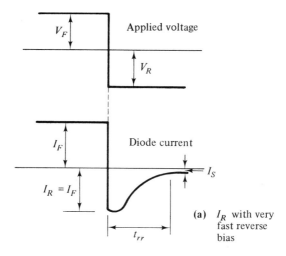

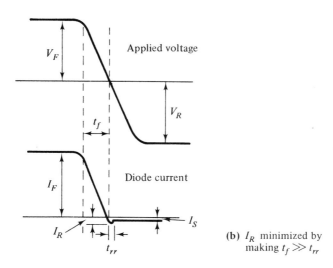

FIGURE 3-3.   Effect of diode reverse recovery time.

minimum reverse current, the fall time of forward current should be much longer than the diode reverse recovery time [see Figure 3-3(b)].

3-2   THE ZENER DIODE

The symbol for and characteristics of a *Zener diode* are shown in Figure 3-4. This device is a semiconductor diode designed to operate in the *reverse breakdown region* of its characteristics. If the reverse current is maintained

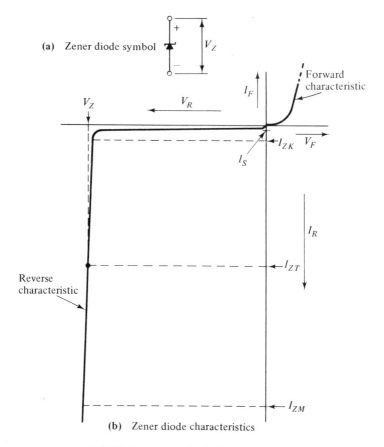

(a) Zener diode symbol

(b) Zener diode characteristics

FIGURE 3-4.   Zener diode characteristic.

within certain limits, the voltage drop across the diode is maintained at a reliable constant level. Thus the Zener diode (also known as an *avalanche diode* or *breakdown diode*) is useful as a voltage reference source.

From Figure 3-4(b), V_Z is the *Zener voltage* measured at *test current* I_{ZT}. The *knee current*, I_{ZK}, is the minimum current that should pass through the device to maintain a constant V_Z. The maximum Zener current that may be passed for the device not to exceed maximum permissible power dissipation is designated as I_{ZM}.

For correct operation a Zener diode must be positively biased on the cathode and negatively biased on the anode; that is, it must be reverse-biased. When the reverse voltage is smaller than V_Z only the normal diode *reverse saturation current* I_S flows. *When the Zener diode is forward-biased it behaves as an ordinary diode.* Thus a large forward current flows, and the diode voltage is typically $V_F = 0.7$ V.

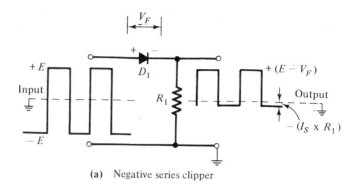

(a)   Negative series clipper

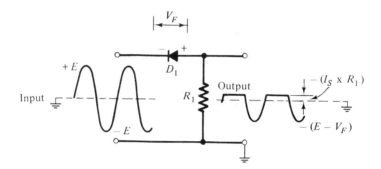

(b)   Positive series clipper

series   limiters

FIGURE 3-5.   Negative and positive series clipping circuits.

3-3   DIODE CLIPPER CIRCUITS

3-3.1   Series Clipper

A *clipper* (or *limiter*) *circuit* is one that clips off a portion of an input waveform. Two clipping circuits are shown in Figure 3-5. In the *negative series clipper*, the diode is forward-biased when the input becomes positive. Thus, the output voltage at this time is the peak input voltage minus the diode voltage drop. When the input becomes negative, the diode is reverse-biased and the reverse saturation current I_S flows through resistor R_1. The output then is a very small negative voltage $-(I_S \times R_1)$. The resultant output waveform from the circuit is essentially the input, with the negative portion clipped off.

The *positive series clipper* operates in the same way as the *negative clipper* except that, in this case, the diode is turned around, and the positive portion of the input waveform is clipped off. The input waveform may be square, sinusoidal, or any other shape.

EXAMPLE 3-1

The negative series clipping circuit in Figure 3-5(a) is to have an input of $E = \pm 50$ V. The output current from the circuit is to be $I_L = 20$ mA, and the negative output voltage $-V_O$ is not to exceed 0.5 V. Calculate the value of R_1. Specify the diode in terms of forward current, power dissipation, and peak reverse voltage.

solution

For a negative input:

$$I_S = 5 \ \mu\text{A} \ (\text{maximum likely; see Appendix 1-1})$$

$$-V_O = I_S \times R_1$$

$$R_1 = \frac{V_O}{I_S} = \frac{0.5 \text{ V}}{5 \ \mu\text{A}} = 100 \text{ k}\Omega$$

Use a 100-kΩ standard value (see Appendix 2).
Diode peak reverse voltage:

$$-E = -50 \text{ V}$$

For a positive input:

$$I_{R1} = \frac{E}{R_1}$$

$$= \frac{50 \text{ V}}{100 \text{ k}\Omega}$$

$$\simeq 0.5 \text{ mA}$$

Power dissipation in R_1:

$$P_{R1} = \frac{E^2}{R} = \frac{(50 \text{ V})^2}{100 \text{ k}\Omega}$$

$$= 25 \text{ mW}$$

Diode forward current:

$$I_F = I_L + I_{R1}$$

$$= 20 \text{ mA} + 0.5 \text{ mA}$$

$$= 20.5 \text{ mA}$$

Diode power dissipation:

$$P_{D1} = V_F \times I_F$$
$$= 0.7 \text{ V} \times 20.5 \text{ mA} \text{ (for a silicon diode)}$$
$$= 14.35 \text{ mW}$$

EXAMPLE 3-2

From the diode data sheets in Appendix 1 select a suitable device for the circuit designed in Example 3-1.

solution

From Example 3-1:

Reverse saturation current $I_S \leqslant 5 \mu$A

Peak reverse voltage $E \geqslant 50$ V

Forward current $I_F \geqslant 20.5$ mA

Therefore, the diode selected must have a reverse current I_R *not greater than* 5 μA, a maximum reverse voltage V_R *not less than* 50 V, and a maximum forward current *not less than* 20.5 mA. The 1N914, 1N915, and 1N916 diodes fulfill all of the required conditions. The 1N917 can take a maximum reverse voltage of only 30 V; therefore it is not suitable.

3-3.2   Series Noise Clipper

Frequently a signal has unwanted voltage fluctuations (called *noise*) which can trigger sensitive circuits. To eliminate noise, a *series noise clipping circuit* may be employed. If the noise is considerably smaller than the normal forward voltage drop of a diode and the signal voltages are larger than V_F, then the diode noise clipper shown in Figure 3-6(a) may be employed. Since the peaks of noise voltage are not large enough to forward-bias either D_1 or D_2, the output during the time between signals is zero. The wanted signals readily forward-bias the diodes, and the output peak voltage is $(E - V_F)$. A *dead zone* of $\pm V_F$ exists around ground level at the output. This simply indicates that for signals to be passed to the output, they must exceed $\pm V_F$.

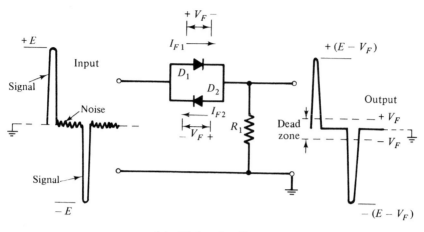

(a) Diode noise clipper

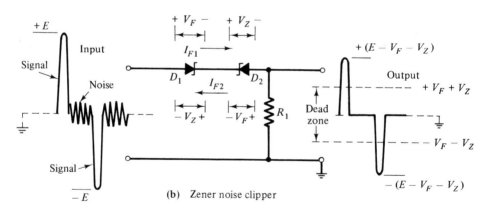

(b) Zener noise clipper

FIGURE 3-6. Series noise clipping circuits.

When noise is too large for ordinary diodes, Zener diodes can be used, as shown in Figure 3-6(b). In this case, when the signal input goes positive, D_1 behaves as an ordinary forward-biased diode, while D_2 goes into breakdown. Similarly, when the signal is negative, D_2 is forward-biased and D_1 is in breakdown. The *dead zone* is now $\pm(V_F + V_Z)$, and only signals greater than this will be passed to the output. That is, the signals must be large enough to drive one diode into breakdown and to forward-bias the other diode. The voltage drop across the two diodes is subtracted from the input signal, so the output peak is $(E - V_F - V_Z)$.

EXAMPLE 3-3 _____

The Zener diode noise clipper in Figure 3-6(b) has input signals of $E = \pm 6$ V. The input noise has an amplitude of ± 2 V. Specify the Zener diodes required and calculate the resistance of R_1. Also calculate the amplitude of the output signals.

solution

$$V_Z > 2 \text{ V}$$

From the Zener diode data sheet in Appendix 1-3, the 1N746 with $V_Z = 3.3$ V is a suitable device. The output signal amplitude is

$$V_O = \pm (E - V_F - V_z)$$
$$= \pm (6 \text{ V} - 0.7 \text{ V} - 3.3 \text{ V})$$
$$= \pm 2 \text{ V}$$

In the absence of a load current, R_1 must pass enough current to keep the diode conducting when the signal is present. From the data sheet, $I_{ZT} = 20$ mA. To ensure that I_{R1} is greater than I_{ZK}, make

$$I_{R1} \simeq \tfrac{1}{4} I_{ZT} = 5 \text{ mA}$$

and

$$V_{R1} = V_O = \pm 2 \text{ V}$$
$$R_1 = \frac{V_{R1}}{I_{R1}} = \frac{2 \text{ V}}{5 \text{ mA}}$$
$$= 400 \text{ } \Omega$$

Use a standard value resistor ($R_1 = 390$ Ω). See Appendix 2.
Power dissipation in R_1 is

$$P_{R1} = \frac{V_O^2}{R_1}$$
$$= \frac{(2 \text{ V})^2}{390 \text{ } \Omega}$$
$$= 10.3 \text{ mW}$$

3-3.3   Shunt Clipper

Negative and positive shunt clipping circuits are shown in Figure 3-7. In each case the clipping circuit is applied to protect the base-emitter junction of a transistor from excessive reverse bias. Most transistors will not survive more than 5 V applied in reverse across the base-emitter junction. Consequently, when input signals are greater than 5 V, some sort of protective circuitry is needed.

    In the circuit of Figure 3-7(a), the negative portion of the input signal is clipped off to protect the *npn* transistor. When the signal becomes positive, D_1 is reverse-biased and all of the current I_O flows through the

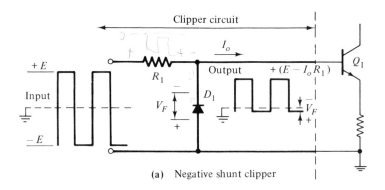

(a)   Negative shunt clipper

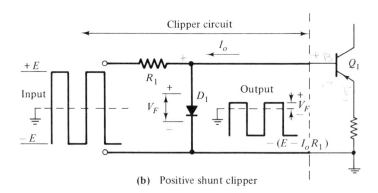

(b)   Positive shunt clipper

FIGURE 3-7.   Negative and positive shunt clipping circuits.

transistor base. The voltage at the transistor base, *i.e.*, the clipper output, is

$$(\text{Input voltage}) - (\text{Voltage drop across } R_1) = E - I_O R_1$$

When the input becomes negative, the transistor base-emitter junction is reverse-biased, and the diode is forward-biased. Since the diode is in parallel with the transistor input, the maximum reverse base-emitter voltage is limited to the diode forward voltage drop V_F.

Diode D_1 in Figure 3-7(b) is forward-biased when the input is positive. Thus, the transistor base-emitter voltage is limited to V_F, positive on the base and negative on the emitter, which is reverse-biased for the *pnp* transistor. When the input becomes negative, D_1 is reverse-biased and I_O flows through the transistor base.

As in the case of the series clippers, shunt clipping circuits can be employed with sine or other input waveforms, as well as with square waves.

EXAMPLE 3-4 _____

The negative shunt clipper circuit in Figure 3-7(a) is to have an output voltage of 9 V, and output current of approximately 1 mA. If the input voltage is ± 10 V, calculate the value of R_1 and the diode forward current.

solution

When the input is $+10$ V:

$$\text{Output} = 9 \text{ V} = E - I_O R_1$$
$$I_O R_1 = E - 9 \text{ V} = 10 \text{ V} - 9 \text{ V} = 1 \text{ V}$$
$$R_1 = \frac{1 \text{ V}}{I_O} = \frac{1 \text{ V}}{1 \text{ mA}} = 1 \text{ k}\Omega$$

When the input is -10 V, D_1 is forward-biased and

$$V_F \simeq 0.7 \text{ V}$$
$$V_F = E - I_F R_1$$
$$I_F = \frac{E - V_F}{R_1} = \frac{10 \text{ V} - 0.7 \text{ V}}{1 \text{ k}\Omega}$$
$$= 9.3 \text{ mA}$$

3-3.4   Biased Shunt Clipper

All the shunt clipping circuits discussed so far clip off either the positive or the negative portion of an input waveform. The unwanted output is limited to a maximum of V_F above or below ground. In the circuit of Figure 3-8(a), diode D_1 has its cathode connected to a bias of $+2$ V, and D_2 has its anode connected to -2 V. In this case the diode D_1 will not be forward-biased while the output of the clipping circuit is below 2 V. The presence of D_1 then limits the positive output to a maximum of $(2\,\text{V} + V_F)$. Similarly, D_2 will be reverse-biased until the output is more negative than -2 V. This limits the negative output to a maximum of $-(2\,\text{V} + V_F)$.

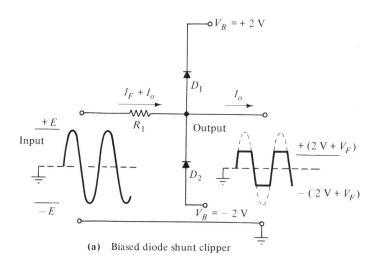

(a)   Biased diode shunt clipper

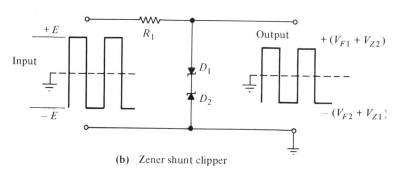

(b)   Zener shunt clipper

FIGURE 3-8.   Biased diode shunt clipper circuit and Zener shunt clipper circuit.

The biased shunt clipper normally is used to protect a device or circuit which has both positive and negative input signals. The bias voltage is selected to prevent the input (either positive or negative) from exceeding a maximum safe level.

The Zener clipper shown in Figure 3-8(b) performs a function similar to that of the circuit of Figure 3-8(a). In this case, however, no separate bias voltage supplies are necessary. When the input signal becomes positive, D_1 operates like an ordinary forward-biased diode while D_2 goes into Zener breakdown. The output voltage at this time is $(V_{F1} + V_{Z2})$. When the input is negative, D_1 is in Zener breakdown and D_2 is forward-biased. The output voltage now is $-(V_{F2} + V_{Z1})$.

EXAMPLE 3-5 _____

A biased shunt clipper circuit [as shown in Figure 3-8(a)] is to be designed to protect a circuit which cannot accept voltages exceeding $V_O = \pm 2.7$ V. The input to the clipper is a ± 8 V square wave, and the output current is to be a maximum of 1 mA. Calculate the value of R_1 and specify the diodes to be used.

solution

To ensure that the diodes become properly forward-biased, take the minimum forward current I_F as 10 mA (see the typical diode characteristics in Figure 3-1).

$$V_O = \text{bias voltage } V_B + \text{diode voltage drop } V_F$$
$$V_B = V_O - V_F$$
$$= 2.7 \text{ V} - 0.7 \text{ V} \quad \text{(using a silicon diode)}$$
$$= 2 \text{ V}$$

Voltage across $R_1 = (I_F + I_O) \times R_1$

$$(I_F + I_O) \times R_1 = E - V_B - V_F$$
$$R_1 = \frac{E - V_B - V_F}{I_F + I_O}$$
$$= \frac{8 \text{ V} - 2 \text{ V} - 0.7 \text{ V}}{10 \text{ mA} + 1 \text{ mA}}$$
$$= 482 \text{ }\Omega \quad \left[\begin{array}{l} \text{use 470 }\Omega\text{ standard value} \\ \text{resistor (see Appendix 2)} \end{array}\right]$$

The diodes selected should be low current devices with $V_F \simeq 0.7$ V at $I_F = 10$ mA, and should have a peak reverse voltage greater than 10 V.

EXAMPLE 3-6

Design a Zener diode shunt clipper circuit [as shown in Figure 3-8(b)] to be connected between a ± 25 V square wave signal and a circuit which cannot accept inputs greater than ± 11 V. The output current is to be a maximum of 1 mA.

solution

Clipper circuit output is $V_O = \pm 11$ V

$$V_O = V_F + V_Z$$
$$V_Z = V_O - V_F$$
$$= 11 \text{ V} - 0.7 \text{ V} = 10.3 \text{ V}$$

Refer to the breakdown diode data sheet in Appendix 1-3;

$$V_Z = 10 \text{ V for the 1N758 device}$$

Use 1N758 breakdown diodes.

$$V_{R1} = E - V_O$$
$$= E - (V_F + V_Z)$$
$$= 25 \text{ V} - 10 \text{ V} - 0.7 \text{ V} = 14.3 \text{ V}$$

To ensure that $I_Z > I_{ZK}$ make

$$I_Z \simeq \tfrac{1}{4} I_{ZT}$$
$$= \tfrac{1}{4} \times 20 \text{ mA} = 5 \text{ mA}$$
$$I_{R1} = I_z + I_O$$
$$= 5 \text{ mA} + 1 \text{ mA} = 6 \text{ mA}$$
$$R_1 = \frac{V_{R1}}{I_{R1}} = \frac{14.3 \text{ V}}{6 \text{ mA}} = 2.38 \text{ k}\Omega \qquad \begin{bmatrix} \text{use a 2.2 k}\Omega \text{ standard value} \\ \text{resistor (see Appendix 2)} \end{bmatrix}$$

3-4   DIODE CLAMPER CIRCUITS

3-4.1   Negative and Positive Clamping Circuits

The *clamper circuit* (also known as a *dc restorer circuit*) changes the dc level but does not affect the shape of a waveform. When the input is positive, in the *negative voltage clamper* circuit of Figure 3-9(a), diode D_1 is forward-biased, and capacitor C_1 charges with the polarity shown. During the positive input peak, the output cannot exceed the diode forward-bias voltage V_F. At this time, therefore, the voltage on the right-hand side (RHS) of the capacitor is $+V_F$, while on the left-hand side (LHS) of the capacitor the voltage is $+E$. Thus, the capacitor is charged to $E - V_F$, positive on the LHS, negative on the RHS, as shown.

When the input becomes negative, the diode is reverse-biased and it has no further effect on the capacitor voltage. Also, resistor R_1 has a very large value and cannot discharge the capacitor significantly during the negative (or positive) portion of the waveform. While the input is negative,

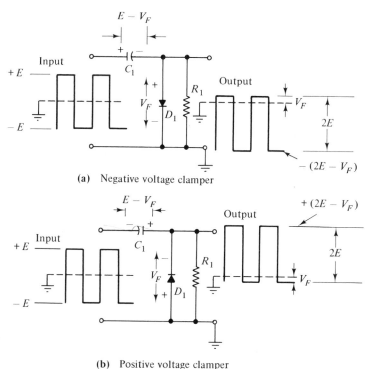

(a)   Negative voltage clamper

(b)   Positive voltage clamper

FIGURE 3-9.   Negative and positive clamping circuits.

the output voltage is the sum of the input voltage and the capacitor voltage. Since the polarity of the capacitor voltage is the same as the (negative) input, the result is a negative output larger than the input voltage.
Thus

$$\text{Negative output} = -E - (E - V_F)$$
$$= -(2E - V_F)$$

The peak-to-peak (p-to-p) output is the difference between the negative and positive peak voltages.

$$p\text{-to-}p \text{ output} = (\text{positive peak}) - (\text{negative peak})$$
$$= V_F - [-(2E - V_F)]$$
$$= 2E$$

It is seen that the amplitude of the output waveform from the *negative voltage clamper* is exactly the same as that of the input. Instead of being symmetrical above and below ground, however, the output positive peak is clamped to a level of V_F above ground. The difference between *clipping* and *clamping* circuits is that while the clipper *clips off* an unwanted portion of the input waveform, the clamper simply *clamps* the maximum positive or negative peak to a desired dc level.

The function of R_1 is to discharge C_1 over several cycles of the input waveform. This would not be necessary if the input signal never changed. However, if the input is made smaller, C_1 must be partially discharged for the positive output peak to rise to V_F once again.

The *positive voltage clamping circuit* [Figure 3-9(b)] functions in exactly the same way as the negative voltage clamper. The diode, connected as shown, clamps the negative output peak at $-V_F$. Capacitor C_1 charges to $E - V_F$ positive on the RHS, negative on the LHS. The positive output then becomes $2E - V_F$.

To design a clamping circuit, C_1 should be selected so that it becomes completely charged during about five cycles of the input waveform. Since a capacitor takes approximately five time constants to become fully charged,

$$5\,CR = 5 \times \text{PW},$$
or
$$CR = \text{PW}$$

where PW is the width of the pulse which forward-biases the diode. In this case, R is the total resistance in series with the capacitor when it is being charged. This is the sum of the source resistance R_S and the diode forward resistance R_F [see Figure 3-10(a)].

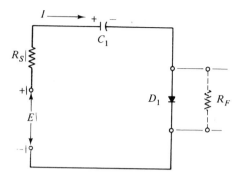

(a)   Charge of C_1 via R_S and R_F

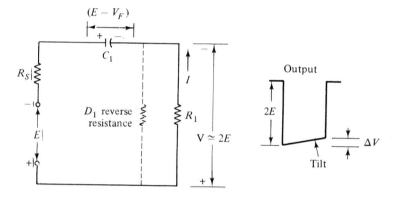

(b)   Discharge of C_1

FIGURE 3-10.   Capacitor charge and discharge circuits for clamping circuit.

$$C(R_s + R_F) = PW$$

Since usually $R_S \gg R_F$,

$$CR_S = PW \qquad (3\text{-}1)$$

When the capacitor partially discharges during the negative input peak (for a *negative* voltage clamper), some tilt appears on the output, as shown in Figure 3-10(b). The acceptable amount of tilt determines the value of the discharge resistor R_1. The voltage across R_1 during this time is approximately $2E$. Therefore, the discharge current is

$$I \simeq \frac{2E}{R_1}$$

The diode reverse resistance is also involved but, since it is in parallel with R_1 and is very much larger than R_1, it may be neglected. The current I remains approximately constant during the discharge time so Equation (2-7), $V = It/C$, may be applied. In this case, a change in capacitor voltage ΔV is involved. Replacing I with $(2E)/R_1$ in the equation gives

$$\Delta V = \frac{2E}{R_1} \times \frac{t}{C}$$

or

$$R_1 = \frac{2Et}{\Delta VC} \tag{3-2}$$

EXAMPLE 3-7

A negative voltage clamper has a 1 kH_z square wave input with an amplitude of ± 10 V. The signal source resistance R_S is 500 Ω, and the tilt on the output waveform is not to exceed 1%. Design a suitable circuit.

solution

For the input,

$$T = \frac{1}{f} = \frac{1}{1 \text{ kHz}} = 1 \text{ ms}$$

and $PW = \frac{1}{2}T = 500$ μs.
From Equation (3-1),

$$C = \frac{PW}{R_S}$$

$$= \frac{500 \ \mu s}{500 \ \Omega} = 1 \ \mu F$$

For 1% tilt on the output,

$$\Delta V = 0.01 \ (2E)$$

From Equation (3-2),

$$R_1 = \frac{2Et}{0.01 \ (2E) \times C}$$

$$= \frac{t}{0.01 \ C}$$

and $t = \mathrm{PW} = 500\ \mu\mathrm{s}$,

$$R_1 = \frac{500\ \mu\mathrm{s}}{0.01 \times 1\ \mu\mathrm{F}} = 50\ \mathrm{k}\Omega \quad (\text{use } 47\ \mathrm{k}\Omega \text{ standard value})$$

3-4.2 Biased Clamping Circuits

In the biased clamping circuit shown in Figure 3-11(a), the cathode of diode D_1 is connected to a 2-V bias level. When the input becomes positive, the output level is clamped to the bias level plus the diode voltage drop, that is, $2\ \mathrm{V} + V_F$. At this time the voltage on the RHS of C_1 is $2\ \mathrm{V} + V_F$, and on the LHS is E. Therefore, C_1 charges to $E - (2\ \mathrm{V} + V_F)$, positive on the LHS and negative on the RHS. When the input goes to

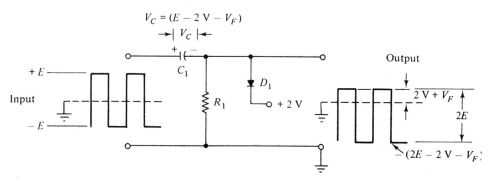

(a) Circuit to clamp output at approximately + 2 V maximum

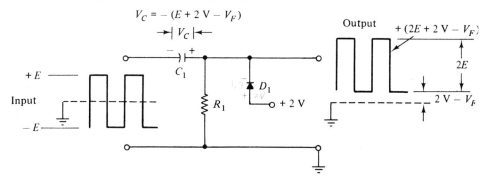

(b) Circuit to clamp output at approximately + 2 V minimum

FIGURE 3-11. Biased clamping circuits.

$-E$, the output becomes $-E$ plus the capacitor voltage. That is,

$$\text{Negative output} = -\left[E + (E - 2\,\text{V} - V_F)\right]$$
$$= -(2E - 2\,\text{V} - V_F)$$

and the peak-to-peak output is equal to (positive peak) $-$ (negative peak), or

$$p\text{-to-}p \text{ output} = (2\,\text{V} + V_F) - \left[-(2E - 2\,\text{V} - V_F)\right]$$
$$= 2E$$

It is seen that, although the output is clamped to a maximum dc level of $2\,\text{V} + V_F$, the wave shape and amplitude is unchanged. As before, the function of R_1 is to partially discharge C_1 over several cycles of the input. The clamper circuit on Figure 3-11(b) is similar to that in Figure 3-11(a) except that the diode is inverted. Since the capacitor charges with a different polarity, C_1 also is inverted from its condition in Figure 3-11(a). The anode is the diode is always at the bias voltage level, which is 2 V in this case. When the diode is forward-biased, its cathode voltage is $2\,\text{V} - V_F$, and cathode cannot go below this level. Therefore, the lowest level of output voltage is $2\,\text{V} - V_F$.

The diode is forward-biased during the time that the input voltage is $-E$. Capacitor C_1 charges, during this time, to $-E$ on the LHS, and to $2\,\text{V} - V_F$ on the RHS.

$$\therefore \quad V_c = \text{Capacitor voltage} = (-E) - (2\,\text{V} - V_F)$$
$$= -(E + 2\,\text{V} - V_F)$$

This is positive on the RHS and negative on the LHS. When the input becomes $+E$, the capacitor voltage and the input have the same polarities and add together to give

$$\text{Output} = E + E + 2\,\text{V} - V_F$$
$$= 2E + 2\,\text{V} - V_F$$

The peak-to-peak value of the output waveform is again $2E$, and its lower level is clamped to $2\,\text{V} - V_F$.

The Zener diode clamping circuits in Figure 3-12 perform the same function as biased clamping circuits. In Figure 3-12(a), the Zener diode behaves like a bias source with a voltage of V_Z. When it is thought of in this way, its operation is seen to be exactly the same as the biased clamper in Figure 3-11(a). The Zener diode circuit in Figure 3-12(b) clamps the negative output at $-(V_Z + V_F)$. The capacitor charge then causes the

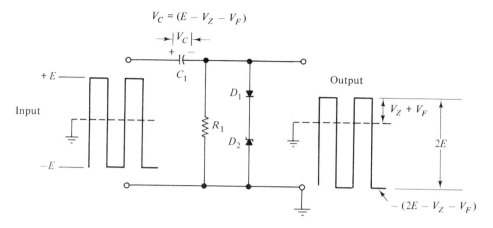

(a) Circuit to clamp output at approximately $+V_Z$ maximum

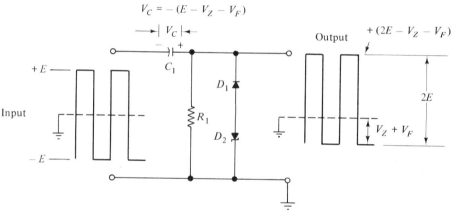

(b) Circuit to clamp output at approximately $-V_Z$ minimum

FIGURE 3-12.   Zener diode clamping circuits.

positive output to be $2E - V_Z - V_F$. As always, the peak-to-peak output voltage is $2E$.

EXAMPLE 3-8

Design the biased positive voltage clamper circuit shown in Figure 3-11(b). The input waveform is ± 20 V with a frequency of 2 kHz, and the tilt on the output is not to exceed 2%. The signal source resistance is 600 Ω.

solution

For the input waveform, $T = 1/f = 1/2$ kHz $= 0.5$ ms.

$$PW = \tfrac{1}{2} T = 250\ \mu s$$

From Equation (3-1),

$$C = \frac{PW}{R_S} = \frac{250\ \mu s}{600\ \Omega} \simeq 0.42\ \mu F \quad \begin{bmatrix} \text{use a 0.5 } \mu F \text{ standard value capacitor} \\ \text{(see Appendix 2)} \end{bmatrix}$$

For 2% tilt, $\Delta V = 0.02 \times 2E$.
  From Equation (3-2),

$$\begin{aligned}
R_1 &= \frac{2E \times PW}{0.02 \times 2E \times C} \\
&= \frac{2E \times 250\ \mu s}{0.02 \times 2E \times 0.5\ \mu F} \\
&= 25\ k\Omega \quad \text{(use 27 } k\Omega \text{ standard value)}
\end{aligned}$$

The bias voltage should be $+2$ V as shown in the figure, or any other desired level. The capacitor voltage should be rated at $V_i + V_B$; i.e., for the circuit designed the capacitor selected should survive at least 22 V.

EXAMPLE 3-9

A square wave having an amplitude of ± 15 V and a source resistance R_s of 1 kΩ is to be clamped to a maximum positive level of approximately 9 V. The square wave frequency ranges from 500 Hz to 5 kHz, and the output tilt is not to exceed 1%. Design a suitable Zener diode clamping circuit.

solution

Maximum tilt occurs when the PW is longest, i.e., when f is a minimum.

$$\text{Maximum } T = \frac{1}{f_{min}} = \frac{1}{500\ Hz} = 2\ ms$$

$$PW = \tfrac{1}{2} T = 1\ ms$$

From Equations (3-1),

$$C = \frac{PW}{R_s} = \frac{1 \text{ ms}}{1 \text{ k}\Omega} = 1 \ \mu F \ \left[\text{standard value (see Appendix 2)}\right]$$

For 1% tilt, $V = 0.01 \times 2E$.
From Equation (3-2),

$$
\begin{aligned}
R_1 &= \frac{2E \times PW}{0.01 \times 2E \times C} \\
&= \frac{2E \times 1 \text{ ms}}{0.01 \times 2E \times 1 \ \mu F} = 100 \text{ k}\Omega \ \left[\text{standard value (see Appendix 2)}\right] \\
V_O &= V_z + V_F = 9 \text{ V} \\
V_Z &= V_O - V_F \\
&= 9 \text{ V} - 0.7 \text{ V} = 8.3 \text{ V}
\end{aligned}
$$

From the regular diode data sheet (Appendix 1-3), the 1N756 has $V_Z = 8.2$ V; therefore, use a 1N756 Zener diode and a low current diode such as a 1N914. The capacitor voltage should be at least $V_i + V_Z$; i.e., minimum capacitor voltage is 23.2 V for this circuit.

3-4.3   Loading Effects on Clamping Circuits

The discussion of loading effects on differentiating and integrating circuits is also relevant to clamping circuits. A known load resistance may be used directly as the capacitor discharge resistance (R_1) if its value is neither too large nor too small. A very large value of load resistance can be ignored if R_1 is much smaller. A very small value of load resistance requires an emitter follower or voltage follower to minimize (or *buffer*) its effect on the clamping circuit.

In the situation where a clamping circuit feeds directly into the base of a transistor, the transistor might be used in place of the diode when the clamped level is to be 0.7 V above or below ground level. When this is not suitable, a resistor R_B must be used (as explained in Section 2-6) to minimize the loading effect and permit the clamper circuit to function correctly. The current flow through the clamper resistance R_1 must be much larger than that into the transistor base.

3-5 VOLTAGE MULTIPLYING CIRCUITS

The *voltage doubling circuit* is simply a two-stage clamper circuit connected as illustrated in Figure 3-13(a). During the positive half-cycle of the input, capacitor C_1 charges to approximately the peak level (E volts) with the polarity illustrated. When the input goes negative, a voltage with a peak value of approximately $-2E$ volts appears at the output of the first stage (i.e., across D_1), as already explained in Section 3-4.1. During the time that $-2E$ is present across D_1, diode D_2 is forward-biased and capacitor C_2 is charged to $2E$ with the polarity illustrated. The dc (output) voltage from the terminals of C_2 is now double the peak value of the input voltage to the circuit.

If further diode-capacitor clamper stages are cascaded, as illustrated in Figure 3-13(b), each capacitor is charged to $2E$ volts. The voltage across C_2 and C_4 is now $4E$ volts. When a large number of stages are employed, high dc voltage levels can be obtained from very low-level supplies.

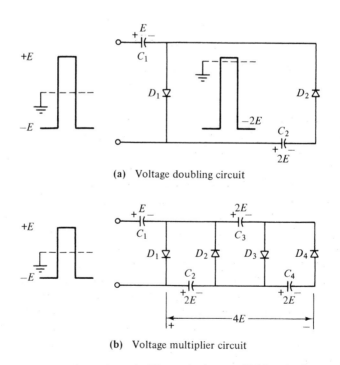

(a) Voltage doubling circuit

(b) Voltage multiplier circuit

FIGURE 3-13. Voltage doubling and voltage multiplying circuits.

REVIEW QUESTIONS AND PROBLEMS

3-1 Sketch typical characteristics for a low-current silicon diode. Briefly explain why the diode can be thought of as a one-way device.

3-2 Sketch ideal diode characteristics, and approximate characteristics for silicon and germanium diodes. Briefly discuss the parameters that should be considered when selecting a diode.

3-3 Explain the origin of *reverse recovery time* for a semiconductor diode. By means of sketches, explain why a large reverse current flows when a very fast reverse bias is applied to a diode. Also show how the reverse current can be minimized.

3-4 Sketch typical characteristics for a Zener diode. Indicate all important quantities related to the characteristics, and define each quantity.

3-5 Sketch the circuit of a positive series clipper, showing the input and output waveforms. Briefly explain its operation.

3-6 Repeat Problem 3-5 for a negative series clipper.

3-7 A negative series clipping circuit employs a diode with $V_F = 0.3$ V and $I_S = 10 \, \mu$A. The input voltage is ± 9 V, and the output current is to be a maximum of 10 mA. Calculate the value of the resistance R_1. Specify the diode in terms of forward current, power dissipation, and peak reverse voltage. The negative output voltage is to be maximum at 0.2 V.

3-8 Design a circuit to clip the positive peaks off a ± 20 V square wave. A silicon diode is available with a maximum reverse leakage current of 10 μA. The positive output voltage is not to exceed 0.5 V. Calculate the amplitude of the negative output peak.

3-9 From the diode data sheets in Appendix 1 select a suitable device for the circuit designed in Problem 3-8.

3-10 Sketch the circuit of a diode noise clipper, showing typical input and output waveforms. Briefly explain how the circuit operates.

3-11 Repeat Problem 3-10 for a Zener diode noise clipper.

3-12 A Zener diode noise clipper has an input pulse signal with an amplitude of ± 7 V and with noise amplitude of ± 3 V. Design a circuit and select suitable Zener diodes and resistance value. Also calculate the amplitude of the output signal.

3-13 A *pnp* transistor which can take a maximum of 5 V in reverse at its base-emitter junction is to be protected from excessive input signal amplitude. Identify the required circuit and sketch the input and output waveforms. Briefly explain the operation of the circuit.

3-14 Repeat Problem 3-13 for an *npn* transistor.

3-15 A negative shunt clipper circuit has a square wave input of ± 15 V. The output voltage is to be 13 V and − 0.7 V, and the output current is to be 250 μA. Calculate the required resistance value, and the diode forward current.

3-16 Sketch the circuit of a biased diode shunt clipper that has an output limited to a maximum of approximately ± 4 V. Explain the operation of the circuit.

3-17 The input to the circuit of Problem 3-16 is ± 16 V, and the output current is to be 500 μA. Determine the required resistance value, allowing the diode forward currents to be 10 mA.

3-18 Sketch a Zener diode shunt clipper circuit, and select suitable diodes which will clip off input peaks greater than approximately 6 V. Explain the operation of the circuit.

3-19 A ± 14 V square wave is applied to the circuit of Problem 3-18. The output current is to be 2 mA maximum. Design a suitable circuit.

3-20 Define the difference between clipping and clamping circuits. A ± 10 V square wave is applied to the input terminals of a negative voltage clamping circuit, and to the input of a negative shunt clipper. Sketch the output waveform that will result in each case.

3-21 Sketch a negative voltage clamping circuit, showing input and output waveforms. Briefly explain the operation of the circuit.

3-22 Repeat Problem 3-21 for a positive voltage clamper.

3-23 A negative voltage clamper has a 5 kHz square wave input with an amplitude of ± 6 V. The signal source resistance is 1 kΩ, and the tilt on the output waveform is not to exceed 1%. Design a suitable circuit.

3-24 Sketch the output waveforms you would expect from each of the circuits shown in Figure 3-14. Assume that the input to each circuit is a ± 12 V square wave.

3-25 Sketch the output waveforms you would expect from each of the circuits shown in Figure 3-15. Assume that the input to each circuit is a ± 9 V square wave.

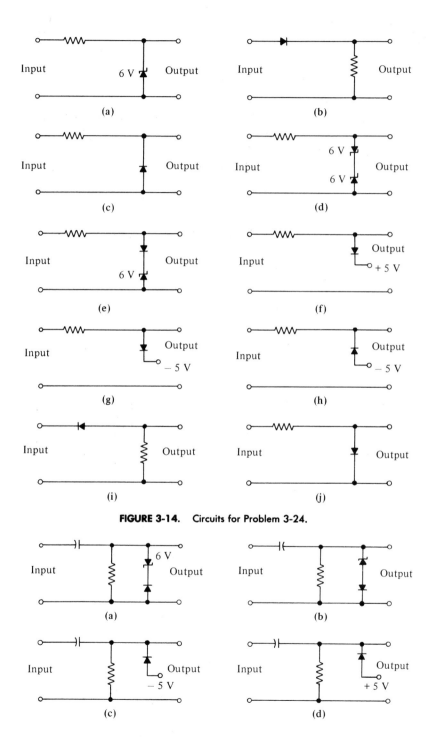

FIGURE 3-14. Circuits for Problem 3-24.

FIGURE 3-15. Circuits for Problem 3-25.

3-26 Design a biased clamper circuit to clamp a ± 12 V square wave to a minimum level of + 3 V. The input waveform has a frequency which ranges from 1 kHz to 10 kHz, and the signal source resistance is 500 Ω. The tilt on the output is not to exceed 1%.

3-27 A square wave having an amplitude of ± 18 V and a source resistance of 700 Ω is to be clamped to a maximum positive level of approximately 10 V. The square wave frequency is 800 Hz, and the output tilt is not to exceed 0.5%. Design a suitable Zener diode clamping circuit.

3-28 Sketch a diode-capacitor *voltage multiplier* circuit, and explain how it operates.

CHAPTER 4

Transistor Switching

INTRODUCTION

A bipolar transistor can be made to approximate an ideal switch. When the transistor is off, a small collector-base *leakage current* flows through the load. When it is on, there is a small collector-emitter *saturation voltage* across the device. A transistor will not switch *on or off* instantaneously. *Turn-on* and *turn-off times* depend upon the device and the circuit conditions. Field effect transistors have several advantages over bipolar transistor switches.

4-1 IDEAL TRANSISTOR SWITCH

Figure 4-1(a) shows a common emitter transistor circuit arranged to function as a switch. A load resistance R_L is connected from the transistor collector to the supply voltage V_{CC}. The emitter terminal of the device is grounded. For the transistor to simulate a switch, the terminals of the switch are the transistor collector and emitter. The input voltage, or controlling voltage, for the transistor switch is the base-emitter voltage

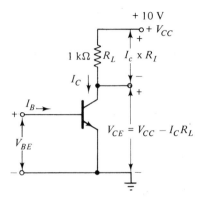

(a)   Common emitter
      circuit

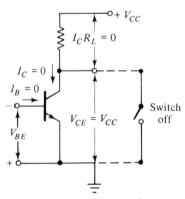

(b) Ideal transistor switch
     in *off* condition

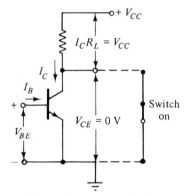

(c) Ideal transistor switch
     in *on* condition

FIGURE 4-1.   Ideal transistor switch.

V_{BE}. The collector-emitter voltage V_{CE} is equal to the supply voltage minus the voltage drop across R_L:

$$V_{CE} = V_{CC} - I_C R_L \qquad (4\text{-}1)$$

When the transistor base-emitter voltage is zero, or reverse-biased, as in Figure 4-1(b), the base current I_B is zero and the collector current I_C is also zero. The transistor switch is now in its *off* condition. Since there is no collector current, there can be no voltage drop across the load resistor. When $I_C = 0$, Equation (4-1) gives

$$V_{CE} = V_{CC} - (0 \times R_L)$$
$$= V_{CC}$$

Thus, when the ideal transistor is *off*, its collector-emitter voltage equals the supply voltage.

When the transistor base is made positive with respect to the emitter [Figure 4-1(c)], a base current I_B flows. The collector current I_C is equal to I_B multiplied by the transistor *common emitter dc current gain* h_{FE}; that is, $I_C = h_{FE} \times I_B$. If I_B is made large enough, $I_C \times R_L$ can become equal to the supply voltage V_{CC}. Then by Equation (4-1).

$$V_{CE} = V_{CC} - V_{CC}$$
$$V_{CE} = 0$$

When the ideal transistor is *on*, its collector-emitter voltage equals zero.

The transistor can also simulate a switch in that, *ideally*, it dissipates zero powder both when *on* and *off*. The only time power is dissipated in the device is when it is switching between *on* and *off*. Transistor power dissipation is given by

$$P_D = I_C \times V_{CE}$$

When *off*,

$$I_C = 0, \qquad P_D = 0 \times V_{CE} = 0$$

When *on*,

$$V_{CE} = 0, \qquad P_D = I_C \times 0 = 0$$

As described above, the transistor can be operated as a switch which is *off* when V_{BE} is zero or negative, and which is *on* when V_{BE} is positive. Ideally, $V_{CE} = V_{CC}$ when the transistor is *off*, and $V_{CE} = 0$ V when the device is *on*. With a practical transistor these conditions are not achieved; however, they can be approximated.

4-2 PRACTICAL TRANSISTOR SWITCH

To understand how a practical transistor switch differs from the ideal case, it is necessary to consider the common emitter characteristics. In Figure 4-2, the dc load line for the circuit of Figure 4-1(a) is drawn on the transistor common emitter characteristics. Using Equation (4-1), the procedure for drawing the load line is as follows: When $I_C = 0$, $V_{CE} = V_{CC} - 0$. For the circuit shown, $V_{CC} = 10$ V. Plot point A on the characteristics at $I_C = 0$, and $V_{CE} = 10$ V.

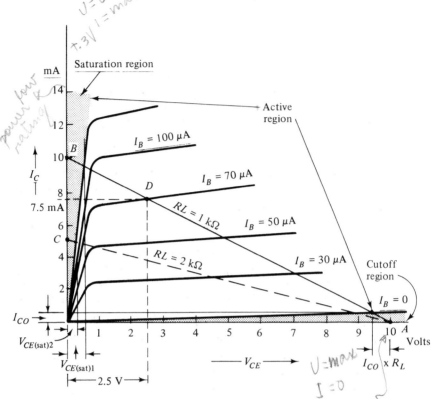

FIGURE 4-2. Characteristics and dc load line for transistor switch.

When $V_{CE}=0$,

$$0 = V_{CC} - I_C R_L$$

$$I_C = \frac{V_{CC}}{R_L}$$

$$= \frac{10 \text{ V}}{1 \text{ k}\Omega} = 10 \text{ mA}$$

[for the circuit of Figure 4-1(a)]

Plot point B on the characteristics at $V_{CE}=0$, $I_C=10$ mA. Draw the dc load line for $R_L=1$ kΩ by joining points A and B together.

The dc load line defines all corresponding current and voltage conditions that can exist in the circuit. For any given level of I_C, a particular V_{CE} is dictated by Equation (4-1) and is illustrated by the load line.

The common emitter characteristics are divided into three regions, as shown in Figure 4-2. The *active region* of the characteristics usually is employed only in amplifier circuits. Here a linear change in base current produces a nearly linear collector-emitter voltage change. When the collector current is so large that V_{CE} is less than approximately 0.7 V, the device is said to be operating in the *saturation region* of the characteristics. The *cutoff region* exists below the level of $I_B = 0$.

Again, with reference to the load line, it is seen that when $I_B = 0$, I_C is *not* zero. Instead, a small current I_{CO} flows. This is the collector-base *reverse saturation current*, or *collector cutoff current*, sometimes designated I_{CBO}. This current is the sum of the minority charge carriers which are crossing the reverse-biased collector-base junction, and leaking along the junction surface. I_{CO} is a very temperature-sensitive quantity. For the most recent transistors, I_{CO} at 25° C is in the nA range, however, at high temperatures this may go into the microamps range.

Refer to the data sheet for 2N3903 and 2N3904 transistors in Appendix 1. The collector cutoff current is designated by I_{CEX}. This is the collector-base leakage current measured under particular conditions specified by the manufacturer. I_{CEX} can be regarded as essentially equal to I_{CO}. From the data sheet, the collector cutoff current for 2N3903 and 2N3904 transistors is 50 nA. The presence of I_{CO} makes V_{CE} slightly less than V_{CC} when the transistor is cut off [see Figure 4-3(a)].

$$V_{CE} = V_{CC} - I_{CO} R_L \qquad (4\text{-}2)$$

For $V_{CC} = 10$ V, $R_L = 1$ kΩ, and $I_{CO} = 1$ μA:

$$V_{CE} = 10 \text{ V} - (1 \ \mu\text{A} \times 1 \text{ k}\Omega)$$
$$= 9.999 \text{ V}$$
$$\simeq V_{CC}$$

When the transistor is in saturation, a small collector-emitter saturation voltage, $V_{CE(\text{sat})}$, exists. Typically about 0.2 V, $V_{CE(\text{sat})}$ largely depends upon I_C and the resistance of the semiconductor material that forms the transistor collector. The load line for $R_L = 2$ kΩ (broken line in Figure 4-2) reveals that when saturation occurs with smaller levels of I_C, then $V_{CE(\text{sat})}$ is reduced. The 2N3903 and 2N3904 data sheet in Appendix 1 specifies $V_{CE(\text{sat})}$ as 0.3 V at $I_C = 50$ mA and 0.2 V at $I_C = 10$ mA.

The saturated transistor circuit in Figure 4-3(b) has typical voltages of $V_{BE} = 0.7$ V and $V_{CE} = 0.2$ V. Thus the base terminal is 0.5 V positive with respect to the collector terminal, and the normally reverse-biased collector-base junction is actually forward-biased. As will be seen later, this forward bias at the collector-base junction limits the switching speed of the transistor.

The forward bias at the collector-base junction when a transistor is saturated reduces the dc current again (h_{FE}). This happens because, to draw the maximum number of charge carriers from emitter to collector, the collector-base junction must be reverse-biased. For saturation to occur, the transistor current gain must have a minimum value, $h_{FE(min)}$, depending upon the circuit conditions. Suppose a transistor has a base current of $I_B = 50$ μA and requires a collector current I_C of 1 mA for saturation. Then, $h_{FE(min)} = I_C/I_B = 1$ mA$/50$ μA$= 20$. If h_{FE} is less than 20 in this case, I_C will be less than 1 mA and saturation will *not* occur. If h_{FE} is greater than 20, I_C will tend to be greater than the required 1 mA, and saturation will occur.

EXAMPLE 4-1

For the circuit of Figure 4-1(a), $I_B = 0.2$ mA. (a) Determine the value of $h_{FE(min)}$ for saturation to occur. (b) If R_L in Figure 4-1(a) is changed to 220 Ω and a 2N3904 transistor is employed, will the transistor be saturated?

solution (a)
For saturation:

$$I_C \simeq \frac{V_{CC}}{R_L}$$

$$= \frac{10 \text{ V}}{1 \text{ k}\Omega} = 10 \text{ mA}$$

$$h_{FE(min)} = \frac{I_C}{I_B}$$

$$= \frac{10 \text{ mA}}{0.2 \text{ mA}} = 50$$

solution (b)
For saturation:

$$I_C \simeq \frac{V_{CC}}{R_L}$$

$$= \frac{10 \text{ V}}{220 \text{ }\Omega} \simeq 45 \text{ mA}$$

$$h_{FE(min)} = \frac{I_C}{I_B}$$

$$= \frac{45 \text{ mA}}{0.2 \text{ mA}}$$

$$= 225$$

From the 2N3904 data sheet in Appendix 1, at $I_C = 50$ mA, $h_{FE(min)} = 60$. Therefore, the transistor will *not* be saturated.

Suppose a 2N3904 transistor is employed in the case of Example 4-1(a). From the 2N3904 data sheet in Appendix 1, at $I_C = 10$ mA, $h_{FE(min)} = 100$ and $h_{FE(max)} = 300$. This suggests that for $I_B = 0.2$ mA, I_C could be any value between 100×0.2 mA and 300×0.2 mA, that is, from 20 mA to 60 mA. In fact, the maximum collector current that can flow is $I_C = V_{CC}/R_L = 10$ mA, as calculated in the example. Thus, with an h_{FE} value greater than 50, more base current flows than is needed to drive the transistor into saturation. The extra base current flows out through the emitter terminal, and in this situation the transistor is said to be *overdriven*.

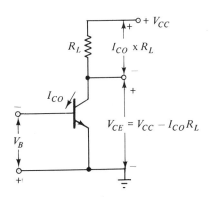

(a) Transistor in cutoff

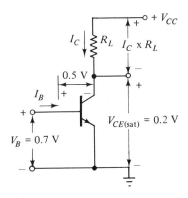

(b) Transistor in saturation

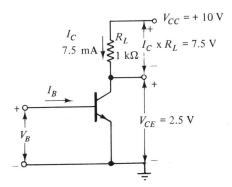

(c) Transistor in active region of characteristics

FIGURE 4-3. Transistors operated in cutoff, saturation, and active region of their characteristics.

Although transistors in switching circuits usually are switched from cutoff to saturation, and *vice versa*, they can also be switched between cutoff and the active region. For example, if the base current is limited to 70 μA for the load line shown in Figure 4-2 (point D), then V_{CE} is 2.5 V. In this case the transistor is referred to as a *nonsaturated switch* [Figure 4-3(c)].

The power dissipation is very small with a practical transistor in saturation or cutoff. For a nonsaturated transistor switch, the power dissipation is much larger than for either the cutoff or saturated cases.

EXAMPLE 4-2

If the circuit of Figure 4-1(a) employs a 2N3904 transistor, calculate the transistor power dissipation (a) at cut-off, (b) at saturation, and (c) when $V_{CE} = 2$ V.

solution (a)
For cutoff:
    From the 2N3904 data sheet, $I_C \simeq I_{CEX} = 50$ nA,

$$P_D \simeq I_C \times V_{CC}$$
$$= 50 \text{ nA} \times 10 \text{ V}$$
$$= 0.5 \ \mu\text{W}$$

solution (b)
For saturation,

$$I_C \simeq \frac{V_{CC}}{R_L}$$
$$= \frac{10 \text{ V}}{1 \text{ k}\Omega} = 10 \text{ mA}$$

from the 2N3904 data sheet; at $I_C = 10$ mA, $V_{CE(\text{sat})} = 0.2$ V

$$P_D = I_C \times V_{CE(\text{sat})}$$
$$= 10 \text{ mA} \times 0.2 \text{ V}$$
$$= 2 \text{ mW}$$

solution (c)

At $V_{CE} = 2$ V:

From Equation (4-1),

$$V_{CE} = V_{CC} - I_C R_L$$

$$I_C = \frac{V_{CC} - V_{CE}}{R_L}$$

$$= \frac{10\text{ V} - 2\text{ V}}{1\text{ k}\Omega}$$

$$= 8\text{ mA}$$

$$P_D = I_C \times V_{CE}$$

$$= 8\text{ mA} \times 2\text{ V}$$

$$= 16\text{ mW}$$

4-3 TRANSISTOR SWITCHING TIMES

One most important characteristic of a switching transistor is the speed with which it can be switched *on* and *off*. Consider Figure 4-4, where the time relationship between collector current and base current is shown. When the input current I_B is applied, the transistor does not switch *on* immediately. The time between application of base current and commencement of collector current is termed the *delay time* t_d (see Figure 4-4). The delay time is defined as the time required for I_C to reach 10% of its final level, after I_B has commenced. Even when the transistor begins to switch *on*, a finite time elapses before I_C reaches its maximum level. The *rise time* t_r is defined as the time it takes for I_C to go from 10% to 90% of its maximum level. The *turn-on time* (t_{on}) for the transistor is the sum of t_d and t_r (see Figure 4-4).

Similarly, a transistor cannot be switched *off* instantaneously. The *turn-off time* t_{off} is composed of a *storage time* t_s and a *fall time* t_f (Figure 4-4). The storage time results from the fact that the collector-base junction is forward-biased when the transistor is in saturation. Charge carriers crossing a forward-biased junction are trapped (or stored) in the depletion region when the junction is reversed. These charge carriers must be withdrawn or made to recombine with charge carriers of an opposite type before the collector current begins to fall. The storage time t_s is defined as

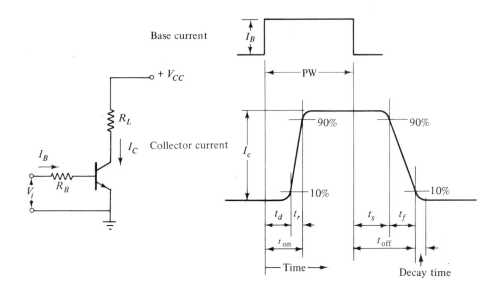

FIGURE 4-4. Time relationships between base current and collector current in a transistor switching circuit.

the time between I_B switch *off* and I_C falling to 90% of its maximum level. The fall time t_f is the time required for I_C to fall from 90% to 10% of its maximum. A further quantity, *the decay time* is sometimes included in the turn-off time. This is the time required for I_C to go from its 10% level to I_{CO}. Usually, this is not an important quantity since the transistor is regarded as being *off* when I_C falls to the 10% level.

Refer to the data sheet for the 2N3904 transistor in Appendix 1-4. The turn-on and turn-off times given are

$$\text{Turn-on time} = t_d + t_r = 35 \text{ ns} + 35 \text{ ns} = 70 \text{ ns}$$
$$\text{Turn-off time} = t_s + t_f = 200 \text{ ns} + 50 \text{ ns} = 250 \text{ ns}$$

In the case of a nonsaturated transistor switch, the collector-base voltage is reverse-biased when the transistor is *on*. Therefore, no storage time is involved, and the turn-off time is not much larger than the fall time. This faster turn-off time is the major advantage of the nonsaturated switch.

Figure 4-5 shows the time relationship of the input and output voltages as well as the I_B and I_C waveforms for the circuit in Figure 4-4. I_B commences almost immediately when V_i is applied. The approximate level of I_B is input voltage V_i divided by base resistor R_B, that is, V_i/R_B (assuming V_{BE} is initially zero). The output voltage at any instant depends upon the instantaneous level of I_C. Thus, V_{CE} is initially $(V_{CC}-I_{CO}R_L)$ and falls to 90% of V_{CC} when I_C becomes 10% of $I_{C(max)}$ after t_d (see Figure 4-5). When I_C is 90% of $I_{C(max)}$, V_{CE} is 10% of V_{CC} and finally falls to $V_{CE(sat)}$ when I_C reaches its maximum level. When I_B goes to zero, the storage time elapses before I_C commences to fall. Then V_{CE} again becomes approximately 0.1 V_{CC} when I_C is 90% of its maximum level, and V_{CE} becomes 0.9 V_{CC} when I_C is 10% of maximum. Finally, V_{CE} returns to $V_{CC}-I_{CO}R_L$ when I_C falls to the level of the reverse saturation current.

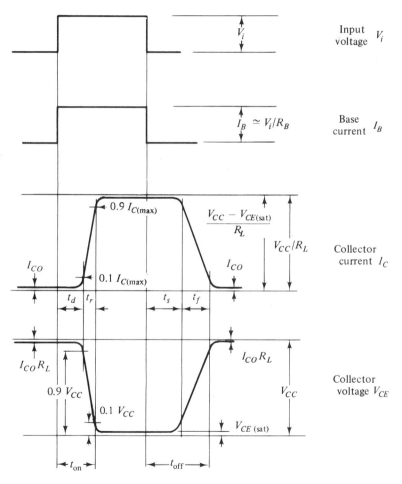

FIGURE 4-5.   Time relationships between voltages and currents in a transistor switching circuit.

EXAMPLE 4-3 _____

The circuit in Figure 4-4 has $V_{CC} = 12$ V and $R_L = 3.3$ kΩ. The transistor employed is a 2N3904, and the input voltage has a PW of 5 μs. Calculate the level of V_{CE} (a) before the input pulse is applied, (b) at the end of the delay time, (c) at the end of the turn-on time. Also determine the time from commencement of the input pulse until the transistor switches *off*.

solution

In the 2N3904 data sheet (Appendix 1-4) the collector cutoff current is defined as $I_{CEX} = 50$ nA. Before the transistor switches *on*,

$$V_{CE} = V_{CC} - I_{CEX} R_L$$
$$= 12 \text{ V} - (50 \text{ nA} \times 3.3 \text{ k}\Omega)$$
$$= 11.9998 \text{ V}$$

At the end of the delay time,

$$V_{CE} = V_{CC} - (0.1 I_{C(\max)} R_L)$$
$$= V_{CC} - \left(0.1 \times \frac{V_{CC}}{R_L} \times R_L\right)$$
$$= 12 \text{ V} - (0.1 \times 12 \text{ V})$$
$$= 10.8 \text{ V}$$

At the end of the turn-on time,

$$V_{CE} = V_{CC} - \left(0.9 \times \frac{V_{CC}}{R_L} \times R_L\right)$$
$$= 12 \text{ V} - (0.9 \times 12 \text{ V})$$
$$= 1.2 \text{ V}$$

For the 2N3904, $t_{\text{off}} = 250$ ns. Time from commencement of input to transistor switching off is $\text{PW} + t_{\text{off}}$.

$$\text{PW} + t_{\text{off}} = 5 \text{ μs} + 250 \text{ ns}$$
$$= 5.25 \text{ μs}$$

4-4 IMPROVING THE SWITCHING TIMES

If the base-emitter of the transistor is reverse-biased before switch-on, the delay time is longer than in the case when V_{BE} is initially zero. This is because the transistor input capacitance is charged to the reverse bias voltage, and must be discharged before V_{BE} can become positive. Therefore, to minimize the turn-on time, V_{BE} should be zero or have a very small reverse bias before switch-on. Both the delay time and the rise time can be reduced if the transistor is *overdriven*, i.e., if I_B is made larger than the minimum required for saturation. With a larger I_B, the junction capacitances are charged faster, thus reducing the turn-on time.

A major disadvantage of overdriving is that the storage time is extended, by the larger current flow across the forward-biased collector-base junction, when the transistor is in saturation. Therefore, although an overdriven transistor will turn on faster, it has a longer turn-off time than a transistor which has just enough base current for saturation. One way to shorten the turn-off time is to provide a large negative input voltage at switch-off. This produces a reverse base current flow which causes the junction capacitance to discharge rapidly. Again, this has a disadvantage in that the turn-on time is increased because of the initial large reverse bias of the base-emitter junction.

Ideally, for fast switching V_{BE} should start at zero volts, and I_B should initially be large at switch-on but should rapidly settle down to the minimum required for saturation. Also, switch-off should be accomplished by a large reverse bias voltage which quickly returns to zero. Exactly these conditions are achieved when a capacitor is connected in parallel with R_B, as shown in Figure 4-6. This capacitor, termed a *speed-up capacitor*, is initially uncharged before the input voltage pulse is applied. When the input voltage rises, the capacitor commences to charge to $(V_i - V_{BE})$ [Figure 4-6(a)]. The capacitor charging current flows into the transistor base terminal. Thus I_B is initially large, but quickly settles down to its minimum dc level as the capacitor becomes charged. At switch-off [Figure 4-6(b)], the capacitor discharge produces a reverse base current which rapidly returns to zero.

The speed-up capacitor tends to reduce t_d and t_s as well as t_r and t_f. However, if C_1 is so small that it becomes completely charged within the delay time, then it will not have a significant effect upon the rise time. Similarly, if C_1 is completely discharged during the storage time, it will not produce a marked improvement in the fall time.

Consider the circuit of Figure 4-7. The settled base current level (*i.e.*, after the capacitor is completely charged) can be calculated by using V_i, R_B, and R_s.

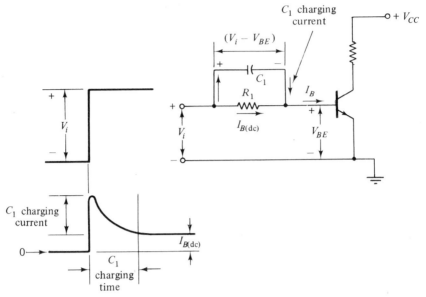

(a) Effect of C_1 charge at switch-on

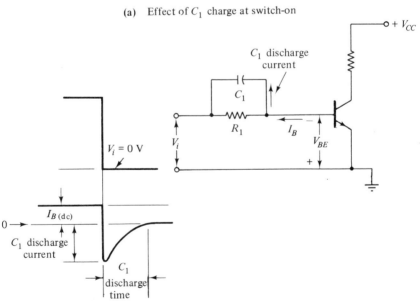

(b) Effect of C_1 discharge at switch-off

FIGURE 4-6. Effect of C_1 charge and discharge when the transistor is switched *on* and *off*.

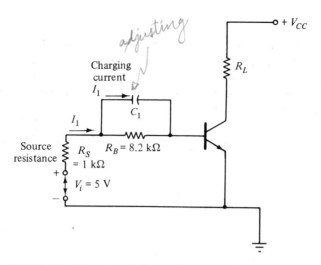

FIGURE 4-7.   Circuit for calculation of initial charging current level.

$$I_B = \frac{V_i - V_{BE}}{R_s + R_B}$$

$$= \frac{5\text{ V} - 0.7\text{ V}}{1\text{ k}\Omega + 8.2\text{ k}\Omega} \simeq 0.5\text{ mA}$$

The initial level of capacitor charging current is approximately the signal voltage divided by the signal source resistance.

$$I_1 \simeq \frac{V_i - V_{BE}}{R_s}$$

$$= \frac{5\text{ V} - 0.7\text{ V}}{1\text{ k}\Omega} = 4.3\text{ mA}$$

This is considerably greater than the dc level of I_B. Therefore, an improvement in switching speed may be expected. For the best possible improvement in switching speed, a speed-up capacitor should be selected that is large enough to maintain the charging current (*i.e.*, base current) nearly constant at its maximum level during the transistor turn-on time. The charging current will drop by only 10% from its maximum level, if the capacitor is allowed to charge by 10% during the turn-on time. C_1 charges by 10% during a time of $0.1\,C_1R_s$ (Chapter 2). Therefore,

$$t_{on} = 0.1\,C_1R_s$$

and

$$C_1 = \frac{t_{on}}{0.1 R_s} \qquad (4\text{-}3)$$

For $t_{on} = 300$ ns, and $R_s = 1$ kΩ:

$$C_1 = \frac{300 \text{ ns}}{0.1 \times 1 \text{ k}\Omega} = 3000 \text{ pF}$$

A larger capacitor than this is not likely to offer any greater improvement in switching time. Also, if a ten times improvement in switching time is achieved, then a capacitor of 300 pF might be almost as effective as one with a value of 3000 pF. This is because t_{on} in the above calculation would be reduced from 300 ns to 30 ns and, consequently, C_1 is calculated as 300 pF. To achieve such an improvement in switching time, however, the transistor must initially operate well below its maximum switching speed. Also, the input pulse must have a rise time very much less than the minimum switching time sought.

The upper limit to the value of C_1 that may be used depends upon the maximum signal frequency. When the transistor is switched *off*, C_1 discharges through R_B. For correct switching, C_1 should be at least 90% discharged during the time interval between transistor *switch-off* and *switch-on*. The time required for the capacitor to return to its discharged condition is variously referred to as the *settling time*, the *resolving time*, or the *recovery time* t_{re} of the circuit. In this case, the transistor is *off* and the capacitor is discharged through R_B. C_1 discharges by 90% in a time $t = 2.3 C_1 R_B$ (Chapter 2).

$$t_{re} = 2.3 \, C_1 R_B$$

or

$$\text{maximum } C_1 = \frac{t_{re}}{2.3 R_B} \qquad (4\text{-}4)$$

EXAMPLE 4-4 _____

The circuit of Figure 4-7 is to have a 50 kHz input square wave. Calculate the maximum value of the speed-up capacitor that may be used.

solution

$$T = \frac{1}{f} = \frac{1}{50 \text{ kHz}} = 20 \text{ } \mu s$$

$$t_{re} \text{ between } \textit{switch-off} \text{ and } \textit{switch-on} = \frac{T}{2} = 10 \text{ } \mu s$$

$$C_{1(\max)} = \frac{t_{re}}{2.3 R_B} = \frac{10 \text{ } \mu s}{2.3 \times 8.2 \text{ k}\Omega}$$

$$= 530 \text{ pF}$$

EXAMPLE 4-5 _____

Determine the maximum input frequency for the circuit of Figure 4-7 when $C_1 = 200$ pF.

solution

$$t_{re} = 2.3 C_1 R_B$$

$$= 2.3 \times 200 \text{ pF} \times 8.2 \text{ k}\Omega$$

$$= 3.772 \text{ } \mu s$$

$$T = 2 t_{re} = 7.544 \text{ } \mu s$$

$$f = \frac{1}{T} = \frac{1}{7.544 \text{ } \mu s} \simeq 133 \text{ kHz}$$

4-5 JUNCTION FIELD EFFECT TRANSISTOR (JFET) SWITCH

An *n*-channel *junction field effect transistor* (JFET), connected in common source configuration, is shown in Figure 4-8(a). The output voltage from the circuit equals the supply voltage minus the voltage drop across R_L.

$$V_0 = V_{DD} - I_D R_L \tag{4-5}$$

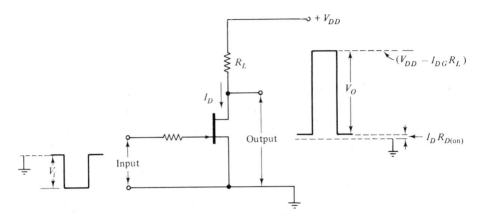

(a)   Common source circuit as switch

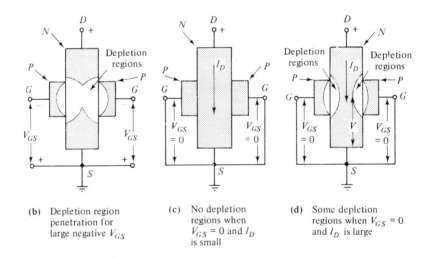

(b)   Depletion region
penetration for
large negative V_{GS}

(c)   No depletion
regions when
$V_{GS} = 0$ and I_D
is small

(d)   Some depletion
regions when $V_{GS} = 0$
and I_D is large

FIGURE 4-8.   JFET switching circuit and effects of V_{GS} and I_D.

Ideally, $V_O = V_{DD}$ when the device is *off*, and $V_O = 0$ when the transistor is *on*. The input signal biases the transistor *off* when V_i is negative. The effect is illustrated in Figure 4-8(b) where the *depletion regions* resulting from the negative bias on the gate are shown to penetrate so deeply into the n-type channel that they meet at the center. The channel is interrupted by the depletion regions, so that the *drain current* I_D cannot flow. When the input signal is at ground level, the *gate* potential equals that at the source terminal. In this case there are no depletion regions, so the drain current easily flows through the channels [Figure 4-8(c)]. If the drain current is small, there will be only a small voltage drop along the channel, due to the small *drain-source on resistance* $R_{D(on)}$.

Actually there could be depletion regions even when the gate is at source potential, if a substantial drain current flows. Figure 4-8(d) illustrates the situation. The drain current I_D causes sufficient voltage drop along the channel so that the source (and, consequently, the gate) becomes negative with respect to part of the channel. This produces depletion

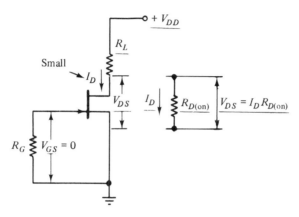

(a) *On*-biased JFET with small I_D

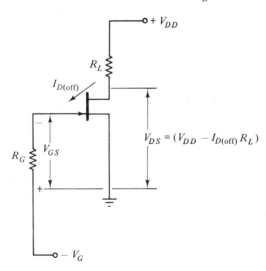

FIGURE 4-9. Drain-source voltage for *on-* and *off-*biased JFETs

regions which again penetrate into the channel. The narrowed channel has an increased drain-source resistance, and thus a relatively large drain-source *on* voltage.

When the JFET is biased *on* [Figure 4-9(a)], the output voltage is

$$V_{D(on)} = I_D \times R_{D(on)} \qquad (4\text{-}6)$$

Typically, $R_{D(on)}$ is 30 Ω or less. With a drain current of 100 μA, the typical level of $V_{D(on)}$ can be quite small:

$$V_{D(on)} = 100\ \mu A \times 30\ \Omega$$
$$= 3\ mV$$

A comparison of $V_{D(on)}$ with the typical $V_{CE(sat)}$ of 0.2 V for a bipolar transistor shows that $V_{D(on)} \ll V_{CE(sat)}$, and this is a major advantage of the JFET switch.

The *off*-biased JFET has a small *drain-gate leakage-current* $I_{D(off)}$ flowing across the reverse-biased gate-channel junctions. As illustrated in Figure 4-9(b), $I_{D(off)}$ causes a very small voltage drop across R_L.

Another advantage of the JFET switch over a bipolar transistor switch, is that the JFET has a much higher input resistance than a bipolar transistor. Thus, the JFET can be easily switched by signals that have large source resistances.

EXAMPLE 4-6

The circuit in Figure 4-8(a) uses a 2N4857 JFET and has $V_{DD} = 15$ V and $R_L = 3.9$ kΩ. Determine the level of the output voltage (a) when the device is cut off, and (b) when the transistor is switched *on*.

solution

From the 2N4857 data sheet in Appendix 1-8, the maximum drain current at cutoff is

$$I_{D(off)} = 0.25\ nA\ (\text{i.e., at } 25°\ C)$$
$$V_O = V_{DD} - I_{D(off)} R_L$$
$$= 15\ V - (0.25\ nA \times 3.9\ k\Omega)$$
$$= 15\ V - 1\ \mu V$$
$$\simeq 15\ V$$

In the data sheet, $R_{D(on)}$ is identified as

$$r_{ds(on)} = 40\ \Omega$$

and when the FET is *on*:

$$I_D \simeq \frac{V_{DD}}{R_L}$$

$$= \frac{15 \text{ V}}{3.9 \text{ k}\Omega} = 3.85 \text{ mA}$$

$$V_O = I_D r_{ds(on)}$$

$$= 3.85 \text{ mA} \times 40 \ \Omega$$

$$= 154 \text{ mV}$$

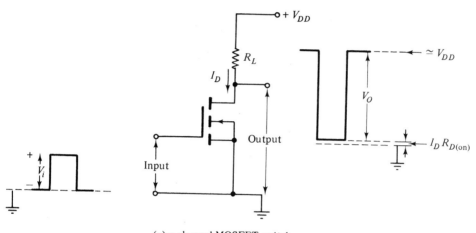

(a) *n*-channel MOSFET switch

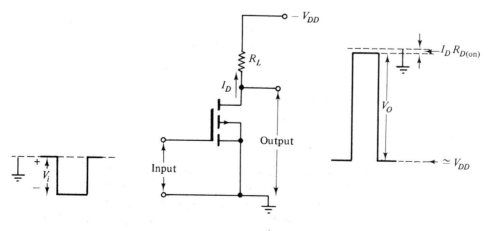

(b) *p*-channel MOSFET switch

FIGURE 4-10.   *n*-channel and *p*-channel MOSFET switches.

4-6 MOSFET SWITCH

N-channel and p-channel *metal oxide semiconductor field effect transistor* (*MOSFET*) switching circuits are shown in Figures 4-10(a) and (b) together with input and output waveforms. In the *enhancement* devices shown no channel exists while the gate is at the same potential as the source. Therefore, these transistors require no external bias voltage to switch them *off;* that is, they can be operated from a single polarity supply. For the n-channel MOSFET a positive input pulse is necessary for switch-on. When the input signal becomes positive, I_D flows, and the output voltage drops from V_{DD} to $I_D R_{D(on)}$. In the case of the p-channel device, the output is $-V_{DD}$ while no drain current is flowing. A negative signal on the gate terminal switches the transistor *on*, causing the output level to change to $-I_D R_{D(on)}$.

Since the gate terminal in a MOSFET is insulated from the channel, there is no drain-gate leakage current. This results in an input resistance even higher than that of a JFET circuit. There is a small drain-source leakage current, which causes some voltage drop along R_L when the MOSFET is *off*.

4-7 CMOS SWITCH

When two devices are identical in every way except for their supply voltage polarities, they are termed *complementary devices*. For example, if two bipolar transistors have identical parameters, but one is an *npn* device and the other is *pnp*, they are complementary. Similarly, two MOSFETS, one of which is p-channel and the other n-channel, can be complementary. When p-channel and n-channel MOSFETS are combined, the resulting circuitry is termed *complementary MOS*, or CMOS.

Figure 4-11 shows the circuit of a CMOS switch. Both devices are enhancement MOSFETS, so that no channel exists until one of them is switched *on*. When the input voltage is zero at the common gate terminal, the p-channel device Q_1 is biased *on* and the n-channel device Q_2 is *off*. In this condition there is only a very small voltage drop from the drain to source for the p-channel device, and the output voltage is very close to V_{DD}. When the input voltage becomes positive, the n-channel transistor is biased *on* and the p-channel device is *off*. There is now only a very small voltage drop along the n-channel device and, consequently, the output voltage is very close to ground.

The major advantage of CMOS circuits is that the power dissipation is extremely small compared to other circuits. The small power dissipation,

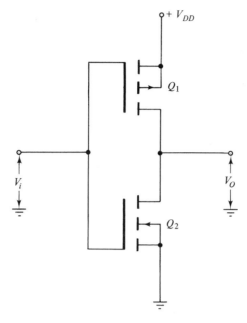

FIGURE 4-11. CMOS switch.

together with the small volt drop across the *on* transistor and the high input impedance, makes the CMOS inverter approach the ideal switch. CMOS is discussed further in Chapter 10.

REVIEW QUESTIONS AND PROBLEMS

4-1 Sketch a circuit to show a bipolar transistor employed as a switch. Compare the transistor to an ideal switch.

4-2 Define the following terms: saturated switch, nonsaturated switch, saturation voltage, collector-base leakage current, and $h_{FE(min)}$. Discuss the importance of the latter three items in relation to a transistor switch.

4-3 Sketch typical transistor common emitter characteristics. Identify the various regions of the characteristics and show how $V_{CE(sat)}$ differs with different transistor load resistances.

4-4 (a) A common emitter transistor circuit has $V_{CC} = 20$ V, $R_L = 2.2$ kΩ and $I_B = 0.3$ mA. Determine $h_{FE(min)}$ if the transistor is to be saturated. (b) If a 2N3903 transistor is used in the above circuit, calculate the minimum I_B level at which the transistor will become saturated.

4-5   A common emitter circuit, using a 2N3904 transistor, has $V_{CC} = 25$ V. The load resistance can be 22 kΩ or 2.2 kΩ. Calculate the minimum level of base current needed to achieve saturation in each case.

4-6   For the circuit described in Problem 4-5, calculate the transistor power dissipation for each load resistance, at both saturation and cutoff. Also calculate the transistor power dissipation for each load resistance, when the collector-emitter voltage is 3 V.

4-7   For a transistor switch, sketch the waveforms of the input voltage, base current, collector current, and collector voltage. Show the various components of transistor turn-on time and turn-off time, and discuss their origins.

4-8   A switching circuit using a 2N4418 transistor (data sheet in Appendix 1-7) has $V_{CC} = 15$ V and $R_L = 2.7$ kΩ. The input pulse width is 2 μs. Calculate the level of V_{CE} (a) before the pulse is applied, (b) at the end of the delay time, (c) at the end of the storage time. Also determine the times from commencement of the input pulse until the transistor switches *on* and until the transistor switches *off*.

4-9   Show how a *speed-up capacitor* may be employed to improve the turn-on and turn-off time of a transistor. Sketch the waveforms of base current with and without the speed-up capacitor. Explain how the capacitor improves switching speed.

4-10   The circuit described in Problem 4-5 has a base resistance of 27 kΩ. (a) If the circuit is to be switched at the maximum frequency of 100 kHz, calculate the maximum size of the speed-up capacitor that should be used. (b) Determine the maximum switching signal frequency when a 100 pF speed-up capacitor is employed.

4-11   Explain how a junction field effect transistor can be employed as a switch. Sketch a circuit which uses a JFET switch. Sketch the input and output waveforms and show how the JFET operates when *on* and when *off*. Compare the JFET switch to a bipolar transistor switch.

4-12   A 2N4856 JFET (data sheet in Appendix 1-8) is connected as a switch, with $V_{DD} = 20$ V and $R_L = 4.7$ kΩ. Determine the level of the output (drain-source) voltage (a) when the device is cut off and (b) when switched *on*.

4-13 Sketch *p*-channel and *n*-channel MOSFET switching circuits. Show input and output waveforms in each case, and discuss the advantages of MOSFET switches.

4-14 Define CMOS. Sketch the basic CMOS inverter circuit and explain how it operates. Describe the advantages and disadvantages of CMOS.

CHAPTER 5

The Inverter Circuit and IC Differentiator

INTRODUCTION

An *inverter circuit* as the name suggests, performs the function of inverting an input signal. When the input goes positive, the output goes negative, and vice versa. Usually, a small input signal can drive the inverter output from one extreme voltage level to the other extreme, but large input signals may also be employed. Bipolar transistor inverters may have the input signal direct-coupled or capacitor-coupled. JFET inverters can also be constructed for direct- or capacitor-coupled signals. An IC operational amplifier without any additional components can be employed as an inverter.

5-1 DIRECT-COUPLED BIPOLAR TRANSISTOR INVERTER

A transistor *inverter circuit* is essentially an overdriven common emitter circuit. The input may be a square wave, pulse waveform, or even a sine wave, provided that the input amplitude is sufficient to drive the transistor into saturation and cutoff. Figure 5-1(a) shows an inverter circuit with a pulse wave input. When the input is zero, there is no collector current and the output is approximately V_{CC}. When the input becomes positive, the transistor switches into saturation, and the output becomes $V_{CE(\text{sat})}$. Thus, a positive-going input produces a negative-going output, and *vice versa*. The output waveform is then the inverse of the input, hence the name *inverter*.

Figure 5-1(b) illustrates the effect of a sine wave input to an inverter. If the amplitude is large enough to switch the transistor rapidly to saturation and cutoff, then an inverted square wave output is the result.

When the input waveform to an inverter has a large amplitude, the base-emitter junction of the transistor may be destroyed by an excessive reverse voltage. Most transistors can take only about 5 V in reverse at the base-emitter junction. To protect the transistor, a diode connected as a negative clipper may be employed (see Figure 3-7). Alternatively, a diode may be connected in series with the transistor emitter terminal, as shown

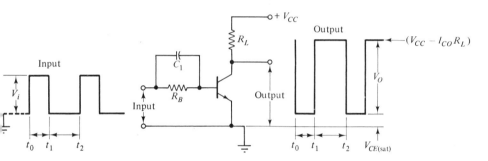

(a)   Inverter with pulse input

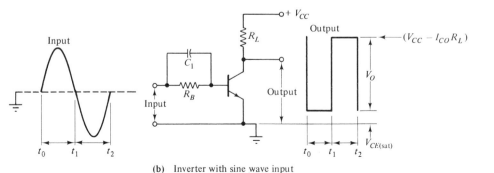

(b)   Inverter with sine wave input

FIGURE 5-1.   Transistor inverter circuits with pulse and sine wave inputs.

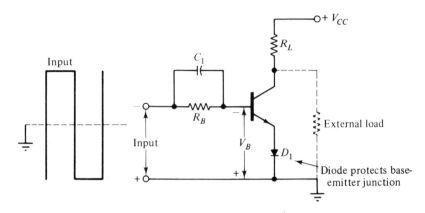

FIGURE 5-2. Use of diode to protect base-emitter junction against excessive reverse voltage.

in Figure 5-2. Since the diode normally can survive reverse bias voltages on the order of at least 50 V, the combined base-emitter junction and diode will withstand a large reverse bias.

The design of a transistor inverter circuit should begin with selection of the load resistance R_L, unless it is already specified. In general, R_L should be much smaller than the external load to be connected to the inverter output. This is to ensure that the external load has no significant effect on the circuit performance. However, R_L should also be made as large as possible in order to keep I_C to a minimum. If this principle is followed in all circuit design, the current demand from power supplies is kept to a minimum. Also, with current maintained as small as possible, power dissipation in all components is minimized. The lower limit for I_C is dependent upon the particular transistor used. The data sheets in Appendix 1-4 and Appendix 1-6 show that for the 2N3904 transistor, $h_{FE(min)}$ is only 40 at $I_C = 0.1$ mA, and $h_{FE(min)} = 100$ at $I_C = 10$ mA. For the 2N930 transistor, $h_{FE(min)} = 100$ at an I_C of only 10 μA, and $h_{FE(min)}$ is 150 at $I_C = 500$ μA. Obviously, the 2N3904 should not be operated with a collector current much below 100 μA, while the 2N930 can easily be operated with I_C as low as 10 μA. One disadvantage of operating a transistor at very low current levels is that the resultant large resistance values make the circuit more susceptible to picking up unwanted signals.

If V_{CC} and I_C are known, R_L is calculated simply as (voltage across R_L) divided by (current through R_L).

$$R_L = \frac{V_{CC} - V_{CE(sat)}}{I_C} \tag{5-1}$$

Usually, the calculated value of R_L is not exactly equal to an available standard resistance value. In this case, the next higher standard value should be selected. The voltage drop across, R_L must at least equal $V_{CC} - V_{CE(sat)}$ for transistor saturation. A value of R_L that is larger than calculated gives saturation with a lower I_C level. If a value of R_L that is smaller than calculated is used, I_C must be increased to ensure saturation. When the design commences with R_L being chosen much smaller than the external load, I_C is calculated from Equation (5-1).

The minimum base current for saturation is calculated as

$$I_{B(min)} = \frac{I_C}{h_{FE(min)}} \qquad (5\text{-}2)$$

Here, the use of $h_{FE(min)}$ is necessary for transistor saturation. If the particular transistor used has a larger than minimum value of h_{FE}, for a given I_B, I_C will tend to be larger than necessary and saturation will be achieved.

The value of R_B is determined by dividing the voltage across R_B by the current through it:

$$R_B = \frac{V_i - V_{BE}}{I_{B(min)}}$$

Again, an available standard resistance must be selected, but this time the next *lower* standard value should be selected. This is because the voltage across R_B is a fixed quantity, $(V_i - V_{BE})$, and

$$I_B = \frac{V_i - V_{BE}}{R_B}$$

If R_B is selected larger than the calculated value, then I_B will be less than the value of $I_{B(min)}$ required to saturate the transistor. If R_B is smaller then calculated, I_B is greater than $I_{B(min)}$ and transistor saturation will occur.

Speed-up capacitor C_1 is calculated for maximum signal frequency, as explained in Sec. 4-4.

EXAMPLE 5-1

Design a transistor inverter circuit using a 2N3904 transistor. The value of V_{CC} is 12 V and the input is a ± 3 V square wave. Use $I_C = 1$ mA.

solution

At saturation,

$$I_C R_L = V_{CC} - V_{CE(\text{sat})}$$

$$R_L = \frac{V_{CC} - V_{CE(\text{sat})}}{I_C} = \frac{12\text{ V} - 0.2\text{ V}}{1\text{ mA}}$$

$$= 11.8\text{ k}\Omega \text{ (use 12 k}\Omega \text{ standard value)}$$

$$I_{B(\text{min})} = \frac{I_C}{h_{FE(\text{min})}}$$

From the 2N3904 data sheet, $h_{FE(\text{min})} = 70$ at $I_C = 1$ mA.

$$I_B = \frac{1\text{mA}}{70} \simeq 14.3\ \mu\text{A}$$

$$R_B = \frac{V_i - V_{BE}}{I_B} = \frac{3\text{ V} - 0.7\text{ V}}{14.3\ \mu\text{A}}$$

$$\simeq 160\text{ k}\Omega \text{ (use 150 k}\Omega \text{ standard value)}$$

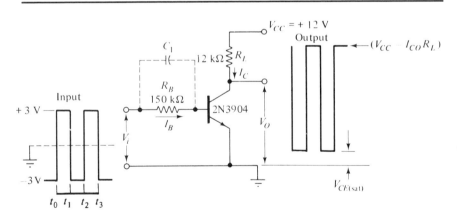

FIGURE 5-3.   Inverter circuit for Example 5-1.

EXAMPLE 5-2

The square wave input to the circuit designed in Example 5-1 has a maximum frequency of 45 kHz. Determine the maximum value of the speed-up capacitor C_1.

solution

C_1 must discharge via R_B by about 90% during the negative (or *off*) portion of the square wave input.

$$\text{Resolving time } t_{re} = \frac{T}{2} = \frac{1}{2f}$$

$$= \frac{1}{2 \times 45 \text{ kHz}} \simeq 11 \text{ } \mu s$$

Recall Equation (4-4):

$$C_{(max)} = \frac{t_{re}}{2.3 \text{ } R}$$

Then, for 90% discharge:

$$C = \frac{t_{re}}{2.3 \text{ } R_B} = \frac{11 \text{ } \mu s}{2.3 \times 150 \text{ k}\Omega}$$

$$\simeq 32 \text{ pF (use 30 pF standard value)}$$

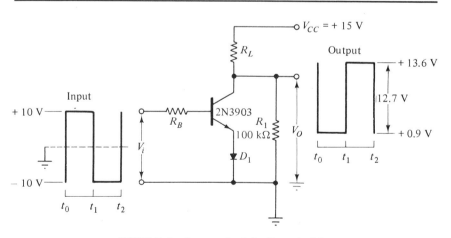

FIGURE 5-4.   Inverter circuit for Example 5-3.

EXAMPLE 5-3

Design a transistor inverter circuit to handle a square wave input of ± 10 V. V_{CC} is 15 V, and the external load has a resistance of 100 kΩ. Use a 2N3903 transistor, and determine the amplitude of the output waveform.

solution

Since the input can be -10 V, diode D_1 (Figure 5-4) is necessary to protect the transistor.

$$R_L \ll R_1$$

Make

$$R_L \simeq \frac{1}{10} R_1$$

$$= \frac{1}{10} \times 100 \text{ k}\Omega = 10 \text{ k}\Omega \text{ (standard value)}$$

At saturation,

$$I_C \simeq \frac{V_{CC} - V_{CE(\text{sat})} - V_{D1}}{R_L}$$

$$= \frac{15 \text{ V} - 0.2 \text{ V} - 0.7 \text{ V}}{10 \text{ k}\Omega}$$

$$= 1.41 \text{ mA}$$

From the 2N3903 data sheet in Appendix 1-4, $h_{FE(\text{min})} = 35$ at $I_C = 1$ mA.

$$I_{B(\text{min})} = \frac{I_C}{h_{FE(\text{min})}}$$

$$= \frac{1.41 \text{ mA}}{35}$$

$$\simeq 40 \text{ } \mu\text{A}$$

and

$$R_B = \frac{V_i - V_{BE} - V_{D1}}{I_{B(\text{min})}}$$

$$= \frac{10 \text{ V} - 0.7 \text{ V} - 0.7 \text{ V}}{40 \text{ } \mu\text{A}} = 215 \text{ k}\Omega \text{ (use 180 k}\Omega \text{ standard value)}$$

D_1 should be a low-current diode with reverse breakdown voltage greater than 10 V. The 2N914 is a suitable device; see the 2N914 data sheet in Appendix 1-1.

When the transistor is saturated:

$$V_o = V_{D1} + V_{CE(\text{sat})}$$
$$= 0.7 \text{ V} + 0.2 \text{ V} = 0.9 \text{ V}$$

At cutoff, R_1 and R_L act as a potential divider, and

$$V_o = \frac{V_{CC} \times R_1}{R_1 + R_L} = \frac{15 \text{ V} \times 100 \text{ k}\Omega}{100 \text{ k}\Omega + 10 \text{ k}\Omega}$$
$$= 13.6 \text{ V}$$

Peak-to-peak output amplitude is equal to 13.6 V − 0.9 V, or 12.7 V.

5-2 CAPACITOR-COUPLED INVERTER CIRCUITS

Sometimes a transistor is required to be biased in the *on* condition, until an input signal is applied to switch it *off*. In this case the signal may be capacitor-coupled. Such a circuit is shown in Figure 5-5. Bias resistor R_B provides base current from the supply to keep the device in saturation. Capacitor C_c couples the negative-going signal to the transistor base. When the signal pulls the base below the emitter voltage level, the transistor

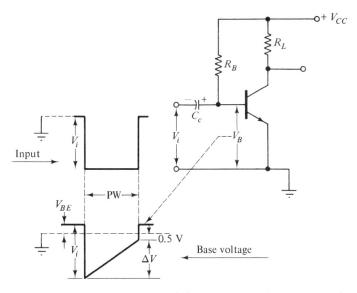

FIGURE 5-5. Normally-*on* capacitor-coupled inverter circuit using an *npn* transistor.

switches *off*. The value of C_c is calculated from a knowledge of the input voltage amplitude and pulse width. For reasons of economy, physical size, and shortest capacitor recharge time, it is best to choose the smallest possible coupling capacitor.

Consider the voltage waveforms shown in Figure 5-5. The circuit input terminal is normally at ground level (*i.e.*, before the signal pulse is applied). The other terminal of C_c (the transistor base terminal) is at V_{BE}. Therefore, the capacitor charge is normally V_{BE}, positive on the RHS, as illustrated on the diagram. When the input pulse with an amplitude of $-V_i$ is applied, the transistor base is pulled down to $-(V_i - V_{BE})$. The transistor switches *off* and the capacitor immediately commences to charge via R_B, so that the negative pulse appearing at the base has tilt, as shown. The tilt must not be so great that V_B rises above ground; otherwise the transistor may switch *on* before the signal pulse has ended. The capacitor charging current is approximately constant, and can be calculated by dividing the voltage across R_B by R_B:

$$I \simeq \frac{V_{CC} - V_i}{R_B}$$

If V_B is allowed to rise to -0.5 V, then the voltage change on the capacitor is $\Delta V = (V_i - V_{BE}) - 0.5$ V. When time t is equal to the pulse width, the simple constant current capacitor equation, Equation (2-8), may be employed. From that equation,

$$C = \frac{It}{\Delta V}$$

EXAMPLE 5-4 _____

The capacitor-coupled inverter circuit of Figure 5-5 has a signal pulse input of -4 V amplitude and PW $= 1$ ms. V_{CC} is to equal 10 V and I_C is to be 10 mA. Using a 2N3904 transistor, design a suitable circuit.

solution

$$R_L = \frac{V_{CC} - V_{CE(sat)}}{I_C} = \frac{10\text{ V} - 0.2\text{ V}}{10\text{ mA}}$$

$$= 980\ \Omega \text{ (use a 1 k}\Omega \text{ standard value)}$$

From the 2N3904 data sheet, $h_{FE(min)} = 100$ at $I_C = 10$ mA:

$$I_{B(min)} = \frac{I_C}{h_{FE(min)}} = \frac{10 \text{ mA}}{100}$$

$$= 100 \ \mu\text{A}$$

$$R_B = \frac{V_{CC} - V_{BE}}{I_{B(min)}} = \frac{10 \text{ V} - 0.7 \text{ V}}{100 \ \mu\text{A}}$$

$$= 93 \text{ k}\Omega \text{ (use 82 k}\Omega \text{ standard value)}$$

$$\Delta V = V_i - V_{BE} - 0.5 \text{ V}$$

$$= 4 \text{ V} - 0.7 \text{ V} - 0.5 \text{ V}$$

$$= 2.8 \text{ V}$$

Capacitor charging current,

$$I \simeq \frac{V_{CC} - V_i}{R_B}$$

$$= \frac{10 \text{ V} - (-4 \text{ V})}{82 \text{ k}\Omega}$$

$$= 0.17 \text{ mA}$$

$$t = PW = 1 \text{ ms}$$

$$C_c = \frac{I \times t}{\Delta V} = \frac{0.17 \text{ mA} \times 1 \text{ ms}}{2.8 \text{ V}} = 0.06 \ \mu\text{F}$$

(use a 0.06 μF standard value)

Although the calculation of C_c in Example 5-4 assumes that the input pulse is negative with respect to ground, actually the pulse could be negative with respect to any other initial level. For example, the pulse could go from +8 V to +4 V, and have exactly the same effect on the inverter circuit as a completely negative pulse. The calculated value of C_c would be the same as in the example.

The circuit in Figure 5-6 is similar to that in Figure 5-5. Since the transistor is now *pnp*, however, all voltage polarities are inverted. For the *pnp* transistor to switch *off*, a positive input pulse must be applied. Also, the voltage at the transistor base should be maintained positive until the end of the pulse width. Design procedure for this circuit is exactly the same as for the *npn* inverter.

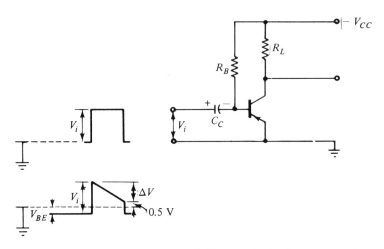

FIGURE 5-6. Normally-*on* capacitor-coupled inverter circuit using a *pnp* transistor.

The converse of the circuit in Figure 5-5 is the normally *off* transistor circuit shown in Figure 5-7. Here R_B connects the base and emitter terminals, and thus keeps V_{BE} equal to zero until a positive-going input pulse is applied. When no input is present, the only current that flows through R_B is the reverse saturation current I_{CO}. The voltage drop across R_B must not be so large as to partially forward-bias the base-emitter

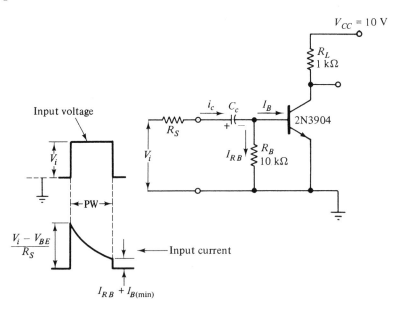

FIGURE 5-7. Normally-*off* capacitor-coupled inverter circuit.

junction; otherwise there may be a small collector current flow. With a maximum I_{CO} of 10 μA (at highest ambient temperature), and $V_{RB} = 0.1$ V, a typical value for R_B is 0.1 V/10 μA = 10 kΩ. Any value below about 22 kΩ is normally suitable for R_B in this circuit. However, normally a resistance less than 1 kΩ should not be used; otherwise the signal source may be overloaded.

In the circuit of Figure 5-7, C_c starts at zero volts and charges via the signal source resistance while the input pulse is present. The charging current i_c flows through R_B and the transistor base terminal. This current begins at $i_c = (V_i - V_{BE})/R_S$, and then falls off as C_c becomes charged. At the end of the input pulse, i_c must still be large enough to provide current through R_B and sufficient base current to saturate the transistor. By use of the equation for capacitor charging current, Equation (2-4) (Chapter 2),

$$i_c = I\epsilon^{-t/CR}$$

an expression for the coupling capacitor can be determined:

$$\frac{I}{i_c} = \epsilon^{t/CR}$$

$$\frac{t}{CR} = \ln\frac{I}{i_c}$$

$$C_c = \frac{t}{R\ln(I/i_c)}$$

In this expression, R is the signal source resistance R_S, t is the input pulse width, I is the initial charging current, and i_c is the charging current at the end of the signal pulse. As in the circuit of Example 5-4, the input pulse does not have to start at ground level.

EXAMPLE 5-5

The inverter circuit in Figure 5-7 has an input pulse of 4 V amplitude and pulse width of 1 ms. The signal source resistance R_S is 1 kΩ. Determine the value of C_c.

solution

$$I_C = \frac{V_{CC} - V_{CE(\text{sst})}}{R_L}$$

$$= \frac{10 \text{ V} - 0.2 \text{ V}}{1 \text{ k}\Omega} = 9.8 \text{ mA}$$

$$I_{B(min)} = \frac{I_C}{h_{FE(min)}} = \frac{9.8 \text{ mA}}{100} = 98 \text{ }\mu\text{A}$$

$$I_{RB} = \frac{V_{BE}}{R_B} = \frac{0.7 \text{ V}}{10 \text{ k}\Omega} = 70 \text{ }\mu\text{A}$$

At the end of the pulse,

$$i_c = I_{B(min)} + I_{RB}$$
$$= 98 \text{ }\mu\text{A} + 70 \text{ }\mu\text{A}$$
$$= 168 \text{ }\mu\text{A}$$

The initial charging current,

$$I = \frac{V_i - V_{BE}}{R_S}$$

$$= \frac{4\text{V} - 0.7 \text{ V}}{1 \text{ k}\Omega} = 3.3 \text{ mA}$$

$$t = PW = 1 \text{ ms}$$

$$C_c = \frac{t}{R_S \ln(I/i_c)} = \frac{1 \text{ ms}}{1 \text{ k}\Omega \ln(3.3 \text{ mA}/168 \text{ }\mu\text{A})}$$

$$= 0.33 \text{ }\mu\text{F (standard value)}$$

Another normally-*off* capacitive coupled inverter circuit is shown in Figure 5-8. In this case the transistor base is biased to a negative voltage $-V_{BB}$, so that the base-emitter junction is reverse-biased. The capacitor C_c now has an initial charge of $E_0 = V_{BB}$, with the polarity shown. When a positive input voltage V_i is applied, the transistor base voltage is

$$V_B = V_i - iR_s - E_O$$

Thus, the transistor will not switch *on* unless V_i is greater than $(iR_s + E_O)$.

When the transistor switches *on*, its base voltage becomes $+V_{BE}$ above ground level. The current through R_B now is

$$I_{RB} = \frac{V_{BE} - V_{BB}}{R_B}$$

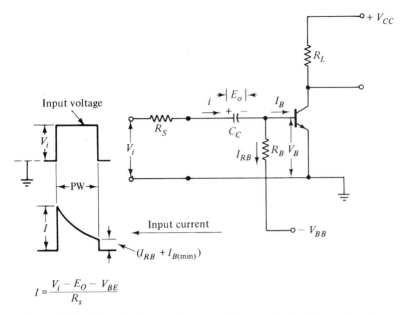

FIGURE 5-8. Normally-*off* capacitor-coupled inverter circuit with negative bias voltage.

Also, the transistor minimum base current remains

$$I_{B(\min)} = \frac{I_C}{h_{FE(\min)}}$$

For the transistor to remain conducting during the input PW, the input (capacitor) current at the end of the PW must be at least

$$i_c = I_{RB} + I_{B(\min)}$$

The initial capacitor charging current for this circuit is

$$I = \frac{\text{initial voltage across } R_s}{R_s}$$

$$= \frac{V_i - E_O - V_{BE}}{R_s}$$

When i_c and I are determined as explained above, the value of C for the circuit in Figure 5-8 can be determined in the same way as in Example 5-5.

5-3 JFET INVERTER CIRCUITS

The direct-coupled inverter circuit illustrated in Figure 5-9(a) has an output which goes approximately from ground to V_{DD}. When the input signal is at ground level, the device is *on*. Drain current I_D flows, causing a voltage drop of almost V_{DD} across R_L. When V_i is negative the device is biased *off* and the output goes to V_{DD}. To ensure that the FET is switched *off*, the negative input pulse must exceed the device *pinchoff voltage V_p*. For a 2N5459 JFET (see data sheet in Appendix 1-9) the *pinchoff* voltage is designated *gate-source cutoff voltage $V_{GS(off)}$*, and is a maximum of 8 V. Therefore, the input pulse to the circuit of Figure 5-9(a) must exceed -8 V, if a 2N5459 is employed. The only function served by R_G in Figure 5-9(a) is to limit the gate current in the event that the input voltage goes positive. Typically R_G is selected as 1 MΩ.

In Figure 5-9(b) a capacitor-coupled JFET inverter circuit is shown. Again, R_G is typically 1 MΩ, this time to present a high input impedance to signals. The circuit shown has the gate biased to the same potential as the source terminal; therefore the FET normally is *on*. To switch the device *off*, the input signal amplitude must exceed V_P. Since the input resistance of the FET circuit is very high, the charging current to the coupling capacitor is extremely small, so C_c can be quite a small capacitor.

A normally *off* JFET inverter circuit is shown in Figure 5-9(c). Here, the gate is biased to a negative dc level to keep the FET *off* when no signal is present. The device pinchoff voltage must be exceeded by V_G. Again, C_c can be a very small capacitor.

5-4 IC OPERATIONAL AMPLIFIER INVERTER

The *operational amplifier* is a very high gain dc amplifier with two input terminals and one output. One input terminal is known as the *inverting input*, because a positive-going signal at this input produces a negative-going output voltage, and *vice versa*. The other input terminal is designated the *noninverting* input. A positive-going signal here produces a positive-going output. The input impedance of the operational amplifier is extremely high, and the output impedance is very low.

The basic circuit symbol for an operational amplifier is shown in Figure 5-10(a). The noninverting and inverting input terminals are identified as + and −, respectively. The output terminal is at the point of the triangle opposite the inputs. Power supply voltages V_{CC} and $-V_{EE}$ normally are symmetrical with respect to ground. A typical supply voltage is

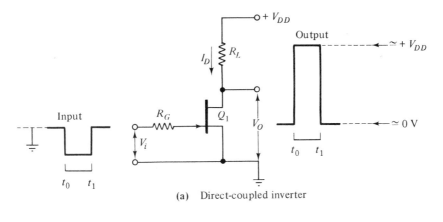

(a) Direct-coupled inverter

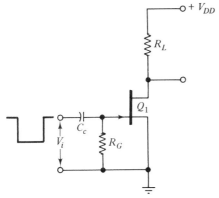

(b) Normally *on* capacitive-coupled inverter

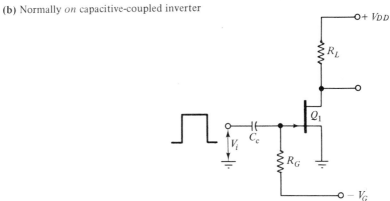

(c) Normally *off* capacitive-coupled circuit

FIGURE 5-9.   JFET inverter circuits.

± 15 V. Frequently, additional terminals are shown for connection of external components. The input terminals usually are biased to ground level.

Consider the data sheet in Appendix 1-11 for the 741 IC operational amplifier. Note that the device parameters are specified for a supply voltage of $V_S = \pm 15$ V, although the supply may be a maximum of ± 22 V. The voltage gain is 50,000 minimum or 200,000 typical. Thus, for $V_o = 10$ V, typical $V_i = 10$ V$/200,000 = 50$ μV. This means that a difference of 50 μV between the inverting and noninverting input terminals will cause the output to go to ± 10 V.

Other important quantities specified are

$$Output\ impedance = 75\ \Omega\ typical$$
$$Input\ impedance = 2\ M\Omega\ typical$$
$$Input\ bias\ current = 80\ nA\ typical$$

The *input bias current* is the current flowing into each of the two input terminals when they are biased to the same voltage level. The *input offset current* is the difference between the two input bias currents. Note that this is typically 20 nA. The *input offset voltage* is the voltage difference that may have to be applied between the two input terminals in order to adjust the output level to exactly zero. For the 741, this quantity is typically 1 mV.

In switching applications it is important to note that the typical *output voltage swing* for the 741 is ± 14 V when a ± 15 V supply is used. Also, note that this output voltage swing may be a minimum of ± 10 V if a 2 kΩ load is connected to the output terminals of the amplifier. An input voltage of approximately $V_i = V_o/$(voltage gain) is required to drive the output to its extreme levels. For a ± 15 V supply, $v_i \simeq 15$ V$/200,000 = 75$ μV. This is the voltage difference between the two input terminals. The actual input voltage can be very much higher than this minimum. The data sheet specifies a typical input voltage range of ± 13 V.

The *slew rate*, particularly important in switching applications, is the rate of change of output voltage, or the speed with which the output changes. For the 741, the slew rate is specified as 0.5 V/μs. Thus, a time of 1 μs is required for the output to change by 0.5 V when a step input is applied. The output moves from -10 V to $+10$ V in a time of 20 V/ 0.5 V$= 40$ μs. In an application for which the slew rate of the 741 is too slow, another amplifier must be selected. The 715, for example, has a slew rate of 100 V/μs.

An IC operational amplifier employed as a direct coupled inverter is shown in Figure 5-10(b). The noninverting terminal is connected directly to ground, and the input pulse is directly connected to the inverting input

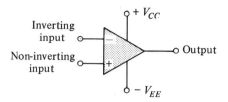

(a)   Circuit symbol for IC operational amplifier

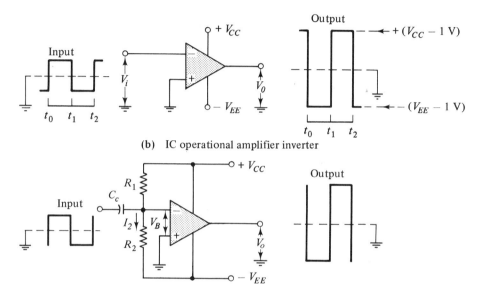

(b)   IC operational amplifier inverter

(c)   Capacitive coupled IC inverter

FIGURE 5-10.   IC operational amplifier circuit symbol and IC inverter circuits.

terminal. When V_i is more than about 75 μV below ground level, the output voltage is approximately $V_{CC} - 1$ V. At $V_i > 75$ μV above ground, the output becomes approximately $-(V_{EE} - 1$ V). Thus an IC operational amplifier can be directly employed as an inverter without using additional components.

Sometimes an IC inverter must have a bias voltage provided at its inverting input terminal, in order to hold the output at its positive or negative extreme when no input signals are present. Such a circuit is shown in Figure 5-10(c). If V_B is positive, the output is negative; if V_B is negative, the output is positive. If the inverting terminal were grounded via a resistance, the dc output voltage cannot be predicted as either positive or negative. Normally, V_B should be about ± 0.5 V, so that a small input signal can easily drive the inverter output from one maximum level to the other.

To design the capacitor-coupled inverter, R_2 should be selected so that the bias current I_2 is very much larger than the input current of the device. Then, R_1 is determined as $R_1 \simeq (V_{CC} - V_B)/I_2$. C_c should be selected for an acceptable level of tilt on the signal at the inverting input terminal (see Sec. 5-2).

EXAMPLE 5-6 _____

Using a 741 IC operational amplifier, design an inverter to provide an output of $V_o \simeq \pm 11$ V. The output normally should be negative when no input is present. The input voltage is a ± 6 V square wave with $f = 1$ kHz. Calculate the rise time of the output voltage.

solution

The circuit is shown in figure 5-10(c). For $V_o \simeq \pm 11$ V, the supply should be ± 12 V.

$$V_{CC} = +12 \text{ V}$$
$$V_{EE} = -12 \text{ V}$$

To maintain a negative dc output, V_B must be positive. Take $V_B \simeq +0.5$ V.

$$V_{R2} = V_B - V_{EE}$$
$$= 0.5 \text{ V} - (-12 \text{ V})$$
$$= 12.5 \text{ V}$$

The maximum input bias current is 500 nA, as taken from the 741 data sheet in Appendix 1-11.

$$I_2 \gg 500 \text{ nA}$$

Make

$$I_2 = 100 \times 500 \text{ nA}$$
$$= 50 \text{ } \mu\text{A}$$
$$R_2 = \frac{V_{R2}}{I_2} = \frac{12.5 \text{ V}}{50 \text{ } \mu\text{A}}$$
$$= 250 \text{ k}\Omega \text{ (use 220 k}\Omega \text{ standard value)}$$

I_2 becomes

$$\frac{V_{R2}}{R_2} = \frac{12.5 \text{ V}}{220 \text{ k}\Omega} = 56.8 \text{ } \mu\text{A}$$

$$R_1 = \frac{V_{CC} - V_B}{I_2}$$

$$= \frac{12 \text{ V} - 0.5 \text{ V}}{56.8 \text{ } \mu\text{A}}$$

$$= 202 \text{ k}\Omega \text{ (use 180 k}\Omega \text{ standard value)}$$

Note that use of a *smaller-than-calculated* value of R_1 ensures that V_B is larger than is required for the output to be negative.

During the pulse time, C_c charges via R_1 and R_2, so the input waveform has *tilt* as it appears at the inverting input terminal. For a ± 6 V input, a tilt of two or three volts will have no effect on the inverter operation.

Let $\Delta V_C = 2$ V.

$$\text{Input resistance} = R_1 \| R_2$$

$$= 180 \text{ k}\Omega \| 220 \text{ k}\Omega$$

$$= 99 \text{ k}\Omega$$

For a ± 6 V input, the initial input current I is calculated as

$$I = \frac{6 \text{ V}}{99 \text{ k}\Omega}$$

$$\simeq 60.6 \text{ } \mu\text{A}$$

$$t = \frac{1}{2f}$$

$$t = \frac{1}{2 \times 1 \text{ kHz}} = 500 \text{ } \mu\text{s}$$

For I assumed constant;

$$C = \frac{It}{\Delta V}$$

Thus,

$$C = \frac{60.6 \text{ } \mu\text{A} \times 500 \text{ } \mu\text{s}}{2 \text{ V}}$$

$$= 1.515 \times 10^{-8}$$

$$\simeq 0.015 \text{ } \mu\text{F (standard capacitance value)}$$

The time taken for the output to go from $+11$ V to -11 V is expressed as:

$$t = \frac{22\ V}{\text{Slew rate}}$$

$$= \frac{22\ V}{0.5\ V/\mu s} = 44\ \mu s$$

5-5 OPERATIONAL AMPLIFIER DIFFERENTIATOR

The CR differentiating circuit discussed in Section 2-5 can be improved upon by the use of an operational amplifier. The operational amplifier circuit can be designed to give an output amplitude larger than the amplitude of the signal being differentiated. This is not the case in the simple CR circuit. A further advantage is that the operational amplifier differentiator has a very low output impedance, whereas, the simple CR circuit can easily be overloaded.

Referring to Figure 5-11, when the input signal is going positive a current I_1 is flowing into C_1, as illustrated. If I_1 is made much larger than the operational amplifier input bias current $I_{B(max)}$ then effectively all of I_1 flows through R_2. The output voltage is

$$V_o = -I_1 R_2$$

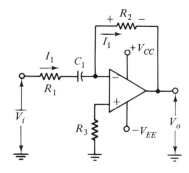

FIGURE 5-11. Operational amplifier differentiating circuit.

This is because the voltage at inverting terminal (in this application) remains close to the level of that at the non-inverting terminal. The non-inverting terminal is grounded, therefore, the inverting terminal voltage is always close to ground, (in fact, in this kind of application, the inverting terminal is termed a *virtual ground* or *virtual earth*). Note that the differentiated output of this circuit is inverted, i.e. when the input is positive-going, the output is negative and vice versa.

The purpose of R_1 is to limit the high frequency gain of the circuit. Without R_1,

$$A_v = \frac{R_2}{X_{C1}}$$

At high frequencies $X_{C1} \ll R_2$, and A_v becomes extremely large. R_1 should be made equal to X_{C1} at a frequency which is approximately ten times the normal operating frequency of the circuit. Thus, R_1 has a negligible effect on the circuit performance, but limits the high frequency gain to $A_V = R_2/R_1$. R_3 is made equal to R_2 to ensure that the dc voltage drop along R_3 due to I_B is equal to $I_B R_2$. This keeps the dc output voltage level close to ground when no input signal is present.

Design of this circuit merely involves: selection of $I_1 \gg I_{B(max)}$, calculation of $R_2 = V_0/I_1$, calculation of C_1 from Equation 2-15, then determining R_1 and R_3 as explained above.

EXAMPLE 5-7 _____

Design a differentiating circuit to give $V_o = 5$ V when the input changes by 1 V over a time period of 10 μ s. Use a 741 operational amplifier.

solution

$$I_1 \gg (I_{B(max)} = 500 \text{ nA})$$

let

$$I_1 = 500 \ \mu A$$

$$R_2 = \frac{V_o}{I_1} = \frac{5 \text{ V}}{500 \ \mu A}$$

$$= 10 \text{ k}\Omega \text{ (standard value)}$$

Eq. 2-15

$$e_R = CR \times \frac{E_{(max)}}{t}$$

or
$$V_o = CR \times \frac{V_i}{t_r}$$

$$5\ V = C \times 10\ k\Omega \times \frac{1\ V}{10\ \mu s}$$

$$C = 5000\ pF\ \text{(standard value)}$$

$$R_3 = R_2 = 10\ k\Omega$$

From Eq. 1-6, the operating frequency of the circuit is

$$f = \frac{0.35}{t_r} = \frac{0.35}{10\ \mu s}$$

$$= 35\ kHz$$

$$R_1 = X_{C1}\ \text{at}\ 10 \times 35\ kHz$$

$$R_1 = \frac{1}{2\pi \times 10 \times 35\ kHz \times 5000\ pF}$$

$$= 90.9\ \Omega\ \text{(use 100 }\Omega\text{ standard value)}$$

Maximum gain of the circuit at high frequency is

$$A_v = \frac{R_2}{R_1} = \frac{10\ k\Omega}{100\ \Omega}$$

$$= 100$$

REVIEW QUESTIONS AND PROBLEMS

5-1 Sketch the complete circuit of a direct-coupled bipolar transistor inverter and explain the function of each component. Show the output waveform when the following inputs are applied: (a) pulse wave, (b) square wave, (c) sine wave.

5-2 Show two methods of protecting a transistor base-emitter voltage against excessive reverse input voltages.

5-3 Design a direct-coupled transistor inverter circuit using a 2N3903 transistor. For this circuit, V_{CC} is 9 V, the input is a ± 5 V square wave, and $I_C = 10$ mA.

5-4 If, in the circuit of Problem 5-3, a 200 pF speed-up capacitor is used, determine the maximum input signal frequency that may be employed.

5-5 A direct-coupled transistor inverter using a 2N3904 transistor has an input square wave of ± 9 V, and $V_{CC} = 20$ V. An external load of 220 kΩ is connected to the inverter output terminals. Design a suitable circuit and determine the amplitude of the output voltage.

5-6 Sketch the circuit of a normally-*on* capacitor-coupled inverter using (a) an *npn* transistor and (b) a *pnp* transistor. In each case show the input voltage and current waveforms, and explain the operation of the circuit.

5-7 Repeat Problem 5-6 for a normally-*off* capacitor-coupled inverter.

5-8 A normally-*on* transistor inverter has a capacitor-coupled pulse input signal with PA = -3 V and PW = 600 μs; $V_{CC} = 12$ V and I_C is to be 1 mA. Design a suitable circuit, using a 2N3904.

5-9 The conditions specified in Problem 5-8 can be applied to a normally-*off* inverter if the pulse amplitude is $+3$ V. Design the circuit, taking the signal source resistance as 1 kΩ.

5-10 A normally-*off* inverter circuit using a 2N4418 transistor has $V_{CC} = 9$ V, and $V_{BB} = -3$ V. I_C is to be approximately 10 mA, and the input pulse has PA = 6 V, PW = 500 μs, and $R_s = 600$ Ω. Design the circuit.

5-11 Sketch the circuits of direct-coupled and capacitor-coupled JFET inverter circuits. Explain the operation of the circuits and discuss their advantages and disadvantages compared to bipolar inverters.

5-12 Using a 741 IC operational amplifier, design an inverter that will provide an output of $V_o \simeq \pm 14$ V. The output normally should be positive when no input is applied. The input voltage is a ± 4 V square wave with $f = 500$ Hz. Calculate the approximate rise time of the output voltage.

5-13 An IC operational amplifier with a maximum input bias current of 750 nA and a slew rate of 10 V/μs is to be used as an inverter. The supply voltage is $V_{CC} = \pm 12$ V, and the inverter output is to be normally positive. Design a suitable circuit if the input voltage has a minimum amplitude of ± 3 V.

5-14 Sketch an operational amplifier differentiating circuit and explain how it operates. Compare this circuit to a simple CR differentiating circuit.

5-15 Using a 741 operational amplifier design a differentiating circuit to give an output of 2 V when the input changes by 1 V/μs.

5-16 Calculate the output voltages when the input waveform in Fig. 2-11 is differentiated by the circuit designed in Problem 5-15.

CHAPTER 6

Schmitt Trigger Circuits and Voltage Comparators

INTRODUCTION

Essentially, a **Schmitt trigger circuit** is a fast-operating voltage level detector. When the input voltage arrives at the upper or lower triggering levels, the output voltage rapidly changes. The circuit operates from almost any input waveform and always gives a pulse-type output. Transistor Schmitt trigger circuits can be designed to trigger at specified upper and lower levels of input voltage. An IC operational amplifier circuit can also be employed as a Schmitt trigger circuit, and several Schmitt trigger circuits are available in a single IC package. **Voltage comparator circuits** produce an output pulse when the input voltage level becomes exactly equal to a reference voltage.

6-1 OPERATION OF SCHMITT TRIGGER CIRCUIT

A transistor Schmitt trigger circuit is shown in Figure 6-1(a). The figure shows that transistors Q_1 and Q_2 have a common emitter resistor R_E. The Q_2 base voltage, V_{B2}, is derived via the potential divider (R_1 and R_2) from

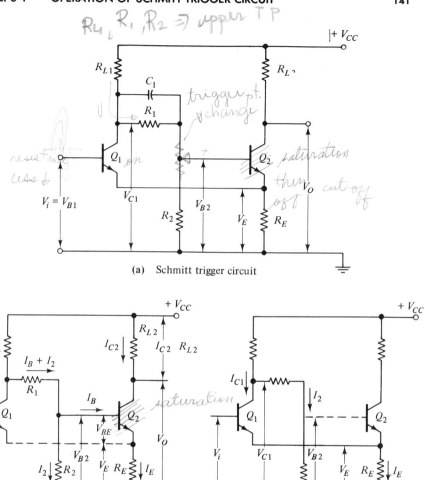

(a)   Schmitt trigger circuit

(b)   dc conditions with Q_2 on

(c)   dc conditions with Q_1 on

FIGURE 6-1.   Schmitt trigger and dc conditions with Q_2 on and with Q_1 on.

the collector of Q_1. Transistors Q_1 and Q_2 have load resistances R_{L1} and R_{L2}, respectively. The arrangement is such that when transistor Q_1 is *on*, transistor Q_2 is *off*, and for Q_2 to switch *on*, Q_1 must switch *off*.

To understand the operation of this type of circuit, first consider the dc conditions when Q_1 is *off*. When it is *off*, Q_1 can be regarded as an open circuit; therefore, it can be left out of the circuit, as is shown in Figure 6-1(b). The Q_2 base voltage now is derived from V_{CC} via a potential divider consisting of R_1, R_2, and R_{L1}. Thus, Q_2 is *on* and a collector current I_{C2}

flows, producing a voltage drop across R_{L2}. The output voltage is $(V_{CC} - I_{C2}R_{L2})$.

If, as in Figure 6-1(c), Q_1 is now triggered *on*, the emitter voltage becomes $V_E = V_i - V_{BE}$. Also, the collector current I_{C1} causes a voltage drop across R_{L1}, in turn causing V_{B2} to fall below the level of V_E. Thus, when Q_1 is *on*, Q_2 is biased *off* and I_{C2} becomes zero. At this point there is no longer any significant drop across R_{L2}, and the output voltage is approximately V_{CC}.

Now, reconsider the conditions illustrated in Figure 6-1 (b). It is seen that with Q_2 *on*, $V_E = V_{B2} - V_{BE}$. This is the voltage at the emitter terminal of both transistors, since they are connected together. Transistor Q_1 will not switch *on* until its base voltage becomes greater than V_E. In fact, Q_1 switches *on* approximately at $V_i = V_E + V_{BE}$. Obviously, if V_i is suddenly made greater than this level, Q_1 would switch *on* rapidly. The lowest level of V_i that causes Q_1 to switch *on* is known as the *upper trigger point* (UTP) for the circuit.

At the moment that Q_1 begins to switch *on* it starts to pull the common emitter voltage up, the flow of I_C causes V_{C1} to fall and, consequently, V_{B2} falls. Thus, as Q_1 switches on, it causes a rapid reduction in V_{BE2}. This effect, known as *regeneration*, produces a very rapid switchover from Q_2 *on* to Q_1 *on*.

Now consider the process of switching from Q_1 *on* to Q_2 *on*. Refer to Figure 6-1(c). With Q_1 *on*, $I_E = (V_i - V_{BE})/R_E$. Thus, a reduction in V_i also reduces I_E, and since $I_C \simeq I_E$, I_{C1} also becomes smaller. The voltage drop across R_{L1} is approximately $(I_{C1}R_{L1})$, and the collector voltage of Q_1 is $V_{C1} \simeq (V_{CC} - I_{C1}R_{L1})$. Therefore, when V_i is reduced, I_{C1} becomes smaller causing V_{C1} to rise. In turn, the base voltage V_{B2} of Q_2 rises. If V_i is reduced by a very small amount, the resultant small increase in V_{B2} may leave Q_2 base still below the level of its emitter voltage. In fact, Q_2 switches *on* again only when V_{B2} and V_i become equal. The input voltage at which this occurs is known as the *lower trigger point* (LTP).

In the changeover from Q_1 *on* to Q_2 *on*, regeneration again occurs. When Q_2 starts to switch *on*, it causes Q_1 to begin to switch *off* because of the rise in the common emitter voltage. Q_1 switching *off* helps Q_2 to turn *on*. Again, the changeover occurs very rapidly.

Speed-up capacitor C_1 [Figure 6-1(a)] is provided solely to improve the circuit switching speed. The effect is essentially as explained in Sec. 4-4. However, it could also be said that during switching, the voltage change at the collector of Q_1 is potentially divided across R_1 and R_2 before being applied to Q_2 base. The presence of C_1 eliminates the potential division (i.e., $\Delta V_{B2} = \Delta V_{C1}$), and thus speeds up the switching process.

Figure 6-2 shows the effects of various input waveforms on the Schmitt trigger circuit. In each case the output is $(V_{CC} - I_{C2}R_{L2})$ until the upper trigger point is reached. Then with Q_2 switched *off*, the output becomes

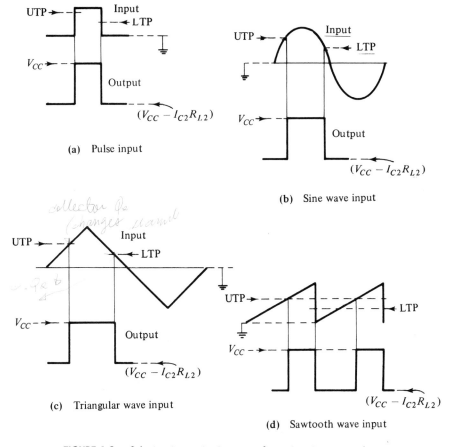

FIGURE 6-2.   Schmitt trigger circuit outputs for various input waveforms.

approximately V_{CC}. When the signal input falls to the LTP, the output again drops to $(V_{CC} - I_{C2}R_{L2})$ as Q_2 switches *on*. The Schmitt trigger circuit is seen to be essentially a voltage level detector, capable of producing a fast-moving output when either a slow- or fast-moving input arrives at the trigger points.

6-2   DESIGNING FOR A GIVEN UPPER TRIGGER POINT

Design of any circuit starts from a specification. For the Schmitt trigger circuit the most important parameters are the upper and lower trigger points, UTP and LTP. A circuit may be designed simply to meet a given

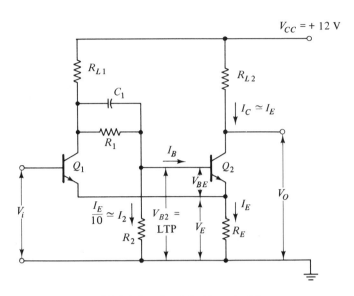

FIGURE 6-3.   Circuit for Example 6-1.

UTP, and to completely ignore the LTP. In this case, the LTP normally is located between the UTP and ground. Such a circuit could function quite satisfactorily with the waveforms illustrated in Figures 6-2. With the pulse or sawtooth input waveforms shown, the LTP can be anywhere between the UTP and ground.

As already explained, the UTP is equal to Q_2 base voltage (V_{B2}) when Q_2 is *on*. Therefore, the circuit is designed to have V_{B2} equal to the specified UTP. Potential divider (R_1 and R_2), together with R_{L1}, must provide a stable bias voltage V_{B2} for Q_2 (Figure 6-3). However, R_1 and R_2 must be large enough to avoid overloading R_{L1}. For example, if R_1 and R_2 were made smaller than R_{L1}, then Q_1 collector current would have very little effect on V_{B2} when Q_1 is switched *on*. If R_1 and R_2 are made very large, I_{B2} will cause a large voltage drop across R_1 when Q_2 is switched *on*. Therefore, R_1 and R_2 must be kept as small as possible, but they must be several times larger than R_{L1}. A good rule of thumb here is to take the value of $I_2 \simeq \frac{1}{10} I_{E2}$. Then calculate R_2 as V_{B2}/I_2. The design procedure is best understood by working through an example.

EXAMPLE 6-1

A Schmitt trigger circuit is to have UTP $= 5$ V. The silicon transistors to be used have $h_{FE(\min)} = 100$, and I_C is to be 2 mA. The available supply is 12 V. Design a suitable circuit.

solution

$$UTP = V_{B2} = 5 \text{ V}$$

$$V_E = V_{B2} - V_{BE} = 5 \text{ V} - 0.7 \text{ V} = 4.3 \text{ V} \qquad \text{(See Figure 6-3.)}$$

$$I_E \simeq I_C = 2 \text{ mA}$$

$$R_E = \frac{V_E}{I_E} = \frac{4.3 \text{ V}}{2 \text{ mA}}$$

$$= 2.15 \text{ k}\Omega \qquad \text{(use 2.2 k}\Omega \text{ standard value)}$$

Taking Q_2 as saturated, $V_{CE(\text{sat})} = 0.2$ V typically. The voltage drop across $R_{L2} = I_C R_{L2}$ and

$$I_C R_{L2} = V_{CC} - V_E - V_{CE(\text{sat})}$$

$$= 12 \text{ V} - 4.3 \text{ V} - 0.2 \text{ V}$$

$$= 7.5 \text{ V}$$

$$R_{L2} = \frac{7.5 \text{ V}}{I_C} = \frac{7.5 \text{ V}}{2 \text{ mA}} = 3.75 \text{ k}\Omega \qquad \text{(use 3.9 k}\Omega \text{ standard value)}$$

$$I_2 \cong \frac{1}{10} I_E = \frac{1}{10} \times 2 \text{ mA} = 0.2 \text{ mA}$$

$$R_2 = \frac{V_{B2}}{I_2} = \frac{5 \text{ V}}{0.2 \text{ mA}} = 25 \text{ k}\Omega \qquad \text{(use 22 k}\Omega \text{ standard value)}$$

I_2 now becomes $\dfrac{5 \text{ V}}{22 \text{ k}\Omega} = 0.227$ mA.

$$I_{B2} = \frac{I_{C2}}{h_{FE(\text{min})}} = \frac{2 \text{ mA}}{100} = 20 \text{ }\mu\text{A}$$

$$I_{B2} + I_2 = 20 \text{ }\mu\text{A} + 0.227 \text{ mA} = 0.247 \text{ mA}$$

$$R_{L1} + R_1 = \frac{V_{CC} - V_{B2}}{I_2 + I_{B2}}$$

$$= \frac{12 \text{ V} - 5 \text{ V}}{0.247 \text{ mA}}$$

$$= 28.3 \text{ k}\Omega$$

When the LTP is not specified, R_{L1} may be made equal to R_{L2}. *This cannot be done when the circuit is to be designed for a given LTP level.*

$$R_{L1} = R_{L2} = 3.9 \text{ k}\Omega$$

$$R_1 = (R_{L1} + R_1) - R_{L1}$$

$$= 28.3 \text{ k}\Omega - 3.9 \text{ k}\Omega$$

$$= 24.4 \text{ k}\Omega \quad \text{(use 22 k}\Omega \text{ standard value)}$$

Since standard value components have been selected in Example 6-1, the UTP will not be exactly as specified. A potentiometer connected between R_1 and R_2 with Q_2 base connected to its moving contact would provide adjustment to obtain a precise level of UTP.

6-3 DESIGNING FOR GIVEN UTP AND LTP

RE fig. 6-1 C

When V_i decreases to LTP, the circuit is about to change state but Q_1 is still *on* and Q_2 is *off*. The situation is illustrated in Figure 6-4. As V_i approaches LTP, V_{B1} is decreasing and V_{B2} is increasing. The Lower Trigger Point occurs when $V_{B2} = V_{B1} = V_i$. The design procedure, when both UTP and LTP are specified, is exactly the same as in Example 6-1, up to the point at which R_{L1} is chosen.

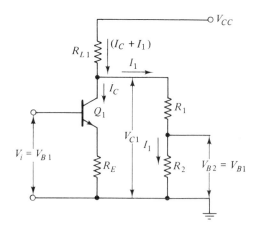

FIGURE 6-4. Determination of LTP.

EXAMPLE 6-2 _____

A Schmitt trigger circuit is to be designed with a UTP of 5 V, and an LTP of 3 V. The silicon transistors employed have $h_{FE(min)} = 100$, and I_C is to be 2 mA. The available supply is 12 V. Design a suitable circuit.

solution

With the exception of the LTP, the circuit is exactly as specified for Example 6-1. Therefore, from Example 6-1,

$$R_{L2} = 3.9 \text{ k}\Omega \qquad R_E = 2.2 \text{ k}\Omega \qquad R_2 = 22 \text{ k}\Omega$$
$$(R_1 + R_{L1}) = 28.3 \text{ k}\Omega$$

Figure 6-4 shows the circuit conditions when Q_1 is *on* and V_i is exactly at the LTP. For $V_i = \text{LTP} = 3$ V, $V_{B2} = \text{LTP} = 3$ V.

$$I_1 = \frac{V_{B2}}{R_2} = \frac{3 \text{ V}}{22 \text{ k}\Omega} = 0.136 \text{ mA}$$

$$I_{C1} \simeq I_E = \frac{V_{B1} - V_{BE}}{R_E} = \frac{3 \text{ V} - 0.7 \text{ V}}{2.2 \text{ k}\Omega} = 1.045 \text{ mA}$$

$$V_{CC} = R_{L1}(I_{C1} + I_1) + I_1(R_1 + R_2)$$

$$12 \text{ V} = R_{L1}(1.045 \text{ mA} + 0.136 \text{ mA}) + 0.136 \text{ mA}(R_1 + 22 \text{ k}\Omega)$$

$$R_1 + R_{L1} = 28.3 \text{ k}\Omega \qquad \text{(from Example 6-1)}$$

$$R_1 = 28.3 \text{ k}\Omega - R_{L1}$$

$$12 \text{ V} = R_{L1}(1.045 \text{ mA} + 0.136 \text{ mA})$$
$$+ 0.136 \text{ mA}(28.3 \text{ k}\Omega - R_{L1} + 22 \text{ k}\Omega)$$

$$= R_{L1}(1.181 \text{ mA}) + 3.85 \text{ V} - R_{L1}(0.136 \text{ mA}) + 2.99 \text{ V}$$

$$12 \text{ V} - 3.85 \text{ V} - 2.99 \text{ V} = R_{L1}(1.181 \text{ mA} - 0.136 \text{ mA})$$

$$R_{L1} = 4.94 \text{ k}\Omega \qquad \text{(use 4.7 k}\Omega \text{ standard value)}$$

$$R_1 = 28.3 \text{ k}\Omega - R_{L1}$$

$$= 28.3 \text{ k}\Omega - 4.7 \text{ k}\Omega$$

$$= 23.6 \text{ k}\Omega \qquad \text{(use 22 k}\Omega \text{ standard value)}$$

6-4   SELECTION OF THE SPEED-UP CAPACITOR

The effects of a speed-up capacitor on transistor switching times were discussed in Sec. 4-4. As already explained, the capacitor should be as large as possible, but must be small enough to allow its voltage to return to normal dc levels between switching. For a Schmitt circuit to trigger at the UTP and LTP levels as designed, the capacitor C_1 voltage must settle to the dc level across R_1 in the time interval between triggering.

EXAMPLE 6-3

The Schmitt trigger circuit designed in Example 6-2 is to be triggered at a maximum frequency of 1 MHz. Determine the largest speed-up capacitor that may be used.

solution

Resistance in parallel with the capacitor terminals when Q_1 is *off*:

$$R = R_1 \| (R_{L1} + R_2)$$

$$= 22 \text{ k}\Omega \| (4.7 \text{ k}\Omega + 22 \text{ k}\Omega)$$

$$\simeq 12 \text{ k}\Omega$$

Actually, the input resistance of Q_2 is in parallel with R_2 but is very large compared to R_2.

$$\text{Resolving time} = t_{re} = \frac{1}{\text{Triggering frequency}}$$

$$= \frac{1}{1 \text{ MHz}} = 1 \text{ } \mu s$$

For C to charge through 90% of its total voltage change:

$$C_{\text{max}} = \frac{t_{re}}{2.3R} \qquad \text{[Equation (4.4)]}$$

$$C = \frac{1 \text{ } \mu s}{2.3 \times 12 \text{ k}\Omega} = 36 \text{ pF} \qquad \text{(use 35 pF standard value)}$$

6-5   OUTPUT/INPUT CHARACTERISTICS

Consider the Schmitt trigger circuit design as finalized in Example 6-2. When Q_2 is *on*, the output voltage is

$$V_o = V_{CC} - I_C R_{L2}$$
$$= 12 \text{ V} - (2 \text{ mA} \times 3.9 \text{ k}\Omega)$$
$$= 4.2 \text{ V}$$

When Q_2 is *off*, the output voltage is

$$V_o = V_{CC} - I_{Co} R_{L2}$$
$$\simeq V_{CC} = 12 \text{ V}$$

The UTP and LTP are approximately, as designed, 5 V and 3 V, respectively. With the triggering levels and the output voltage levels known, a graph showing output voltage *versus* input voltage may be plotted.

When the input voltage is zero, Q_1 is *off* and Q_2 is *on*. Therefore, $V_o = 4.2$ V. This may be plotted as point A on the output/input characteristics in Figure 6-5(a). As V_i is increased above zero volts, Q_1 remains *off* and Q_2 remains *on* until V_i becomes equal to the UTP, which for this particular circuit is at 5 V. Hence V_o remains at 4.2 V from $V_i = 0$ to $V_i = 5$ V. Point B is plotted at $V_o = 4.2$ V and $V_i = 5$ V.

When the UTP is reached, Q_1 switches *on* and Q_2 switches *off*. Thus, V_o changes from 4.2 V to 12 V. Point C is plotted at $V_o = V_{CC}$ and $V_i = \text{UTP}$. Any further increase in V_i has no effect on V_o. The horizontal line from point C to point D in Figure 6-5(a) shows that V_o remains equal to V_{CC} as V_i increases above the UTP.

Now, consider the effect of reducing V_i from a level greater than the UTP. The output voltage V_o remains equal to 12 V until V_i becomes equal to the LTP, at point E on Figure 6-5(b). At the LTP, Q_1 switches *off* and Q_2 switches *on*, returning V_o to 4.2 V. Point F is plotted at $V_o = 4.2$ V and $V_i = \text{LTP} = 3$ V. As V_i is reduced below the LTP, V_o remains at 4.2 V, shown by the line from point F to point A in Figure 6-5(b). The two graphs taken together in Figure 6-5(c) give the complete output/input characteristics for the circuit.

The difference between the UTP and LTP levels is termed the *hysteresis* of the circuit. For many circuit applications, the hysteresis is not very important. In some circumstances, however, circuits with the least possible hysteresis are desirable. Zero hysteresis occurs when the upper and lower trigger points are equal.

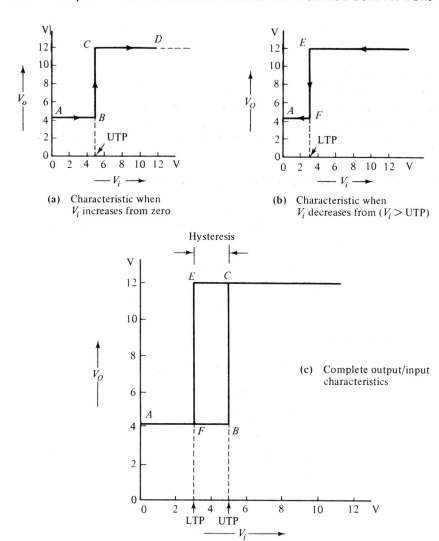

(a) Characteristic when V_i increases from zero

(b) Characteristic when V_i decreases from $(V_i > \text{UTP})$

(c) Complete output/input characteristics

FIGURE 6-5. Output/input characteristics for Schmitt trigger circuit.

The hysteresis can be adjusted by altering the ratio of R_1 and R_2 or by adjusting the value of R_{L1}. The *loop gain* or stage gain of Q_1 (when Q_1 is *on*) determines the amount of hysteresis in the circuit. With a large loop gain, a very small value of input voltage is sufficient to produce a large (phase inverted) output to Q_2 base, which keeps Q_2 biased *off*. Consequently, the lower trigger point is found at a very low level of input voltage. Thus, it is seen that the largest loop gain produces greatest hysteresis.

6-6 THE IC OPERATIONAL AMPLIFIER AS A SCHMITT TRIGGER CIRCUIT

6-6.1 Inverting Schmitt Circuit

The IC operational amplifier, previously treated in Sec. 5-4, may be employed as a Schmitt trigger circuit. The design of such a circuit is quite simple. Consider Figure 6-6 which shows a circuit in which the input triggering voltage is applied to the inverting input terminal. The noninverting terminal is connected to the junction of resistors R_1 and R_2; these resistors operate as a potential divider from output to ground. The voltage at the noninverting terminal is the voltage across R_2, as shown in the figure.

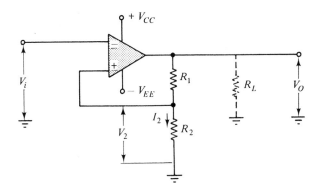

FIGURE 6-6.   IC operational amplifier as Schmitt trigger circuit.

When V_i is less than V_2, the noninverting terminal input voltage V_2 is greater than the inverting input. Therefore, the output is positive. In this case, $V_o \simeq +V_{CC} - 1$ V if the load resistance is 10 kΩ or greater (see the 741 data sheet in Appendix 1-11). The voltage at the noninverting input terminal is calculated by using the output voltage V_o and R_1 and R_2:

$$V_2 = \frac{R_2}{R_1 + R_2}(V_{CC} - 1 \text{ V})$$

When the input voltage is raised to the level of V_2, the output begins to go negative. This causes V_2 to fall; thus the noninverting input terminal rapidly becomes negative with respect to the inverting input terminal. When this occurs the output changes over very rapidly from approximately $(V_{CC} - 1$ V) to approximately $(V_{EE} + 1$ V). It is seen that the upper trigger point is equal to V_2 when the output is positive.

When the output is $(V_{EE}+1 \text{ V})$, V_2 becomes:

$$V_2 = \frac{R_2}{R_1 + R_2}(V_{EE}+1 \text{ V})$$

Since V_{EE} is negative, V_2 is a negative voltage, and the output V_o remains negative until the voltage at the inverting input terminal is reduced to the new (negative) level of V_2.

From the above discussion, it can be seen that

$$\text{UTP} \simeq \frac{R_2}{R_1 + R_2}(V_{CC}-1 \text{ V}) \tag{6-1}$$

and

$$\text{LTP} \simeq \frac{R_2}{R_1 + R_2}(V_{EE}+1 \text{ V}) \tag{6-2}$$

EXAMPLE 6-4

A 741 operational amplifier is to be employed as a Schmitt trigger circuit with a UTP of 3 V. Design a suitable circuit and calculate the actual UTP and LTP when resistors with standard values are selected. Take R_L as 10 kΩ and use a supply voltage of ± 15 V.

solution

From the 741 data sheet in Appendix 1-11:

$$I_{B(\text{max})} = 500 \text{ nA}$$

For a stable level of V_2, $I_2 \gg I_{B(\text{max})}$.

$$I_2 = 100 \times 500 \text{ nA}$$

$$= 50 \ \mu\text{A}$$

$$R_2 = \frac{\text{UTP}}{I_2} = \frac{3 \text{ V}}{50 \ \mu\text{A}}$$

$$= 60 \text{ k}\Omega \qquad \text{(use 56 k}\Omega \text{ standard value)}$$

Thus, I_2 becomes 3 V/56 kΩ, which is equal to 53.57 μA.

$$V_{R1} = V_o - V_2$$
$$= (V_{CC} - 1 \text{ V}) - V_2$$
$$= 15 \text{ V} - 1 \text{ V} - 3 \text{ V}$$
$$= 11 \text{ V}$$

$$R_1 = \frac{V_{R1}}{I_2} = \frac{11 \text{ V}}{53.57 \ \mu\text{A}}$$
$$= 205 \text{ k}\Omega \qquad \text{(use 220 k}\Omega \text{ standard value)}$$

$$\text{Actual UTP} \simeq \frac{V_o R_2}{R_1 + R_2}$$
$$= \frac{14 \text{ V} \times 56 \text{ k}\Omega}{220 \text{ k}\Omega + 56 \text{ k}\Omega} = 2.8 \text{ V}$$

$$\text{LTP} \simeq -2.8 \text{ V}$$

The operational amplifier *slew rate* (discussed in Sec. 5-4) imposes limitations on the performance of the Schmitt circuit. The 741 has a typical slew rate of 0.5 V/μs (see Appendix 1). For the output voltage to go from -14 V to $+14$ V requires a time of

$$t = \frac{28 \text{ V}}{0.5 \text{ V}/\mu\text{s}} = 56 \ \mu\text{s}$$

If this time is allowed to occupy 10% of the output pulse width, (see the discussion in Sec. 1-6), then minimum pulse width is

$$\text{PW} = 10 \times t = 560 \ \mu\text{s}$$

If it is assumed that the output is a square wave, the maximum satisfactory operating frequency is

$$f = \frac{1}{2 \times \text{PW}} = 893 \text{ Hz}$$

For higher-frequency performance, an operational amplifier with a faster slew rate must be employed. The 715, for example, has a slew rate of 100

V/μs. Going through the same calculations as above, the maximum operating frequency for a Schmitt trigger using a 715 is approximately 179 kHz.

6-6.2   Adjusting the Trigger Points

A Schmitt trigger circuit in which the lower trigger point is clamped to -0.7 V is shown in Figure 6-7(a). When the output voltage becomes negative, diode D_1 is forward-biased. Thus, D_1 holds the noninverting input to 0.7 V below ground. When V_i (at the inverting input) drops below -0.7 V, the output again becomes positive. The LTP is now -0.7 V. The UTP is unaffected by the diode, since D_1 is reverse-biased when the output is positive.

The circuit can be designed for any desired UTP. Then, by use of a diode and potential divider, as shown in Figure 6-7(b), the LTP can be fixed at any desired level. A potentiometer placed between R_3 and R_4 [Figure 6-7(c)] provides an adjustable LTP. A diode connected in series with R_1, as illustrated in Figure 6-7(d), gives an LTP which is very close to ground. When the output is negative, D_1 is reverse-biased and only the diode reverse leakage current I_R flows. The LTP now becomes $V_2 = -I_R R_2$. Since I_R normally is very small, and R_2 can be selected as low as a few kilohms, the LTP can be only millivolts from ground. By reversal of D_1 the UTP can be brought close to ground. Then, the LTP is specified by V_o, R_1, and R_2.

Figure 6-7(e) shows an arrangement by which the UTP and LTP can be made completely independent of each other at the design stage. When the output is positive, D_2 is reverse-biased and D_1 is forward-biased. The UTP of the circuit is determined by R_1 and R_2. When the output of the circuit is negative, D_1 is reverse-biased while D_2 is forward-biased. The LTP is now determined by R_1' and R_2.

EXAMPLE 6-5 _____

The Schmitt trigger circuit designed in Example 6-4 is to have the LTP adjustable over the range from 2 V to -2 V. Design a suitable circuit.

solution

Consider the circuit of Figure 6-7(c). For a stable bias voltage V_A, set $I_4 \gg I_2$. Let

$$I_4 = 100 \times I_2 = 100 \times 50 \; \mu A$$
$$= 5 \; mA$$

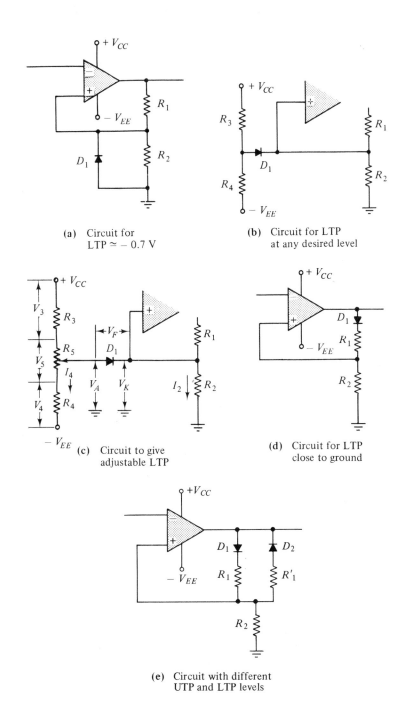

(a) Circuit for
LTP $\simeq -0.7$ V

(b) Circuit for LTP
at any desired level

(c) Circuit to give
adjustable LTP

(d) Circuit for LTP
close to ground

(e) Circuit with different
UTP and LTP levels

FIGURE 6-7. Operational amplifier Schmitt trigger circuits with various methods of
setting LTP.

For

$$V_{K1} = -2\ V$$

$$V_{A1} = V_{K1} + V_F$$

$$= -2\ V + 0.7\ V$$

$$= -1.3\ V$$

$$V_4 = -1.3\ V - (-15\ V) = 13.7\ V$$

$$R_4 = \frac{V_4}{I_4} = \frac{13.7\ V}{5\ mA}$$

$$= 2.72\ k\Omega \qquad \text{(use 2.7 k}\Omega\text{ standard value)}$$

$$R_5 = \frac{V_5}{I_4} = \frac{4\ V}{5\ mA}$$

$$= 800\ \Omega \qquad \text{(use 1 k}\Omega\text{ standard potentiometer value)}$$

$$V_{A2} = V_{K2} + V_F$$

$$= 2\ V + 0.7\ V = 2.7\ V$$

$$R_3 = \frac{V_3}{I_4} = \frac{15\ V - 2.7\ V}{5\ mA}$$

$$= 2.46\ k\Omega \qquad \text{(use 2.7 k}\Omega\text{ standard value)}$$

6-6.3 Noninverting Schmitt Circuit

In the circuit shown in Figure 6-8(a) the input voltage (V_i) is applied to the noninverting input terminal of the operational amplifier. This means that the circuit output goes positive when V_i is increased to the UTP, and negative when V_i is lowered to the LTP level. Assume that the output voltage of the circuit is negative at a level of approximately $-(V_{EE} - 1\ V)$. If the input voltage is zero, then the voltage across R_1 is

$$V_{R1} = \frac{R_1}{R_1 + R_2} \left[-(V_{EE} - 1\ V) \right]$$

[see Figure 6-8(b)]. Thus, the voltage at the noninverting input is a negative quantity, and this keeps the output at its negative saturation level.

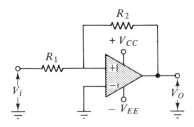

(a) Non-inverting Schmitt trigger circuit

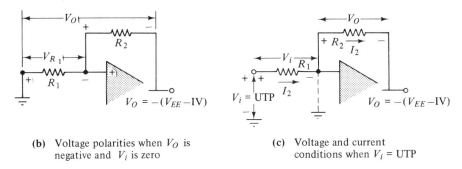

(b) Voltage polarities when V_O is
negative and V_i is zero

(c) Voltage and current
conditions when V_i = UTP

FIGURE 6-8. Noninverting IC operational amplifier Schmitt trigger circuit.

For the output of the circuit to go positive, V_i must be raised until the voltage at the noninverting terminal goes slightly above ground level (i.e., above the level of the inverting input terminal). When this occurs, the voltage at the right-hand side of R_2 is $-V_o$ and that at the left-hand side is 0 V; i.e., the voltage across R_2 is V_o [see Figure 6-8(c)]. Therefore, the current flowing through R_2 is

$$I_2 = V_o / R_2$$

Since the current flowing into the input terminal of the operational amplifier is negligibly small, the current through R_1 is also I_2. The voltage at the left-hand side of R_1 is V_i, and that at the right-hand side is 0 V. Thus,

$$V_{R1} = V_i = I_2 \times R_1 = \text{UTP}$$

The circuit is designed very simply by first selecting I_2 very much larger than the maximum input current to the amplifier. Then, R_1 is calculated as UTP/I_2, and R_2 as V_o/I_2. The LTP is numerically equal to the UTP but with reversed polarity.

The modifications to the noninverting Schmitt circuit shown in Figure 6-9 are self-explanatory. Figure 6-9(a) shows a circuit in which the UTP is a positive quantity, but the LTP is very close to ground level. The circuit in Figure 6-9(b) has (numerically) different UTP and LTP values, while that in Figure 6-9(c) has an adjustment to raise or lower the level of the trigger points.

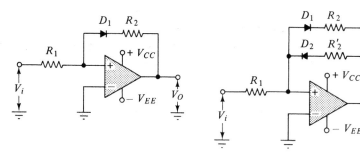

(a) Schmitt circuit with
     UTP positive and LTP $\simeq 0\,V$

(b) Schmitt circuit with
     numerically different
     UTP and LTP levels

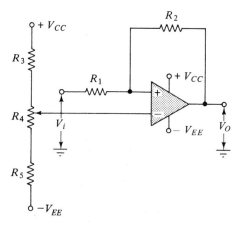

(c) Schmitt circuit with
     trigger level adjustment

FIGURE 6-9. Operational amplifier noninverting Schmitt trigger circuits with various methods of setting the trigger levels.

EXAMPLE 6-6 _____

A noninverting Schmitt trigger circuit is to be designed to have trigger points of ± 2 V. The available supply is ± 15 V. Using a 741 operational amplifier, design a suitable circuit.

solution

Use the circuit as in Figure 6-8(a). From the 741 data sheet (Appendix 1-11).

$$I_{B(max)} = 500 \text{ nA}$$

$$I_2 \gg I_{B(max)}$$

Make

$$I_2 = 100 \times I_{B(max)}$$
$$= 100 \times 500 \text{ nA} = 50 \ \mu\text{A}$$
$$V_o \simeq \pm (V_{cc} - 1 \text{ V})$$
$$\simeq \pm (15 \text{ V} - 1 \text{ V}) = \pm 14 \text{ V}$$
$$R_2 = V_o / I_2 = 14 \text{ V} / 50 \ \mu\text{A}$$
$$= 280 \text{ k}\Omega \qquad \text{(use 270 k}\Omega \text{ standard value)}$$

I_2 now becomes $\simeq 14$ V$/270$ k$\Omega = 51.9 \ \mu$A

$$R_1 = (\text{UTP or LTP}) / I_2$$
$$= 2 \text{ V} / 51.9 \ \mu\text{A}$$
$$= 38.5 \text{ k}\Omega \qquad \text{(use 39 k}\Omega \text{ standard value)}$$

6-7   IC SCHMITT TRIGGER CIRCUITS

Schmitt trigger circuits are available as integrated circuits, i.e. without any external components required. The 7413, for example (see Appendix 1-14), is described as a *Dual 4-Input Schmitt Trigger*. This means that each IC package contains two complete Schmitt trigger circuits, and each circuit has four input terminals.

The UTP and LTP of the 7413 are identified on the device data sheet as V_{T+} (*positive-going threshold*) and V_{T-} (*negative-going threshold*), respectively. With $V_{CC} = +5$ V, the typical triggering levels are UTP = 1.7 V and LTP = 0.9 V. The circuit output switches (typically) between a low level of 0.2 V and a high output voltage of 3.4 V. The switching time between levels is 15 to 18 ns. Input current might be as high as 1 mA, and the output can handle as much as 55 mA. The input terminals of the 7413 are connected in a way that permits the circuit to function as a *nand gate* (see Sec. 10-4).

Obviously, the 7413 switches very much faster than an IC operational amplifier or discrete component Schmitt trigger circuit. It is also much less expensive than such other circuits, if only because no external components are required. However, its trigger points are fixed, and where different UTP and LTP levels are required it is necessary to use a discrete component or IC operational amplifier circuit.

6-8   IC VOLTAGE COMPARATORS

The output of a *voltage comparator* changes rapidly from one level to another when the input arrives at a predetermined voltage. In this respect the voltage comparator is similar to a Schmitt trigger circuit. Unlike a Schmitt circuit, a voltage comparator has two inputs. The change in output level occurs at the instant that the two input voltages become equal. The input voltages are *compared*, hence the name *comparator*.

Since an IC operational amplifier has two input terminals, it can be employed as a voltage comparator. But operational amplifiers have much slower response times than circuits designed as comparators (i.e., they are relatively slow in switching from *low* output to *high*, and *vice versa*). This is because the circuitry of a comparator prevents its transistors from going into saturation. In an operational amplifier the transistors do saturate when the output is at one extreme or the other. Voltage comparators can be thought of as fast-switching operational amplifiers not usually intended for amplifier applications.

In Appendix 1-19 and 1-20 partial data sheets are shown for two IC voltage comparators, the 710 and the 311. The 710 is described as a high-speed comparator. Its typical response time is listed as 40 ns. The 311 is much slower with a response time of 200 ns. (Considering the slew rate of a 741 operational amplifier, the response time could be as large as 50 μs).

The 311 takes an *input bias current* of only 60 nA typically, while the 710 requires 13 μA. Consequently, the 311 offers a much higher input resistance to signals than does the 710. The *input offset voltage* is the voltage difference between input terminals when the comparator detects equality. For the 710 and 311, this quantity is 0.6 mV and 0.7 mV, respectively. The *differential input voltage*, the maximum allowable voltage difference between input terminals, is ± 5 V for the 710 and ± 14 V for the 311.

The 710 switches between a *low* output of -0.5 V and a *high* output level of $+3.2$ V. The 311 output can switch between ground and a load voltage (separate from the circuit supply) as high as 40 V. The 710 requires supply voltages of $V_{CC} = +12$ V to $+14$ V and $V_{EE} = -6$ V to -7 V. The

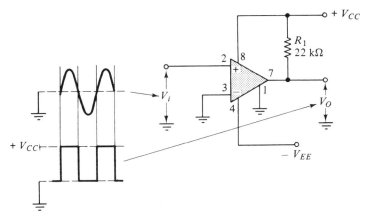

(a)   Zero-crossing detector

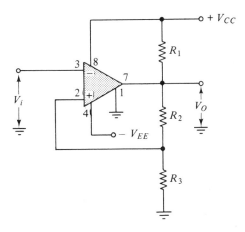

(b) Comparator as a Schmitt trigger

FIGURE 6-10.   Voltage comparator as a zero-crossing detector and as a Schmitt trigger circuit.

311 can operate from a single polarity supply ranging from 5 V to 36 V, or can use the typically available supply of ±15 V.

It is seen that with the single exception of response time, the performance of the 311 comparator is superior to the 710.

Figure 6-10(a) shows a 311 comparator employed as a *zero-crossing detector*. The inverting input is grounded, and the waveform to be monitored is directly connected to the noninverting input. At the instant that the input voltage rises above ground, the output voltage goes positive. When the input drops below ground again, the comparator output goes to zero. The output is a square wave with its leading and lagging edges occurring exactly at the instants that the input waveform crosses the zero level. The zero-crossing detector could be described as a Schmitt trigger with UTP=LTP=0 V. Since UTP=LTP, the circuit is said to have zero hysteresis.

A 311 comparator employed as an inverting-type Schmitt trigger is shown in Figure 6-10(b). The upper trigger point for the circuit is

$$UTP = \frac{V_{CC} \times R_3}{R_1 + R_2 + R_3}$$

and LTP=0 V, because the *low* output level of the circuit is ground.

Other very important applications of voltage comparators are shown in Secs. 14-3, 14-5, and 14-6.

REVIEW QUESTIONS AND PROBLEMS

6-1   Sketch a transistor Schmitt trigger circuit, and briefly explain its operation.

6-2   Define the terms *upper trigger point*, *lower trigger point*, *hysteresis*, and *regeneration*.

6-3   Design a transistor Schmitt trigger circuit with UTP=6 V. Use 2N3904 transistors with $I_C = 1$ mA. The available supply is 15 V. Use standard value resistors.

6-4   The circuit of Problem 6-3 is to have an LTP of 5 V. Make the necessary design modifications and select suitable standard value resistors.

6-5   Plot the output/input characteristics for the Schmitt trigger circuit designed in Problem 6-4.

6-6   (a) Briefly explain how a *speed-up capacitor* improves the switching time of a Schmitt trigger circuit. (b) The Schmitt trigger circuit designed in Problem 6-4 is to be triggered at a maximum frequency of 800 kHz. Determine the maximum size of the speed-up capacitor that may be employed.

6-7 (a) Sketch the circuit of an operational amplifier employed as a Schmitt trigger circuit. Briefly explain how it functions. (b) Show how this circuit could be modified so that (1) LTP$\simeq -0.7$ V, (2) LTP$\simeq 0$ V, (3) UTP$\simeq 0$ V.

6-8 Design inverting and noninverting Schmitt trigger circuits using a 741 operational amplifier. The supply voltage is to be ± 12 V, and the trigger points ± 2 V. Select standard value resistors and calculate the actual triggering levels.

6-9 Plot the output/input characteristics for the Schmitt trigger circuits designed in Problem 6-8.

6-10 The Schmitt trigger circuits designed in Problem 6-8 are to have an LTP adjustable over the range ± 1 V. Suitably modify each circuit.

6-11 The circuits of Problem 6-8 have the following input waveforms:
(a) A triangular waveform with an amplitude of ± 5 V;
(b) A square wave with an amplitude of 3 V;
(c) A square wave with an amplitude of ± 4;
(d) A sine wave with an amplitude of ± 10 V;
(e) A sawtooth wave with an amplitude of ± 6 V.
Sketch the above waveforms and the resultant output wave from the Schmitt trigger circuit for each case.

6-12 Using a 741 IC operational amplifier, design inverting and noninverting Schmitt trigger circuits to have LTP$= -2$ V and UTP$=0$ V. The available supply is ± 9 V.

6-13 Design an inverting Schmitt trigger circuit to have UTP$= +4$ V, LTP$= -3$ V. Use a 741 operational amplifier with $V_{CC}= \pm 18$ V.

6-14 Explain the operation of a voltage comparator circuit, and discuss how it differs from an operational amplifier.

6-15 Sketch the circuit of a voltage comparator employed as a zero-crossing detector. Also sketch the input and output waveforms. Briefly explain.

6-16 A voltage comparator is to be employed as a Schmitt trigger circuit with UTP$=2.5$ V. The available supply is ± 18 V and the external load (connected between $+V_{CC}$ and the output terminal) is 12 kΩ. Sketch the circuit and determine the values of the other components.

CHAPTER 7

Ramp Generators and Integrators

INTRODUCTION

A simple ramp generator circuit can be constructed using a capacitor charged via a resistance, in conjunction with a discharge transistor. To improve the ramp output linearity, a transistor **constant current circuit** can be employed. When the discharge is replaced by a **unijunction** transistor, the circuit becomes a **relaxation oscillator**. The **bootstrap ramp generator**, which produces a closely linear ramp, can be constructed using a transistor or an IC operational amplifier. An IC operational amplifier can also be employed in a **Miller integrator**.

7-1 CR RAMP GENERATOR

The simplest ramp generator circuit is a capacitor charged via a series resistance. A transistor must be connected in parallel with the capacitor to provide a discharge path, as shown in the circuit of Figure 7-1(a). Capaci-

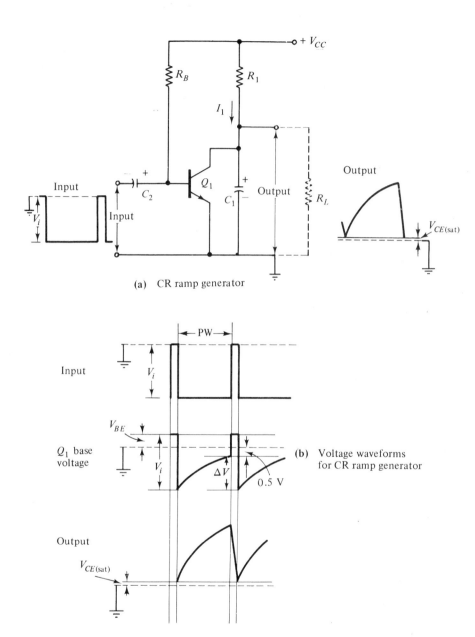

FIGURE 7-1. CR ramp generator circuit and voltage waveforms.

tor C_1 is charged from V_{CC} via R_1. Q_1 is biased *on* via R_B so the capacitor is normally in a discharged state. When a negative-going input pulse is coupled by C_2 to Q_1 base, the transistor switches *off*. Then, C_1 begins to charge; this provides an approximate ramp output until the input pulse ends [see Figure 7-1(b)]. At this point, Q_1 switches *on* again, and rapidly discharges the capacitor.

The output from a simple CR circuit is exponential rather than linear. For voltages very much less than the supply voltage, however, the output is approximately linear. When the transistor is *on*, the capacitor is discharged to $V_{CE(sat)}$. Hence, $V_{CE(sat)}$ is the starting level of the output ramp. Output amplitude control can be provided by making the charging resistance (R_1) adjustable.

Capacitor C_2, which couples the input pulse to the transistor base, should be selected as small as possible, both for minimum cost and smallest possible physical size. The minimum suitable size can be determined by allowing the base voltage of Q_1 to rise during the input pulse time, as shown in Figure 7-1(b). The base voltage starts approximately at 0.7 V when Q_1 is *on*. Then, V_{B2} is pulled negative by the input pulse, but starts to rise again as C_2 is charged through R_B. To ensure that Q_1 is still *off* at the end of the pulse time, V_{B2} should not rise above -0.5 V. This approach to coupling capacitor selection is outlined in Sec. 5-2.

EXAMPLE 7-1

Design a simple CR ramp generator to give an output that peaks at 5 V. The supply voltage is 15 V, and the load to be connected at the output is 100 kΩ. The ramp is to be triggered by a negative-going pulse with an amplitude of 3 V, PW = 1 ms, and the time interval between pulses is 0.1 ms. Take the transistor $h_{FE(min)}$ as 50.

solution

This circuit is shown in Figure 7-1(a). The maximum output current is

$$I_{L(max)} = \frac{V_P}{R_L}$$

$$= \frac{5 \text{ V}}{100 \text{ k}\Omega} = 50 \text{ } \mu\text{A}$$

Select the minimum capacitor charging current $I_1 \gg I_{L(max)}$. At peak output voltage, let

$$I_1 = 100 \times I_{L(max)}$$
$$= 100 \times 50 \ \mu A = 5 \ mA$$
$$R_1 = \frac{V_{CC} - V_P}{I_1}$$
$$= \frac{15 \ V - 5 \ V}{5 \ mA} = 2 \ k\Omega \qquad \text{(use 2.2 k}\Omega \text{ standard value)}$$

The voltages for capacitor C_1 are

$$\text{Initial voltage} = E_o \simeq 0$$
$$\text{Final voltage} = e_c = 5 \ V$$
$$\text{Charging voltage} = E = V_{CC} = 15 \ V$$
$$e_c = E - (E - E_o)\epsilon^{-t/CR} \quad [\text{Equation (2-2)}]$$
$$C_1 = \frac{t}{R \ln \dfrac{E - E_o}{E - e_c}}$$
$$= \frac{1 \ ms}{2.2 \ k\Omega \ \ln \dfrac{15 \ V - 0}{15 \ V - 5 \ V}}$$
$$\simeq 1 \ \mu F$$

The discharge time for C_1 is 0.1 ms, which is one-tenth of the charging time. For Q_1 to discharge C_1 in 1/10 of the charging time,

$$I_C \simeq 10 \times (C_1 \text{ charging current})$$
$$= 10 I_1 = 50 \ mA$$
$$I_B = \frac{I_C}{h_{FE(min)}}$$
$$= \frac{50 \ mA}{50} = 1 \ mA$$
$$R_B = \frac{V_{CC} - V_{BE}}{I_B}$$
$$= \frac{15 \ V - 0.7 \ V}{1 \ mA}$$
$$= 14.3 \ k\Omega \qquad \text{(use 12 k}\Omega \text{ standard value)}$$

For Q_1 to remain biased *off* at the end of the input pulse, let $V_B = -0.5$ V.

$$\therefore \quad \Delta V = V_i - V_{BE} - V_B \quad [\text{See Figure 7-1(b)}]$$
$$= 3\text{ V} - 0.7\text{ V} - 0.5\text{ V} = 1.8\text{ V}$$

The charging current for C_2 is equal to the current through R_B when Q_1 is *off*:

$$I \simeq \frac{V_{CC} - V_i}{R_B} = \frac{15\text{ V} - (-3\text{ V})}{12\text{ k}\Omega}$$
$$= 1.5\text{ mA}$$

From Equation (2-7):

$$C_2 = \frac{It}{\Delta V} = \frac{1.5\text{ mA} \times 1\text{ ms}}{1.8\text{ V}}$$
$$= 0.83\ \mu\text{F} \qquad (\text{use }1\ \mu\text{F standard value})$$

7-2 CONSTANT CURRENT RAMP GENERATORS

7-2.1 Bipolar Transistor Constant Current Circuits

The major disadvantage of the single CR ramp generator is its nonlinearity. To produce a linear ramp, the capacitor charging current must be held constant. This can be achieved by replacing the charging resistance with a *constant current circuit*.

A basic transistor constant current circuit is shown in Figure 7-2(a). The potential divider (R_1 and R_2) provides a fixed voltage V_1 at the base of *pnp* transistor Q_2. The voltage across the emitter resistor R_3 remains constant at $(V_1 - V_{BE})$. Thus, the emitter current is also constant: $I_E = (V_1 - V_{BE})/R_3$. Since $I_C \simeq I_E$, the collector current remains constant. Figure 7-2(b) shows an arrangement that allows the level of constant current to be adjusted. R_4 provides adjustment of V_B. Since $V_3 = (V_B - V_{BE})$, V_3 also is adjustable by R_4, and I_E can be set to any desired level over a range dependent upon R_4.

Figure 7-3(a) shows a ramp generator that employs the constant current circuit. Note that because I_C of Q_2 is a constant charging current

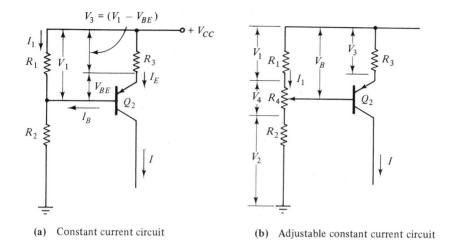

(a) Constant current circuit   (b) Adjustable constant current circuit

FIGURE 7-2.   Transistor fixed and adjustable constant current circuits.

for C_1; the capacitor voltage V_O grows linearly. The simpler capacitor-charging equation, Equation (2-7), may now be used for C_1 calculations. The circuit of Figure 7-3(a) functions like the simple CR ramp generator, with R_1 replaced by the constant current circuit.

The output voltage from the constant current ramp generator remains linear only if a sufficient voltage is maintained across Q_2 for it to operate in the active region of its characteristics. If Q_2 reaches saturation, the output stops at a constant level. Therefore, V_{CE2} should not fall below about 3 V. Because of this and the constant voltage V_3 across resistor R_3, the maximum ramp output voltage obtainable from the circuit of Figure 7-3 is approximately $V_O = V_{CC} - V_3 - 3$ V.

In the circuit of Figure 7-3(b) the input pulse is directly connected to the base of transistor Q_1. When the input is at ground level, Q_1 is *off* and capacitor C_1 charges via Q_2. When a positive input is applied, Q_1 is switched *on* and C_1 is rapidly discharged. Q_1 remains *on* during the positive input pulse; thus C_1 is held in a discharged condition, and the ramp generator output voltage remains at the $V_{CE(sat)}$ of Q_1.

EXAMPLE 7-2

Using a constant current circuit, modify the ramp generator designed in Example 7-1 to produce a linear ramp output.

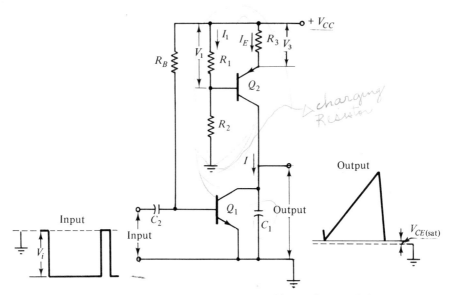

(a) Constant current ramp generator with capacitor-coupled input

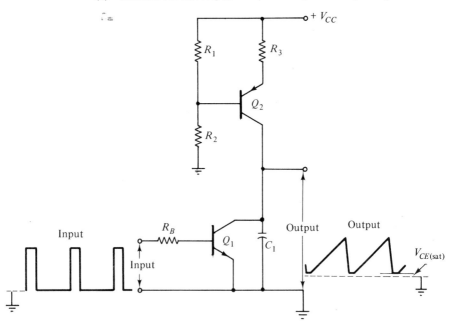

(b) Constant current ramp generator with direct-coupled input

FIGURE 7-3. Constant current ramp generators.

solution

Refer to the circuit of Figure 7-3(a).

$$V_{O(max)} = 5 \text{ V}$$
$$V_3 + V_{CE2} = V_{CC} - 5 \text{ V} = 10 \text{ V}$$

Let

$$V_{CE2} = 3 \text{ V, minimum}$$

then

$$V_3 = 10 \text{ V} - 3 \text{ V} = 7 \text{ V}$$

To maintain a constant level of I_E (and I_C), the voltage across R_3 should be several times larger than the base emitter voltage V_{BE}. This ensures that changes in V_{BE} do not significantly affect I_E.

$$C_1 = \frac{It}{\Delta V} \quad \therefore \quad I = \frac{C_1 \Delta V}{t}$$

For $C_1 = 1 \ \mu\text{F}$, $V = 5$ V, and $t = 1$ ms,

$$I = \frac{1 \ \mu\text{F} \times 5 \text{ V}}{1 \text{ ms}} = 5 \text{ mA}$$

$$R_3 \simeq \frac{7 \text{ V}}{5 \text{ mA}} = 1.4 \text{ k}\Omega \qquad \text{(use 1.2 k}\Omega \text{ standard value)}$$

For $I_E = 5$ mA,

$$V_3 = 5 \text{ mA} \times 1.2 \text{ k}\Omega$$
$$= 6 \text{ V}$$
$$V_1 = V_3 + V_{BE2} = 6 \text{ V} + 0.7 \text{ V}$$
$$= 6.7 \text{ V}$$

V_1 must be a stable bias voltage unaffected by I_{B2}. Make $I_1 \simeq I_E = 5$ mA.

$$R_1 = \frac{V_1}{I_1}$$
$$= \frac{6.7 \text{ V}}{5 \text{ mA}}$$
$$= 1.34 \text{ k}\Omega \qquad \text{(use 1.2 k}\Omega \text{ standard value)}$$

Then I_1 becomes

$$I_1 = \frac{6.7 \text{ V}}{1.2 \text{ k}\Omega} = 5.58 \text{ mA}$$

$$V_2 = V_{CC} - V_1 = 15 \text{ V} - 6.7 \text{ V} = 8.3 \text{ V}$$

$$R_2 \simeq \frac{V_2}{I_1} = \frac{8.3 \text{ V}}{5.58 \text{ mA}}$$

$$= 1.49 \ k\Omega \qquad \text{(use 1.5 k}\Omega \text{ standard value)}$$

EXAMPLE 7-3

Redesign the circuit of Example 7-2 to make the ramp amplitude adjustable from 3 V to 5 V.

solution

The circuit modification is shown in Figure 7-2(b). The charging current, with $\Delta V = 3$ V, is

$$I = \frac{C_1 \Delta V}{t}$$

$$= \frac{1 \ \mu\text{F} \times 3 \text{ V}}{1 \text{ ms}} = 3 \text{ mA}$$

For $\Delta V = 5$ V,

$$I = \frac{1 \ \mu\text{F} \times 5 \text{ V}}{1 \text{ ms}} = 5 \text{ mA}$$

For $I = 3$ mA, $I_E \simeq 3$ mA and

$$V_3 = I_E \times R_3 = 3 \text{ mA} \times 1.2 \text{ k}\Omega$$

$$= 3.6 \text{ V}$$

$$V_B = V_1 = V_3 + V_{BE} = 3.6 \text{ V} + 0.7 \text{ V}$$

$$= 4.3 \text{ V}$$

(At this point, the moving contact on the potentiometer is at the upper end.)

For $R_1 = 1.2$ kΩ,

$$I_1 = \frac{4.3 \text{ V}}{1.2 \text{ k}\Omega} \simeq 3.6 \text{ mA}$$

For $I = 5$ mA,

$$V_3 = 5 \text{ mA} \times 1.2 \text{ k}\Omega$$
$$= 6 \text{ V}$$

and

$$V_B = 6.7 \text{ V}$$

(At this point the potentiometer moving contact is at the lower end.)

$$V_B = V_1 + V_4$$
$$V_4 = 6.7 \text{ V} - 3.6 \text{ V} = 3.1 \text{ V}$$

and

$$R_4 = \frac{V_4}{I_1} = \frac{3.1 \text{ V}}{3.6 \text{ mA}}$$
$$= 0.86 \text{ k}\Omega \qquad \text{(use a 1 k}\Omega \text{ standard potentiometer value)}$$

Then V_4 becomes

$$V_4 = I_1 R_4 = 3.6 \text{ mA} \times 1 \text{ k}\Omega = 3.6 \text{ V}$$

and

$$V_2 = V_{CC} - V_1 - V_4$$
$$= 15 \text{ V} - 4.3 \text{ V} - 3.6 \text{ V}$$
$$= 7.1 \text{ V}$$
$$R_2 = \frac{V_2}{I_1}$$
$$= \frac{7.1 \text{ V}}{3.6 \text{ mA}}$$
$$= 1.97 \text{ k}\Omega \qquad \text{(use 2.2 k}\Omega \text{ standard value)}$$

7-2.2   FET Constant Current Circuits

A field effect transistor with a single source resistance can function as a constant current circuit. A p-channel FET is shown in Figure 7-4(a) with a resistor connected between the source terminal and V_{CC}. With the gate terminal also connected to V_{CC}, the gate-source voltage is the voltage drop across R_S, which is $I_S R_S$ or $I_D R_S$. Referring to the FET transconductance characteristics in Figure 7-4(b), the desired drain/source current (I_S) can be selected and the corresponding gate-source voltage (V_{RS}) determined as illustrated. Then,

$$R_S = \frac{V_{RS}}{I_S} \tag{7-1}$$

This approach is satisfactory only when the transconductance characteristic for the particular FET has been plotted. For any given FET type, there are two possible extreme characteristics as shown in Figure 7-4(c). These occur because of the spread in values of *drain-source saturation current* [$I_{DSS(\max)}$ and $I_{DSS(\min)}$], and *pinch-off voltage* [$V_{P(\max)}$ and $V_{P(\min)}$]. In this case it is necessary to draw a *bias line* for each possible value of source resistance. This is done simply by using Equation (7-1) to determine two convenient corresponding values of I_D and V_{GS}:
When $V_{GS} = 0$,

$$I_D = V_{RS}/R_S = 0$$

Plot point A on Figure 7-4(c) at $V_{GS} = 0$ and $I_D = 0$.
When $V_{RS} = 6$ V and $R_S = 3.3$ kΩ,

$$I_D = 6 \text{ V}/3.3 \text{ k}\Omega = 1.8 \text{ mA}$$

Plot point B on Figure 7-4(c) at $V_{GS} = 6$ V and $I_D = 1.8$ mA.
Draw the bias line through points A and B. The maximum and minimum source current levels that can flow are now shown at the intersections of the bias lines and the characteristics. When it is desired to set I_S to a precise level R_S must be made adjustable.

One important caution that must be observed when using a FET constant current circuit is that the drain-source voltage V_{DS} must not be allowed to fall below the maximum value of pinch-off voltage $V_{P(\max)}$. Just as a bipolar transistor cannot be expected to function linearly if its collector-base voltage approaches the saturation level, so too a FET will not function correctly in a linear circuit if its drain-source voltage falls below the pinch-off level.

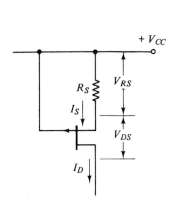

(a) FET constant current circuit

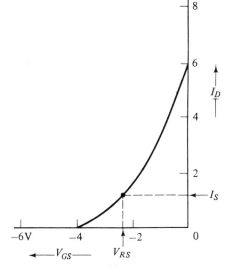

(b) FET transconductance characteristics

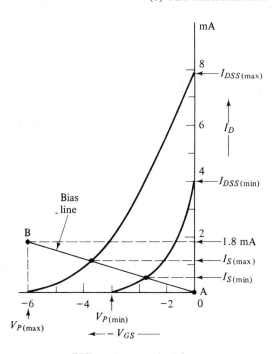

(c) FET maximum and minimum
transconductance characteristics

FIGURE 7-4. **FET constant current circuit and transconductance characteristics.**

A *constant current diode* (or *field effect diode*) is essentially a FET and a resistor connected as illustrated in Figure 7-4(a), and contained in a single package. These devices can be purchased with various constant current levels.

7-3   UJT RELAXATION OSCILLATORS

7-3.1   The Unijunction Transistor

The basic construction of a unijunction transistor (UJT) and its equivalent circuit are shown in Figure 7-5. The device can be thought of as a bar of lightly doped *n*-type silicon with a small piece of heavily doped *p*-type joined to one side [see Figure 7-5(a)]. The *p*-type is named the *emitter*, while the two end terminals of the bar are designated bases 1 and 2 (B_1 and B_2), as shown. In the equivalent circuit of Figure 7-5(b), the silicon bar is represented as two resistors, r_{B1} and r_{B2} while the *pn* junction formed by the emitter and the bar is represented by a diode.

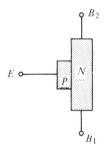

(a) Basic construction

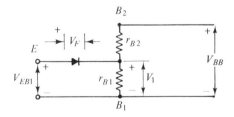

(b) Equivalent circuit

FIGURE 7-5.   Basic construction and equivalent circuit of unijunction transistor.

The ratio, $r_{B1}/(r_{B1}+r_{B2})$ is termed the *intrinsic standoff ratio* of the UJT, and is designated η. Thus the voltage across r_{B1} is given by

$$V_1 = V_{BB}\frac{r_{B1}}{r_{B1}+r_{B2}}$$

or

$$V_1 = V_{BB}\eta \qquad\qquad (7\text{-}2)$$

The *pn* junction becomes forward-biased at a peak voltage, $V_P = V_{EB1} = V_1 + V_F$. When this peak is reached, the flow of charge carriers through r_{B1} causes its resistance to fall. Thus, a capacitor connected across E and B_1 is rapidly discharged. The flow of current into the emitter terminal continues until V_E falls to the *emitter saturation voltage $V_{EB1(sat)}$*, at which time the device switches *off*.

Two more important parameters for the UJT are *peak point current I_P* and the *valley point current I_V*. The peak point current is the minimum emitter current that must flow for the UJT to switch *on* or *fire*. This current occurs when V_E is at the *firing voltage*, that is, at peak point V_P. The valley point current is the emitter current that flows when V_E is at the emitter saturation voltage, $V_{EB1(sat)}$.

7-3.2 UJT Relaxation Oscillator

A unijunction transistor can be used in conjunction with a capacitor and a charging circuit, to construct an oscillator with an approximate ramp-type output. Figure 7-6(a) shows the simplest form of such a circuit, which is called a *UJT relaxation oscillator*. The UJT remains *off* until its emitter voltage V_{EB1} approaches the firing voltage V_P for the particular device. At this point, the UJT switches *on* and a large emitter current I_E flows. This causes capacitor C_1 to discharge rapidly. When the capacitor voltage falls to the emitter saturation level, the UJT switches *off*, allowing C_1 to begin to charge again.

The frequency of a relaxation oscillator can be made variable by switched selection of capacitors and/or by adjustment of the charging resistance [see Figure 7-6(b)]. The resistance R_2, in series with UJT terminal B_1, allows synchronizing input pulses to be applied. When an input pulse pulls B_1 negative, V_{EB1} is increased to the level at which the UJT fires. Once the UJT fires it will not switch *off* again until the capacitor is discharged.

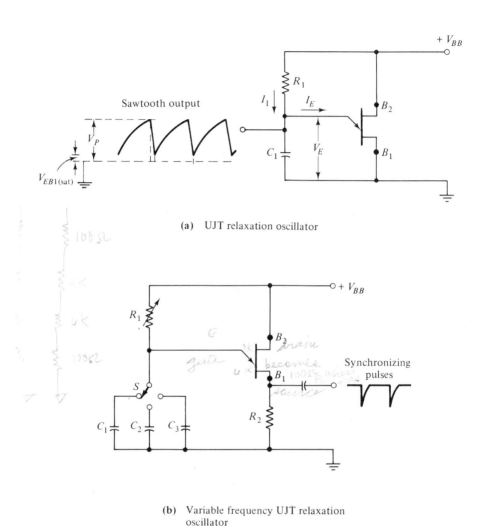

(a)   UJT relaxation oscillator

(b)   Variable frequency UJT relaxation
      oscillator

FIGURE 7-6. Basic UJT relaxation oscillator and variable frequency circuit.

In the design of a UJT relaxation oscillator, the charging resistance R_1 must be selected between certain upper and lower limits. Resistance R_1 must not be so large that the emitter current is less than the peak point current when V_{EB1} is at the firing voltage; otherwise, the device may not switch *on*. If R_1 is very small, then when V_{EB1} is at the emitter saturation level, a current greater than the valley point current might flow into the emitter terminal. In this case, the UJT may not switch *off*. Thus, for correct UJT operation, R_1 must be selected between two limits that allow the emitter current to be a minimum of I_P and a maximum of I_V.

The UJT oscillator circuits shown in Figures 7-6(a) and (b) will produce exponential output waveforms because the capacitors are charged by resistances. Constant current circuits could be used here to generate linear ramp output waveforms.

EXAMPLE 7-4

The circuit of Figure 7-6(a) is to use a 2N3980 UJT. The supply voltage V_{BB} is 20 V, and output frequency is to be 5 kHz. Design a suitable circuit, and calculate the output amplitude.

solution

Capacitor C_1 charges from $V_{EB1(sat)}$ to the firing voltage, $V_P = V_F + \eta V_{BB}$. The data sheet for the 2N3980 (Appendix 1-12) gives the following specifications:

$$V_{EB1(sat)} = 3 \text{ V maximum}, \qquad I_p = 2 \ \mu\text{A}, \qquad I_V = 1 \text{ mA}$$

and

$$\eta = 0.68 \text{ to } 0.82$$
$$\simeq 0.75 \text{ average}$$
$$V_P = 0.7 + (0.75 \times 20 \text{ V})$$
$$= 15.7 \text{ V}$$

Therefore, for the capacitor,

$$E = \text{Supply voltage} = V_{BB} = 20 \text{ V}$$
$$E_o = \text{Initial charge} = V_{EB1(sat)} = 3 \text{ V}$$
$$e_c = \text{Final charge} = V_P = 15.7 \text{ V}$$

Now, to select R_1:

$$R_{1(max)} = \frac{V_{BB} - V_P}{I_P}$$
$$= \frac{20 \text{ V} - 15.7 \text{ V}}{2 \ \mu\text{A}} \simeq 2.15 \text{ M}\Omega$$

$$R_{1(\min)} = \frac{V_{BB} - V_{EB1(sat)}}{I_V}$$

$$= \frac{20\,\text{V} - 3\,\text{V}}{1\,\text{mA}} \simeq 17\,\text{k}\Omega$$

So R_1 must be in the range 17 kΩ to 2.15 MΩ. If R_1 is very large, C_1 must be a very small capacitor. Let R_1 have a value of 22 kΩ; then from Equation (2-2)

$$C_1 = \frac{t}{R_1 \ln\left(\dfrac{E - E_o}{E - e_c}\right)}$$

$$t = \frac{1}{\text{Output frequency}} = \frac{1}{5\,\text{kHz}} = 200\,\mu\text{s}$$

then

$$C_1 = \frac{200\,\mu\text{s}}{22\,\text{k}\Omega \ln\left(\dfrac{20\,\text{V} - 3\,\text{V}}{20\,\text{V} - 15.7\,\text{V}}\right)}$$

$$= 6600\,\text{pF} \qquad \begin{bmatrix} \text{use 6800 pF standard capacitor} \\ (\textit{see Appendix } 2\text{-}2) \end{bmatrix}$$

$$\text{Output amplitude} = V_P - V_{EB1(sat)}$$
$$= 15.7\,\text{V} - 3\,\text{V}$$
$$= 12.7\,\text{V}$$

7-4 FOUR-LAYER DIODE RELAXATION OSCILLATOR

A very simple relaxation oscillator can be constructed using a *four-layer diode* or *Shockley diode*.

The theory of operation of the four-layer diode is illustrated in Figure 7-7. The device consists of four semiconductor layers p_1, n_1, p_2, and n_2, with a connecting terminal to p_1 identified as the *anode* and that to n_2 as

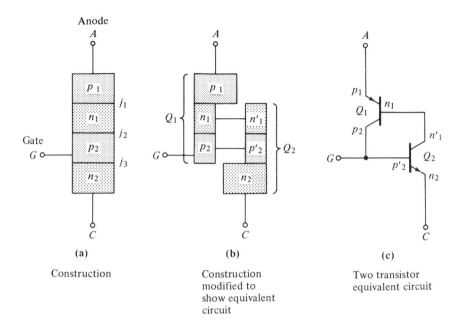

Anode
A

Gate
G

(a)
Construction

(b)
Construction
modified to
show equivalent
circuit

(c)
Two transistor
equivalent circuit

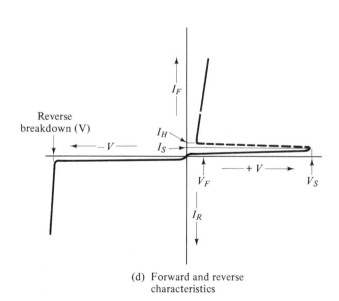

(d) Forward and reverse
characteristics

FIGURE 7-7. Four-layer diode: construction, equivalent circuit, and characteristics.

the *cathode* [Figure 7-7(a)]. To help understand the operation of the device, Figure 7-7(b) shows layers n_1 and p_2 split into sections n_1, n'_1, p_2 and p'_2. Since these sections are connected together, there is no real change. However, p_1, n_1, and p_2 can now be thought of as a *pnp* transistor, and n'_1, p'_2 and n_2 can be considered an *npn* transistor. This gives the two-transistor equivalent circuit in Figure 7-7(c). Now return to Figure 7-7(a) and note that when the anode is biased positively with respect to the cathode (forward bias), junctions j_1 and j_3 are forward-biased while junction j_2 is reverse-biased. Thus, at small forward bias voltages only a very low leakage current flows. When the forward bias is increased to the breakdown voltage of junction j_2, a large forward current flows.

Going back to Figure 7-7(c), it is seen that when a substantial current flows into the emitter of transistor Q_1, an equally large collector current flows from the collector of Q_1 into the base of Q_2. Similarly, Q_2 has substantial emitter and collector currents, and the collector current of Q_2 provides base current for Q_1. The result is that both transistors are switched *on* into saturation, and the total anode-to-cathode voltage V_{AK} is around 0.9 V.

The device forward characteristics shown in Figure 7-7(d) can now be understood. When $+V$ is small only a low level leakage current flows. When the breakdown or *switching voltage* V_s is reached, j_2 breaks down, the two transistors switch *on* into saturation, and the device voltage rapidly falls to a low level V_F. Any further increase in forward current now causes only a slight increase in V_F. The device reverse characteristics are similar to those of a reverse-biased diode, except that two junctions j_1 and j_3 must break down before the four-layer diode goes into reverse breakdown.

The circuit of a four-layer diode relaxation oscillator is shown in Figure 7-8. Note the device circuit symbol. Capacitor C_1 is charged via resistor R_1 until the diode switching voltage is reached. Then D_1 rapidly switches to the low level *on* voltage V_F, discharging C_1 in the process. D_1 continues conducting until its current falls below the minimum level that can maintain conduction. This level is known as the holding current I_H [see Figure 7-7(d)]. Once D_1 ceases to conduct, current flows into the capacitor again (via R_1), charging it up to V_s. The cycle is repetitive, generating an output waveform as illustrated.

In designing a relaxation oscillator using a four-layer diode the considerations are similar to those that apply for the UJT. Resistor R_1 must not be so large that the current flowing through it is less than the diode switching current I_S. Neither can it be so small that its current is greater than the diode holding current I_H. Thus R_1 is selected between these limitations, and C_1 is determined exactly as in the case of the UJT relaxation oscillator.

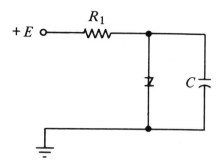

Relaxation oscillator using four-layer diode.

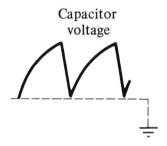

FIGURE 7-8.   Relaxation oscillator using four-layer diode.

7-5 PROGRAMMABLE UJT RELAXATION OSCILLATORS

The *programmable unijunction transistor* (PUT) is a four-layer device used in a particular way to simulate a UJT. The interbase resistances r_{B1} and r_{B2} and the intrinsic standoff ratio η may be programmed to any desired values by selecting two resistors. This means that the device firing voltage V_P can also be programmed.

Consider Figures 7-9(a) and (b). The *gate* of the *pnpn* device is connected to the junction of resistors R_1 and R_2. The gate voltage is

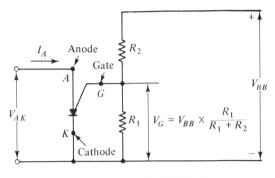

(a) Programmable UJT circuit

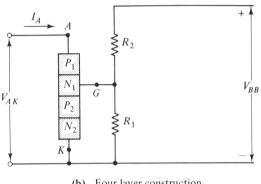

(b) Four layer construction
of programmable UJT

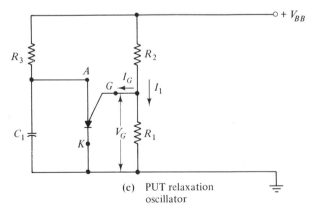

(c) PUT relaxation
oscillator

FIGURE 7-9. Programmable UJT, four-layer construction, and PUT relaxation oscillator circuit.

$V_G = V_{BB} R_1/(R_1 + R_2)$. The device will trigger *on* when the input voltage V_{AK} makes the anode (layer P_1) positive with respect to the gate (layer N_1). [This forward-biases the base-emitter junction of Q_1 in Figure 7-7(c).] When this occurs, the anode-to-cathode voltage rapidly drops to a low level, and the device conducts heavily from anode to cathode. This situation continues until the current becomes too low to sustain conduction. With the anode used as an emitter terminal, and with R_1 and R_2 substituted for r_{B1} and r_{B2}, the circuit action simulates a UJT. Figure 7-9(c) shows the PUT employed in a relaxation oscillator.

A data sheet for 2N6027 and 2N6028 PUT devices is included in Appendix 1-13. For the 2N6027, the value of I_P is given as 1.25 μA typical, and I_V as 18 μA typical. The offset voltage, which is equivalent to $V_{EB1(\text{sat})}$, is typically 0.7 V.

EXAMPLE 7-5

Design a relaxation oscillator using a 2N6027 PUT. The supply voltage is 15 V, and the output is to be 5 V peak at 1 kHz.

solution

The circuit is as shown in Figure 7-9(c).

$$V_P = V_G + (p_1 n_1 \text{ junction voltage drop})$$
$$5 \text{ V} = V_G + 0.7 \text{ V}$$
$$V_G = 5 \text{ V} - 0.7 \text{ V} = 4.3 \text{ V}$$

To provide a stable gate bias voltage the current through the potential divider (R_1 and R_2) must be much larger than the gate current at switch-*on:*

$$I_1 \gg I_G$$

Since $I_G \simeq 5$ μA (typical), let

$$I_1 = 100 \times I_G$$
$$= 100 \times 5 \text{ }\mu\text{A} = 0.5 \text{ mA}$$
$$R_1 = \frac{V_G}{I_1} = \frac{4.3 \text{ V}}{0.5 \text{ mA}}$$
$$= 8.6 \text{ k}\Omega \qquad (\text{use } 8.2 \text{ k}\Omega \text{ standard})$$

Now, I_1 becomes

$$I_1 = \frac{4.3 \text{ V}}{8.2 \text{ k}\Omega} = 524 \text{ }\mu\text{A}$$

$$V_{R2} = V_{BB} - V_G$$

$$= 15 \text{ V} - 4.3 \text{ V} = 10.7 \text{ V}$$

$$R_2 = \frac{V_{R2}}{I_1} = \frac{10.7 \text{ V}}{524 \text{ }\mu\text{A}}$$

$$= 20.4 \text{ k}\Omega \qquad \text{(use 18 k}\Omega \text{ standard)}$$

Now V_G becomes

$$V_G = \frac{15 \text{ V} \times 8.2 \text{ k}\Omega}{18 \text{ k}\Omega + 8.2 \text{ k}\Omega}$$

$$= 4.69 \text{ V} \qquad (i.e., \text{ instead of 4.3 V})$$

and

$$V_P = V_G + 0.7 \text{ V}$$

$$= 4.69 \text{ V} + 0.7 \text{ V} = 5.39 \text{ V}$$

The valley voltage V_V is 0.7 V.
For the capacitor C_1;

$$E = \text{Supply voltage} = V_{BB} = 15 \text{ V}$$

$$E_O = \text{Initial charge} = V_V = 0.7 \text{ V}$$

$$e_c = \text{Final charge} = V_P = 5.39 \text{ V}$$

For selection of R_3:

$$R_{3(max)} = \frac{V_{BB} - V_P}{I_P}$$

$$= \frac{15 \text{ V} - 5.39 \text{ V}}{1.25 \text{ }\mu\text{A}} = 7.7 \text{ M}\Omega$$

$$R_{3(min)} = \frac{V_{BB} - V_V}{I_V}$$

$$= \frac{15 \text{ V} - 0.7 \text{ V}}{18 \text{ }\mu\text{A}} = 790 \text{ k}\Omega$$

Thus, R_3 must be in the range from 790 kΩ to 7.7 MΩ. Let $R_3 = 1$ MΩ.

$$t = \frac{1}{\text{Output frequency}} = \frac{1}{1 \text{ kHz}} = 1 \text{ ms}$$

and from Equation (2-2)

$$C_1 = \frac{t}{R_3 \ln\left(\dfrac{E - E_O}{E - e_c}\right)}$$

$$= \frac{1 \text{ ms}}{1 \text{ M}\Omega \ln\left(\dfrac{15 \text{ V} - 0.7 \text{ V}}{15 \text{ V} - 5.39 \text{ V}}\right)}$$

$$= 0.0025 \ \mu\text{F} \qquad \text{(standard value)}$$

7-6 TRANSISTOR BOOTSTRAP RAMP GENERATOR

The circuit of a transistor *bootstrap ramp generator* is shown in Figure 7-10(a). The ramp is generated across capacitor C_1, which is charged via resistance R_1. The discharge transistor Q_1 holds the capacitor voltage V_1 down to $V_{CE(\text{sat})}$ until a negative input pulse is applied. Transistor Q_2 is an emitter follower that provides a low-output impedance. The emitter resistor R_E is connected to a negative supply level, rather than to ground. This is to ensure that Q_2 remains conducting when its base voltage V_1 is close to ground. Capacitor C_3, known as the *bootstrapping capacitor*, has a much larger capacitance than C_1. The function of C_3, as will be shown, is to maintain a constant voltage across R_1 and thus maintain the charging current constant.

To understand the operation of the bootstrap ramp generator, first consider the dc voltage levels before an input signal is applied. Transistor Q_1 is *on*, and its voltage is $V_{CE(\text{sat})}$, which is typically 0.2 V. Thus $V_1 = 0.2$ V. This level is indicated as point A on the graph of voltage V_1 in Figure 7-10(b). The emitter of Q_2 is now at $(V_1 - V_{BE2})$, which is also the output voltage V_O (point B on the V_O graph). At this time, the voltage at the cathode of diode D_1 is $V_K = V_{CC} - V_{D1}$, where V_{D1} is the diode forward voltage drop. The voltage, $V_{CC} - V_{D1}$ is shown at point C on the graph of V_K [Figure 7-10(b)]. The voltage across capacitor C_3 is the difference between V_K and V_O.

When Q_1 is switched *off* by an input pulse, C_1 starts to charge via R_1. Voltage V_1 now increases, and the emitter voltage V_O of Q_2 (the emitter follower) also increases. Thus, as V_1 grows V_O also grows, remaining only V_{BE} below V_1 [see Figure 7-10(b)]. As V_O increases, the lower terminal of

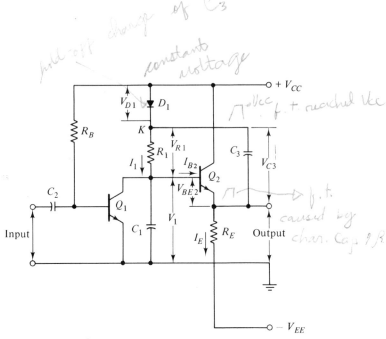

(a) Bootstrap ramp generator

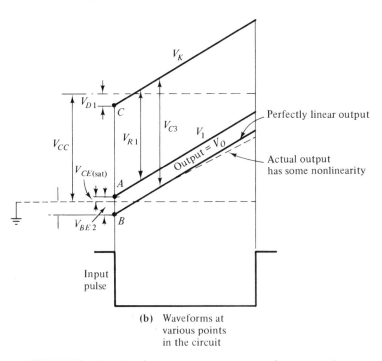

(b) Waveforms at
various points
in the circuit

FIGURE 7-10. Transistor bootstrap ramp generator and circuit waveforms.

C_3 is *pulled up*. Because C_3 is a large capacitor it retains its charge, and as V_O increases the voltage at the upper terminal of C_3 also increases. Thus, V_K increases as V_O increases, and V_K remains V_{C3} volts above V_O. In fact, V_K goes above the level of V_{CC}, and D_1 is reverse-biased. The constant voltage across C_3 maintains the voltage V_{R1} constant across R_1. Therefore, the charging current through R_1 is held constant, and the capacitor charges linearly, giving a linear output ramp.

During the ramp time D_1 is reverse-biased as already explained, and the charging current through R_1 is provided by capacitor C_3. If C_3 is very large and I_1 is small, then C_3 will discharge by only a very small amount. When the input pulse is removed and C_1 is discharged rapidly by Q_1, V_O drops to its initial level. Also, V_K drops, allowing D_1 to become forward-biased. At this time a current pulse through D_1 replaces the small charge lost from C_3. The circuit is then ready to generate another output ramp.

In addition to producing a very linear output ramp, another advantage of the bootstrap generator is that the amplitude of the ramp can approach the level of the supply voltage. Note that the output ramp amplitude may be made adjustable over a fixed time period by making R_1 adjustable.

The broken line on the graph of output voltage [Figure 7-10(b)] shows that the output, instead of being perfectly linear, may be slightly nonlinear. If the difference between the actual output and the ideal output is 1% of the output peak voltage, then the ramp can be said to have 1% nonlinearity. Some nonlinearity results from the slight discharge of C_3 that occurs during the ramp time. Another source of nonlinearity is the base current I_{B2}. As the capacitor voltage grows, I_{B2} increases. Since I_{B2} is part of I_1, the capacitor charging current decreases slightly as I_{B2} increases. Thus, the charging current does not remain perfectly constant, and the ramp is not perfectly linear. The design of a bootstrap ramp generator begins with a specification of ramp linearity. This dictates the charging current and the capacitance of C_3. The percentage of nonlinearity usually is allocated in equal parts to ΔI_{B2} and ΔV_{C3}.

EXAMPLE 7-6

Design a transistor bootstrap ramp generator to provide an output amplitude of 8 V over a time period of 1 ms. The ramp is to be triggered by a negative-going pulse with an amplitude of 3 V, a pulse width of 1 ms, and a time interval between pulses of 1 ms. The load resistor to be supplied has a value of 1 kΩ and the ramp is to be linear within 2%. The supply voltage is to be ±15 V. Take $h_{FE(min)} = 100$.

solution

The circuit is shown in Figure 7-10(a).

$$R_E = R_L = 1 \text{ k}\Omega$$

When $V_O = 0$,

$$I_E \simeq \frac{V_{EE}}{R_E}$$

$$\frac{15 \text{ V}}{1 \text{ k}\Omega} = 15 \text{ mA}$$

When $V_O = V_P$,

$$I_E \simeq \frac{V_{EE} + V_P}{R_E}$$

$$= \frac{15 \text{ V} + 8 \text{ V}}{1 \text{ k}\Omega} = 23 \text{ mA}$$

$$I_{B2} \simeq \frac{I_{E2}}{h_{FE}}$$

At $V_O = 0$,

$$I_{B2} = \frac{15 \text{ mA}}{100} = 0.15 \text{ mA}$$

At $V_O = V_P$,

$$I_{B2} = \frac{23 \text{ mA}}{100} = 0.23 \text{ mA}$$

$$\Delta I_{B2} = 0.23 \text{ mA} - 0.15 \text{ mA} = 80 \text{ }\mu\text{A}$$

Allow 1% nonlinearity due to ΔI_{B2}, (that is, ΔI_{B2} represents a loss of charging current to C_1):

$$I_1 = 100 \times \Delta I_{B2}$$
$$= 100 \times 80 \text{ }\mu\text{A}$$
$$= 8 \text{ mA}$$
$$C_1 = \frac{I_1 t}{\Delta V} = \frac{I_1 \times (\text{Ramp time})}{V_P}$$
$$= \frac{8 \text{ mA} \times 1 \text{ ms}}{8 \text{ V}}$$

$$= 1 \; \mu F \qquad \text{(standard capacitor value)}$$

$$V_{R1} = V_{CC} - V_{D1} - V_{CE(\text{sat})}$$

$$= 15 \text{ V} - 0.7 \text{ V} - 0.2 \text{ V}$$

$$= 14.1 \text{ V}$$

$$R_1 = \frac{V_{R1}}{I_1} = \frac{14.1 \text{ V}}{8 \text{ mA}}$$

$$= 1.76 \text{ k}\Omega \qquad \text{(use 1.8 k}\Omega \text{ standard value)}$$

For 1% nonlinearity due to C_3 discharge,

$$\Delta V_{C3} = 1\% \text{ of initial } V_{C3} \text{ level}$$

$$V_{C3} \simeq V_{CC} = 15 \text{ V}$$

$$\Delta V_{C3} = \frac{15 \text{ V}}{100} = 0.15 \text{ V}$$

and C_3 discharge current is equal to $I_1 = 8$ mA.

$$C_3 = \frac{I_1 t}{\Delta V_{C3}} = \frac{8 \text{ mA} \times 1 \text{ ms}}{0.15 \text{ V}}$$

$$= 53 \; \mu F \qquad \text{(use 56 } \mu F \text{ standard capacitance value)}$$

R_B and C_2 are calculated in the same way as for Example 7-1.

Note that the recharge path for C_3 is via D_1 and R_E in the circuit of Figure 7-10(a). Using Equation (2-2) and the component values from Example 7-6, it is found that the time required to recharge C_3 by ΔV_{C3} of 0.15 V is approximately 0.6 ms. This means that the time interval between ramp outputs (and between input pulses) should be not less than 0.6 ms. Where Q_2 is replaced by a *complementary emitter follower* or *voltage follower* (see Sec. 7-7), the recharge time for C_3 is usually small enough to be ignored.

7-7 IC BOOTSTRAP RAMP GENERATOR

An IC operational amplifier (see Sec. 5-4) connected as a *voltage follower* forms part of the bootstrap ramp generator in Figure 7-11. When an operational amplifier is used as a voltage follower, the inverting input

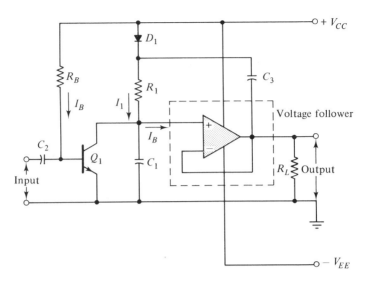

FIGURE 7-11.   Bootstrap ramp generator using an IC operational amplifier.

terminal is connected directly to the output. The input signal is applied at the noninverting input.

The operation of the voltage follower can best be understood if it is assumed that both input terminals are initially at ground level. The output is also at ground level at this time. *Note that the output from an operational amplifier is the amplified voltage difference between the two input terminals.* Now, suppose an input of 1 V is applied at the noninverting terminal. Since the amplifier has a very large gain, the output tends to move positively towards the saturation level. However, as the output increases positively, the voltage at the inverting terminal also increases positively. When the inverting terminal voltage equals the noninverting terminal voltage (*i.e.,* 1 V), there is no longer any voltage difference between the two input terminals. Consequently, there is no longer an input signal, and the output voltage ceases to increase. Thus, the output voltage *follows* the input very closely.

Actually, there is a small voltage difference between the input terminals of a voltage follower. This difference is equal to the output voltage divided by the amplifier gain. For a 741 with an output of 10 V, the input difference would be typically:

$$\frac{10 \text{ V}}{200,000} = 50 \ \mu\text{V}$$

This means that the output voltage is only 50 μV behind the input voltage. This is a big improvement on the transistor emitter follower, where V_O is typically 0.7 V behind V_i.

It is seen that the voltage follower is an amplifier with a gain of 1, and that the output closely follows the input. The voltage follower also has the high input impedance and low output impedance characteristic of the IC operational amplifier.

The circuit of the IC operational amplifier bootstrap generator is almost exactly like that of the transistor bootstrap circuit. The voltage follower takes the place of the emitter follower. Note that although a $\pm$ V supply is still required, the load resistance R_L now can be grounded. Also note that the output ramp starts at $V_{CE(\text{sat})}$ instead of at $V_{CE(\text{sat})} - V_{BE}$. The low input current to the operational amplifier has an almost negligible effect on the charging current to C_1 in the IC bootstrap circuit of Figure 7-11. In fact, the reverse leakage current of D_1 (when it is reverse-biased) is much more significant than the input bias current of the amplifier. Using a 1N914 diode (Appendix 1), I_R is typically 3 μA. For the 741, the maximum input bias current is 500 nA. (Note that for the transistor bootstrap circuit, I_R of D_1 is very much smaller than I_B of transistor Q_2.) The leakage current of D_1 can be the starting point for the IC bootstrap circuit design. This results in a lower charging current to C_1 and in smaller values of C_1, C_2, and C_3.

If D_1 leakage current is extremely small, the above approach may result in a very small charging current and consequently in a very small capacitance value for C_1. The typical input capacitance for an oscilloscope is 30 pF. So C_1 should not be made small enough that the circuit performance is affected when an oscilloscope is connected to any part of it. As a minimum, C_1 should be selected approximately 1000 times greater than the typical 30 pF C_{in} of an oscilloscope. This will also ensure that C_1 is not affected by the *stray capacitance* of wiring, etc.

EXAMPLE 7-7

Design a bootstrap ramp generator using a 741 operational amplifier. The specifications for the circuit are the same as those for the circuit of Example 7-6, with the exception that the time interval between input pulses is 0.1 ms.

solution

The circuit is shown in Figure 7-11.

$$R_L = 1\ k\Omega$$

$$I_R = 3\ \mu A \qquad \text{(when } D_1 \text{ is reverse-biased)}$$

Allow 1% nonlinearity due to I_R:

$$I_1 = 100 \times I_R$$

$$= 100 \times 3 \ \mu A = 300 \ \mu A$$

$$C_1 = \frac{I_1 t}{\Delta V} = \frac{I_1 \times (\text{Ramp time})}{V_P}$$

$$= \frac{300 \ \mu A \times 1 \ ms}{8 \ V}$$

$$= 0.0375 \ \mu F \qquad (\text{use } 0.039 \ \mu F \text{ standard value})$$

$$V_{R1} = V_{CC} - V_{D1} - V_{CE(\text{sat})}$$

$$= 15 \ V - 0.7 \ V - 0.2 \ V$$

$$= 14.1 \ V$$

$$R_1 = \frac{V_{R1}}{I_1} = \frac{14.1 \ V}{300 \ \mu A}$$

$$= 47 \ k\Omega \qquad (\text{standard value})$$

For 1% nonlinearity due to C_3 discharge:

$$\Delta V_{C3} = 1\% \text{ of initial } V_{C3}$$

$$V_{C3} \simeq V_{CC} = 15 \ V$$

$$\Delta V_{C3} = \frac{15 \ V}{100} = 0.15 \ V$$

C_3 discharge current $= I_1 = 300 \ \mu A$

$$C_3 = \frac{I_1 t}{\Delta V_{C3}} = \frac{300 \ \mu A \times 1 \ ms}{0.15 \ V} = 2 \ \mu F \qquad (\text{standard value})$$

Compare this to $C_3 = 56 \ \mu F$ for the transistor circuit of Example 7-6. The discharge time of C_1 is equal to one-tenth of the charge time. Therefore, the discharge current of C_1 is ten times greater than the charge current.

Minimum I_C of $Q_1 = 10 \times I_1$

$$= 10 \times 300 \ \mu A = 3 \ mA$$

$$I_B = \frac{I_C}{h_{FE}} = \frac{3 \ mA}{100}$$

$$= 30 \ \mu A$$

$$R_B = \frac{V_{CC} - V_{BE}}{I_B}$$

$$= \frac{15 \ V - 0.7 \ V}{30 \ \mu A}$$

$$= 477 \ k\Omega \qquad (\text{use } 470 \ k\Omega \text{ standard value})$$

During the input pulse, $\Delta V_{C2} = 1.8$ V (see Example 7-1) and the charging current of C_2 can be expressed by

$$I = \frac{V_{CC} - V_i}{R_B} = \frac{15 \text{ V} - (-3 \text{ V})}{470 \text{ k}\Omega}$$

$$= 38 \ \mu\text{A}$$

Thus,

$$C_2 = \frac{It}{\Delta V} = \frac{38 \ \mu\text{A} \times 1 \text{ ms}}{1.8 \text{ V}}$$

$$= 0.02 \ \mu\text{F} \qquad \text{(standard value)}$$

7-8 FREE-RUNNING RAMP GENERATOR

A bootstrap ramp generator may be made free-running by employing a Schmitt circuit to detect the output peak level and generate a capacitor discharge pulse. In the circuit shown in Figure 7-12(a) *pnp* transistor Q_1 discharges C_1 when the Schmitt circuit output is negative. Diode D_2 protects the base-emitter junction of Q_1 against excessive reverse bias when the Schmitt output is positive.

Consider the circuit waveforms shown in Figure 7-12(b). During the time that the Schmitt circuit output is positive, Q_1 remains *off* and C_1 charges; this provides a positive-going ramp output. When the ramp amplitude arrives at the UTP of the Schmitt circuit, the Schmitt output becomes negative. This causes I_{B1} to flow, biasing Q_1 *on* and discharging C_1. As the voltage of capacitor C_1 falls, the ramp output also falls rapidly, and this continues until the Schmitt LTP is reached. The presence of D_3 makes the Schmitt circuit have an LTP close to ground (see Sec. 6-7.2). Therefore, when the ramp output falls to ground level, the Schmitt output goes positive again, switching Q_1 *off* and allowing ramp generation to commence again.

The free-running ramp generator can be synchronized with another waveform by means of negative pulses coupled via capacitor C_3. The presence of the negative pulse lowers the UTP of the Schmitt circuit, so that the Schmitt output becomes negative, causing the ramp to go to zero when the synchronizing pulse is applied.

Potentiometer R_7 [Figure 7-12(c)] allows the charging current to C_1 to be adjusted, thus controlling the ramp length and the output frequency. In Figure 7-12(d) R_6 affords adjustment of the Schmitt UTP. This provides control of the ramp amplitude.

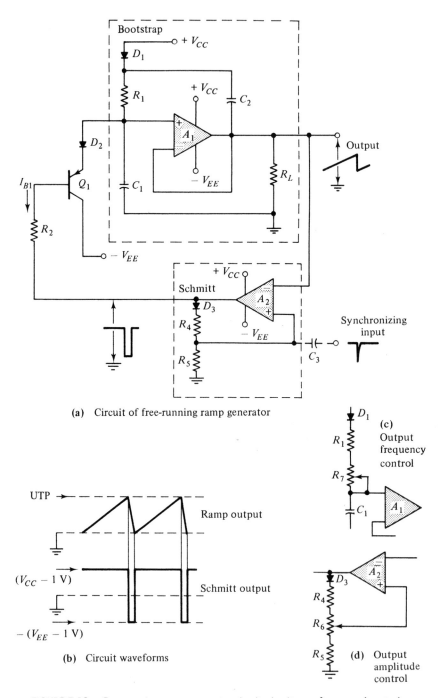

(a) Circuit of free-running ramp generator

(c) Output frequency control

(b) Circuit waveforms

(d) Output amplitude control

FIGURE 7-12. Free-running ramp generator circuit, circuit waveforms, and controls.

EXAMPLE 7-8 _____

Design a free-running ramp generator with an output frequency of 1 kHz and an output amplitude in the range 0 V to 8 V. Use 741 operational amplifiers and a supply voltage of ± 15 V.

solution

Schmitt circuit. For an output of 0 V to 8 V, the Schmitt circuit must have an LTP of 0 V and a UTP of 8 V. Design the Schmitt circuit as explained in Section 6-7.

Bootstrap circuit. The bootstrap output should go from 0 V to 8 V over a time period of 1/1 kHz. (*i.e.*, 1 ms). Design the circuit as in Example 7-7, substituting a *pnp* transistor for Q_1.

_____

7-9   MILLER INTEGRATOR CIRCUIT

7-9.1   Miller Effect

Consider the circuit of Figure 7-13 in which an operational amplifier is connected as an *inverting amplifier*. Let the amplifier voltage gain be $-A_V$. Then,

$$V_O = -A_V V_i$$

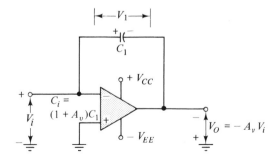

FIGURE 7-13.   Miller effect or amplification of capacitance by inverting amplifier.

Note that because of the amplifier phase shift, the voltage at the left-hand terminal of C_1 increases by V_i, while that at the right-hand terminal of the capacitor decreases by $A_V V_i$ when V_i is positive. This results in a total capacitor voltage change of

$$\Delta V_1 = V_i + A_V V_i$$

$$= V_i(1 + A_V)$$

Using the equation $Q = C \times \Delta V$, the charge supplied to the capacitor is

$$Q = C_1 \times \Delta V_1$$

$$= C_1 \times V_i(1 + A_V)$$

or

$$Q = (1 + A_V)C_1 \times V_i$$

Thus it appears that the input has supplied a charge to a capacitor with a value of $(1 + A_V)C_1$, instead of C_1 alone. Capacitance C_1 is said to have been *amplified* by a factor of $(1 + A_V)$. This is known as the *Miller effect*.

7-9.2 Miller Integrator

The *Miller integrator* utilizes the Miller effect to generate a linear ramp. In the circuit of Figure 7-14(a), a square wave input supplies charging current, alternatively positive and negative, to C_1. The noninverting input terminal is grounded by a resistance R_2 equal to the resistance R_1 at the inverting input terminal. This is to ensure that the small bias currents cause equal voltage drops at each input terminal. Recall that, because of the very large gain of the operational amplifier, the voltage difference between the two input terminals is never greater than about 50 μV. Thus, it can be said that the inverting input terminal is always very close to ground level. The inverting input terminal of an inverting amplifier is frequently termed a *virtual ground*, or *virtual earth*. Thus, the input voltage appears across R_1 and the input current is simply V_i/R_1, which remains constant.

If the input current I_1 is much greater than the input bias current of the amplifier, then I_1 will not flow into the amplifier. Instead, effectively all of I_1 flows through capacitor C_1. For a positive input voltage, I_1 flows into C_1, charging it positively on the left-hand side and negatively on the right-hand side [Figure 7-14(b)]. In this case the output voltage becomes negative, because the positive terminal, that is, the left-hand terminal, of the capacitor is held at the virtual ground level of the inverting input. A

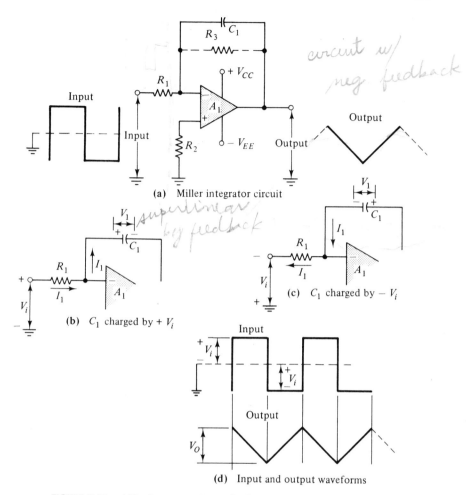

circuit w/ neg feedback

superlinear by feedback

(a) Miller integrator circuit

(b) C_1 charged by $+ V_i$

(c) C_1 charged by $- V_i$

(d) Input and output waveforms

FIGURE 7-14.   Miller integrator circuit, C_1 charging action, and waveforms.

negative input voltage produces a flow of current out of C_1 [Figure 7-14(c)]. Thus the capacitor is charged negatively on the left-hand side and positively on the right-hand side. Now the output becomes positive, because the negative terminal of the capacitor is held at virtual ground.

Since I_1 is a constant ($+$ or $-$) quantity, and since effectively all of I_1 flows through the capacitor, C_1 is charged linearly. Thus the output voltage changes linearly, providing either a positive or negative ramp. When the input voltage is positive, the output is a negative-going ramp. When the input is negative, a positive-going output ramp is generated. Therefore, when the input is a square wave, the output waveform is triangular. This is illustrated in Figure 7-14(d).

Consider the Miller circuit of Figure 7-14(a). If the input is supposed to be zero but is, say, 20 μV away from ground level, then the output voltage could be $(A_v \times 20 \ \mu V) = \pm(200,000 \times 20 \ \mu V) = \pm 4$ V. In this case the output is said to have *drifted* from its zero level. Even when the input terminal is maintained exactly at ground level, there could be a slight difference in the voltage at the amplifier inputs, due to small differences in the resistances of R_1 and R_2, for example. Thus, because of the very high gain of the operational amplifier, its output voltage is very likely to drift from the zero level. The output voltage drift produces a charge on capacitor C_1; this charge gives the output an *offset* so that it is not symmetrical above and below ground (see Figure 7-15).

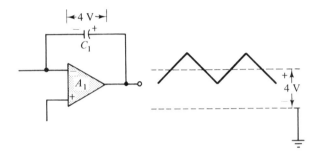

FIGURE 7-15.   Effect of charge on C_1 due to output drift.

To minimize the output voltage drift, a large resistance [R_3 in Figure 7-14(a)] is connected between the output and the inverting input terminals. The effect of this resistance is to *cut down* the dc gain of the amplifier. When $R_3/R_1 = 10$, for example, the output drift will be only 10 times the input voltage difference. A ratio of 10:1 is typical for R_3/R_1.

The presence of R_3 has the disadvantage that it affects the performance of the integrator at low frequencies. If the input frequency is so low that the capacitance impedance is very much larger than R_3, then the capacitor has a negligible effect and the circuit will not function as an integrator. Therefore, C_1 should be selected so that

$$X_{C1} \ll R_3$$

As a lower limit, $X_{C1} = R_3/10$, so

$$\frac{1}{2\pi f C_1} = \frac{R_3}{10}$$

The lowest operating frequency of the integrator is

$$f = \frac{10}{2\pi C_1 R_3} \qquad (7\text{-}3)$$

The design of a Miller integrator circuit begins with selection of the input current I_1 very much larger than the amplifier bias current. Then, R_1 is calculated as V_i/I_1. From $C = It/V$, C_1 is determined using the desired output voltage, the time period, and the input current.

EXAMPLE 7-9 _____

Design a Miller integrator circuit to produce a triangular waveform output with a peak-to-peak amplitude of 4 V. The input is a ± 10 V square wave with a frequency of 250 Hz. Use a 741 operational amplifier with a supply of ± 15 V. Calculate the lowest operating frequency for the integrator.

solution

The circuit is shown in Figure 7-14(a). The 741 data sheet in Appendix 1-11 gives the input bias current as

$$I_B = 500 \text{ nA, maximum} \qquad I_1 \gg I_B$$

Let

$$I_1 = 1 \text{ mA}$$

$$R_1 = \frac{V_i}{I_1} = \frac{10 \text{ V}}{1 \text{ mA}}$$

$$= 10 \text{ k}\Omega$$

Let

$$R_3 = 10 \, R_1 = 100 \text{ k}\Omega$$

$$R_2 = R_3 \| R_1 \simeq 10 \text{ k}\Omega$$

The ramp length is equal to one-half of the time period of the input, which is $1/(2f)$, or

$$t = \frac{1}{2 \times 250 \text{ Hz}} = 2 \text{ ms}$$

The ramp amplitude is equal to the peak-to-peak voltage output, which is 4 V.

$$C_1 = \frac{It}{\Delta V} = \frac{1 \text{ mA} \times 2 \text{ ms}}{4 \text{ V}} = 0.5 \ \mu\text{F}$$

Thus from Equation (7-3) the lowest operating frequency is

$$f = \frac{10}{2\pi \times 0.5 \ \mu\text{F} \times 100 \text{ k}\Omega}$$

$$= 32 \text{ Hz}$$

The circuit in Figure 7-16 shows a Miller integrator operating as a ramp generator. The negative-going pulse generates the positive ramp by producing current I_1 in the direction shown. At this time n-channel FET (Q_1) is biased *off* by the negative input pulse. When the input goes to ground level, I_1 goes to zero and Q_1 is switched *on*. Q_1 discharges C_1 and keeps it discharged until the input becomes negative again. If C_1 is to be discharged in one-tenth of the charge time, then Q_1 must be able to pass a

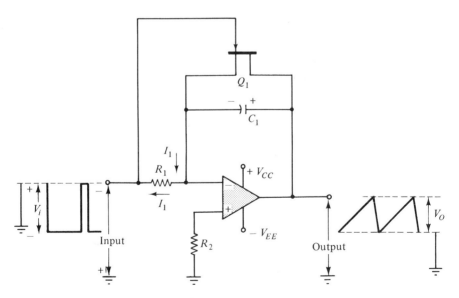

FIGURE 7-16.   Miller integrator circuit as ramp generator.

current ten times greater than the charge current I_1. To ensure that Q_1 is biased *off* when the input pulse is present, the input pulse must have a negative amplitude greater than the FET pinchoff voltage. Because the capacitor is completely discharged by the action of the FET, there is no need to include resistor R_3 [Figure 7-14(a)] in this circuit.

7-10   TRIANGULAR WAVEFORM GENERATOR

A free-running triangular waveform generator can be constructed, using the output of the Miller circuit in Figure 7-14(a) to generate its own square wave input. Consider the circuit in Figure 7-17(a). The Miller integrator circuit used is exactly as discussed in the last section. The output of the Miller circuit is fed directly to a noninverting Schmitt trigger circuit. The Schmitt is designed to have a positive UTP and a negative LTP (see Sec. 6-7.3), and its output is applied as an input to the Miller circuit.

Operation of the circuit is easily understood by considering the waveforms in Figure 7-17(b). At time t_1 the integrator output has reached the UTP (a positive voltage) and the noninverting Schmitt circuit output is positive at approximately $-(V_{EE}-1$ V$)$. The positive voltage from the Schmitt causes current I_1 to flow into the Miller circuit, charging C_1 positive on the left-hand side. As C_1 charges in this direction, the integrator output is a negative-going ramp. The integrator continues to produce a negative-going ramp while its input is a positive voltage. At time t_2, the integrator output arrives at the LTP (negative voltage). The Schmitt trigger circuit output now becomes negative and reverses the direction of I_1. Thus the integrator output becomes a positive-going ramp. This positive-going ramp generation continues until the integrator output arrives at the UTP of the Schmitt circuit once again. Synchronizing pulses may be applied via C_2 to lower the trigger point of the Schmitt circuit, causing it to trigger before the ramp arrives at its normal peak level.

The circuit described above generates a triangular waveform with a constant peak-to-peak output amplitude and a constant frequency. The modifications shown in Figures 7-17(c) and (d) allow both frequency and amplitude adjustment. R_5 adjusts the UTP and LTP of the Schmitt circuit, and thus controls the peak-to-peak output amplitudes. R_6 affords adjustment of the input current to the integrator, and therefore controls the rate of charge of C_1. This means that the ramp time period is controlled by adjusting R_6.

In the design of a triangular waveform generator, each section must be treated separately.

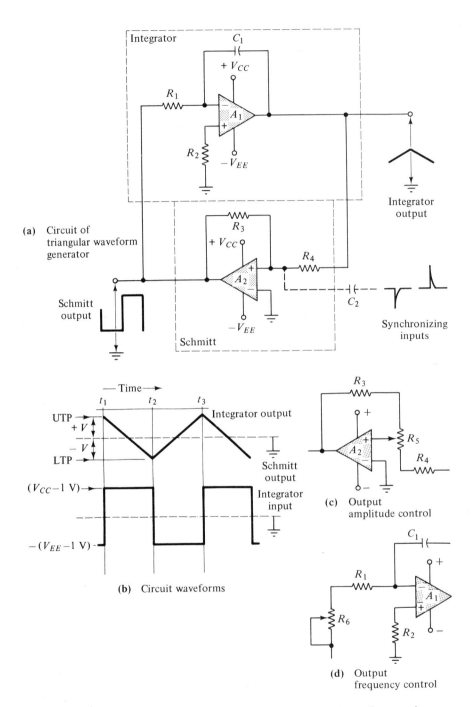

FIGURE 7-17. Triangular waveform generator circuit, circuit waveforms, and controls.

(a) Circuit of triangular waveform generator

(b) Circuit waveforms

(c) Output amplitude control

(d) Output frequency control

EXAMPLE 7-10 _____

Design a free-running triangular waveform generator to have a peak-to-peak output of 4 V at a frequency of 250 Hz. Use 741 operational amplifiers and a supply voltage of ± 15 V.

solution

Schmitt circuit. For 4 V, p-to-p, the Schmitt circuit UTP = 2 V and LTP = −2 V. A Schmitt circuit can be designed as in Example 6-6 to give these desired trigger points.

Miller integrator circuit. The input to the Miller circuit is the Schmitt output, that is, ≃ ± 14 V.

$$\text{Ramp amplitude} = 4 \text{ V}$$
$$\text{Ramp time period} = \frac{1}{2f} = \frac{1}{2 \times 250 \text{ Hz}}$$
$$= 2 \text{ ms}$$

Design the Miller circuit as in Example 7-9.

_____

When amplitude and frequency controls are employed with the circuit of Figure 7-17, a disadvantage is that the amplitude control affects the frequency. The frequency control Figure 7-17(d) sets the capacitor charging current through R_1 and R_6 at a constant level. Variation of the UTP and LTP of the Schmitt circuit has no effect on the rate of charge. Consequently, when the output amplitude is adjusted, for example from ±2 V to ±4 V, the time required for the capacitor to charge from −4 V to +4 V is twice that to charge from −2 V to +2 V. This means that the output frequency has been halved. The problem is solved by replacing the noninverting Schmitt with an inverting Schmitt trigger circuit and an inverting amplifier, as illustrated in Figure 7-18.

The amplifier has a gain of 1 when $R_7 = R_8$, and with the amplifier input terminal connected to moving contact of R_5, the capacitor charging current is always proportional to the Schmitt trigger voltage. Thus, if the output voltage is doubled, the capacitor charging current is doubled, and the frequency is unaffected. In designing this circuit, the current through R_3, R_4 and R_5 must be made much larger than that through R_7 and R_8.

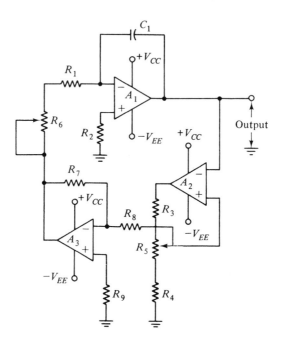

FIGURE 7-18.   Triangular waveform generator with independent frequency and amplitude controls.

7-11   CRT TIME BASE

The need to synchronize a ramp generator is best understood by considering the time base for a cathode-ray oscilloscope. While the signal to be displayed provides vertical deflection of the electron beam, a ramp generator provides horizontal deflection. To obtain a correctly displayed signal, the electron beam must start at the left-hand side of the tube at the same instant that the input signal is going positive. Thus, the ramp must commence at this instant.

Figure 7-19 illustrates the process of obtaining synchronism of the ramp and the input signal. The signal normally is applied to a *vertical amplifier*, which controls the voltage on the vertical deflecting plates. This amplifier produces two equal output voltages, which are opposite in polarity, to the deflecting plates. One output is also fed to a Schmitt trigger which simply converts it to a square wave. The square wave is then *differentiated* to obtain a spike waveform, and the negative spikes are clipped off. This results in a series of positive spikes, each of which occurs exactly at the instant that the signal is entering its positive half-cycle.

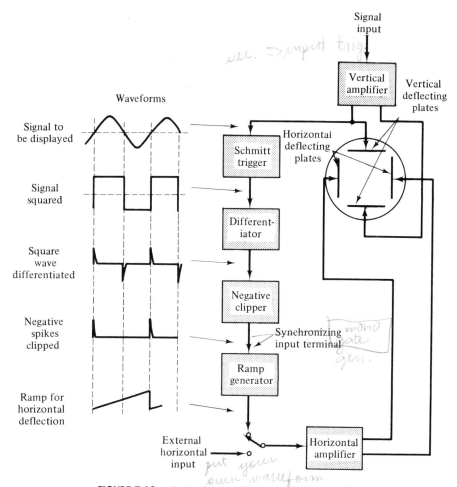

FIGURE 7-19. Automatic time base for cathode ray tube.

These spikes are fed to the ramp generator synchronizing input terminal, causing the ramp to return instantaneously to its starting level.

REVIEW QUESTIONS AND PROBLEMS

7-1 Sketch the circuit of a simple CR ramp generator. Briefly explain its operation and its limitations. Also sketch the typical input and output waveforms.

7-2 Design a CR ramp generator to give an output of 3 V peak. The supply voltage is 20 V, and the load to be connected at the output is

330 kΩ. The ramp is to be triggered by a negative-going pulse with an amplitude of 4 V, PW = 3 ms, and time interval between pulses of 0.3 ms. Take the transistor $h_{FE(min)} = 70$.

7-3   Sketch the circuit of a CR ramp generator using a transistor constant current circuit. Briefly explain how the circuit operates. Also sketch typical input and output waveforms.

7-4   Design a transistor constant current circuit for the CR ramp generator designed in Problem 7-2.

7-5   Redesign the constant current circuit of problem 7-4 to make the ramp amplitude adjustable from 2 V to 4 V.

7-6   Sketch a FET constant current circuit. Explain how the circuit operates and discuss the necessary current and voltage levels.

7-7   A FET with the transconductance characteristics shown in Figure 7-4(c) has $R_S = 2.2$ kΩ. Draw the bias line and determine the maximum and minimum levels of drain current that could flow.

7-8   Sketch the circuit of a UJT relaxation oscillator with adjustable output frequency. Sketch the output waveform, and show how the circuit can be synchronized by external pulses. Briefly explain the operation of the circuit.

7-9   Design a relaxation oscillator using a 2N3980 UJT. The supply voltage is 25 V and the output frequency is to be 2 kHz. Calculate the amplitude of the output waveform.

7-10 Sketch the four-layer construction, two-transistor equivalent circuit, and characteristics of a four-layer diode. Explain how the device operates.

7-11 The four-layer diode employed in the circuit of Figure 7-8 has $V_S = 10$ V and $V_F = 1$ V. If $E = 30$ V, $R_1 = 4.7$ kΩ, and $C = 0.5$ μF, determine the amplitude and frequency of the output waveform.

7-12 Using a 2N6027 PUT, design a relaxation oscillator to operate from a supply of 20 V. The output is to be 7 V peak at a frequency of 3 kHz.

7-13 Sketch the circuit of a transistor bootstrap ramp generator. Show the waveforms, and explain the operation of the circuit.

7-14 A transistor bootstrap generator is to produce an output of 7 V, with a time period of 2.5 ms. The load resistor is to be 1.2 kΩ, and the ramp is to be linear to within 3%. Design a suitable circuit using transistors with $h_{FE(min)} = 120$ and $V_{CC} = \pm 20$ V. The input pulse has: P.A. $= -5$ V, P.W. $= 2.5$ ms, and (space width) $= 1$ ms.

7-15 Sketch the circuit of a bootstrap ramp generator using an IC operational amplifier. Briefly explain the operation of the circuit, drawing a comparison between it and the transistor bootstrap circuit.

7-16 Design a bootstrap generator using a 741 operational amplifier. The circuit specification is the same as for the circuit in problem 7-14.

7-17 Sketch the circuit of a free-running bootstrap ramp generator. Show the waveforms and carefully explain the operation of the circuit. Also, show how the input frequency and amplitude may be controlled.

7-18 Design a free-running bootstrap ramp generator using 741 IC operational amplifiers. The output ramp is to have an amplitude of ± 3 V and frequency of 2 kHz. Use a supply voltage of ± 12 V.

7-19 Sketch the circuit of a Miller integrator, and explain its operation. Show output and input waveforms.

7-20 Design a Miller integrator circuit to produce a triangular output waveform with a peak-to-peak amplitude of 3 V. The input is a ± 8 V square wave with a frequency of 750 Hz. Use a 741 operational amplifier with a supply of ± 12 V. Calculate the lowest operating frequency for the integrator.

7-21 Sketch a Miller integrator circuit connected to operate as a ramp generator. Show the input and output waveforms, and explain the circuit operation.

7-22 Sketch the circuit of a free-running triangular waveform generator using IC operational amplifiers. Show all the waveforms in the circuit, and carefully explain the overall circuit operation. Also, show how the output amplitude and frequency may be controlled.

7-23 Design a free-running triangular waveform generator to have an output of ± 2.5 V at a frequency of 500 Hz. Use 741 operational amplifiers and a supply of ± 12 V.

7-24 The circuit designed for problem 7-23 is to have its output amplitude and frequency adjustable by $\pm 20\%$. Make the necessary design modifications.

7-25 Design a free running triangular waveform generator in which the adjustable output amplitude does not affect the frequency. Use a 741 operational amplifier with $V_{CC} = \pm 18$ V. Make $V_o = (\pm 2$ V to ± 6 V), and $f = 300$ Hz.

7-26 Sketch the block diagram of an automatic time base for a cathode-ray tube. Show the waveforms at the various points in the diagram, and explain the operation of the system.

CHAPTER 8

Monostable and Astable Multivibrators and the 555 Timer

INTRODUCTION

The *monostable multivibrator* has one stable state, and may be triggered temporarily into another state. When triggered, the circuit generates an output pulse of constant width and amplitude. In the collector-coupled monostable circuit, the transistors are switched into saturation. The emitter-coupled circuit may be designed for saturated or unsaturated operation. An IC operational amplifier may be employed as a monostable multivibrator by the external connection of appropriate resistances and capacitances.

The *astable multivibrator* has no stable state; that is, the circuit oscillates between two temporary states. Astable circuits normally are designed to operate as square wave generators. The *555 IC Timer* can be connected to function as a monostable circuit or as an astable multivibrator.

8-1 COLLECTOR-COUPLED MONOSTABLE MULTIVIBRATOR

The monostable multivibrator (also known as a *one-shot multivibrator*) has a single stable condition. One transistor is normally *on* and the other transistor is normally *off*. The condition can be reversed by application of a triggering pulse, which turns *on* the normally *off* transistor and switches *off* the normally *on* transistor. The reversed condition lasts only for a brief time period, dependent upon the circuit components.

Consider the *collector-coupled* monostable circuit shown in Figure 8-1. The circuit is described as collector-coupled because the collector terminal of Q_2 is coupled via R_1 and R_2 to the base terminal of Q_1. In the normal dc condition of the circuit, base current I_{B2} is provided from V_{CC} to Q_2 via resistance R_B. Thus transistor Q_2 is normally *on*. At this time, diode D_1 is forward-biased and has no significant effect on Q_2. The function of D_1 will become apparent later. With Q_2 *on* in saturation, the collector voltage of Q_2 is $(V_{CE(sat)} + V_{D1})$ above ground level. The base voltage V_{B1} of Q_1 is determined by the negative supply voltage V_{BB} and by R_1 and R_2, as well as the collector voltage of Q_2. With Q_2 collector near ground level, V_{B1} is

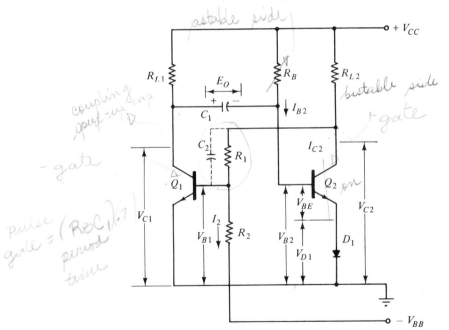

FIGURE 8-1.   Collector-coupled monostable multivibrator circuit.

likely to be negative (*i.e.*, Q_1 base is biased below its grounded emitter). Therefore, with Q_2 normally *on*, Q_1 is normally *off*.

When Q_1 is *off*, its collector current is zero. Therefore, there is no voltage drop across R_{L1}, and the collector of Q_1 is at the supply voltage level V_{CC}. Also, with Q_2 on, the base voltage of Q_2 is $V_{B2} = V_{BE} + V_{D1}$. On the right-hand terminal of capacitor C_1 the voltage is V_{B2}, and on the left-hand terminal it is V_{CC}. Hence the capacitor voltage is $E_o = V_{CC} - V_{B2}$, positive on the left-hand side as shown in Figure 8-1.

Now consider what would occur if Q_1 were triggered *on* to saturation for a brief instant. (This could be made to occur by, for example, capacitor-coupling a positive-going spike to the base of Q_1). The collector voltage of Q_1 drops almost to ground level. Capacitor C_1 will not lose its charge E_o instantaneously; therefore, when the left-hand terminal of C_1 drops to $V_{CE(sat)}$ the right-hand terminal will drop to $(V_{CE(sat)} - E_o)$. Consequently, Q_2 base voltage goes to $(V_{CE(sat)} - E_o)$—*i.e.*, Q_2 is biased *off*. With Q_2 *off*, there is no longer a collector current to produce a voltage drop across R_{L2}. Thus, V_{C2} rises, Q_1 base is biased above ground level, and Q_1 remains *on*. It is seen that when Q_1 is triggered *on* briefly, Q_2 goes *off* and Q_1 remains *on*. As will be seen Q_1 stays *on* only for a brief time.

The transistor switching process is illustrated by the waveforms in Figure 8-2. Prior to Q_1 being triggered on, the voltages are: $V_{B1} = -V$, $V_{C1} = V_{CC}$, $V_{B2} \simeq 1.4$ V, $V_{C2} \simeq 0.9$ V. When Q_1 is triggered *on*, $V_{B1} \simeq 0.7$ V, $V_{C1} \simeq 0.2$ V, $V_{B2} = (V_{C1} - E_o) \simeq -E_o$, $V_{C2} \simeq V_{CC}$.

With the exception of V_{B2} all the above voltages remain constant while Q_2 stays biased *off*. V_{B2} does not remain constant because C_1 discharges via R_B (see Figure 8-3). Voltage E_o across C_1 initially is positive on the left-hand side and negative on the right-hand side. Current I flowing into the right-hand side of C_1 will tend to discharge C_1 and then recharge it with reversed polarity. Thus V_{B2} begins to rise toward ground level. When C_1 is discharged to $e_{c1} \simeq 0$ V, Q_2 base-emitter and D_1 begin to be forward-biased again. At this point I_{C2} again begins to flow and V_{C2} starts to fall (see Figure 8-2). When V_{C2} falls, it causes V_{B1} to fall; consequently V_{C1} rises and causes V_{B2} to rise. The result of this is that Q_1 rapidly switches *off* and Q_2 rapidly comes *on* again, when C_1 is discharged to approximately zero volts. At this time, the negative spike at Q_1 base is due to *speed-up capacitor* C_2 (Figure 8-1) transmitting all the Q_2 collector voltage change to the base of Q_1 and then discharging. When Q_1 switches *off* and Q_2 switches *on* again, C_1 is rapidly recharged to E_o via R_{L1} and Q_2 base. The circuit has now returned to its normal stable state and remains in this condition until Q_1 is triggered *on* again.

Refer again to the waveforms in Figure 8-2. It is seen that when the voltage at Q_2 collector is a positive-going pulse, that at Q_1 collector is a

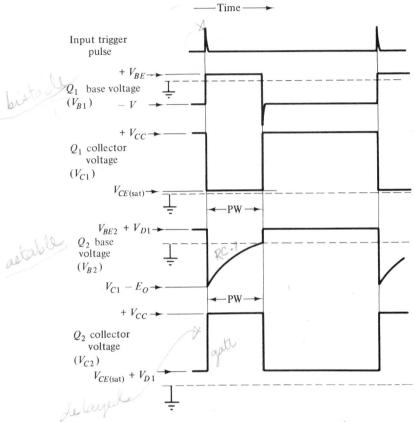

FIGURE 8-2. Monostable multivibrator circuit waveforms.

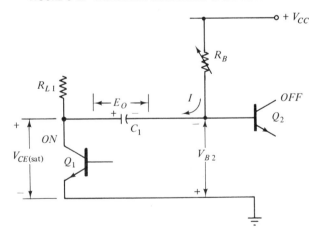

FIGURE 8-3. Negatively charged capacitor C_1 discharged via R_B when Q_1 is *on* and Q_2 is *off*.

negative-going pulse. These two pulses are equal in width and either or both may be taken as output from the circuit. The pulse width of the output depends upon the values of C_1 and R_B. If R_B is made variable the output pulse width may be adjusted.

The monostable multivibrator now can be described as a circuit with one stable state, capable of producing an output pulse when triggered. The output pulse width is constant, and can be made adjustable by making R_B adjustable (see Figure 8-3).

The purpose of D_1 is to protect the base-emitter voltage of transistor Q_2 against excessive reverse bias. When Q_1 is triggered *on*, V_{B2} falls to approximately $-V_{CC}$. Most transistors will not survive a reverse base-emitter voltage of more than 5 V, while most diodes might easily take reverse bias of 50 V without breaking down. Thus diode D_1, in series with Q_2 emitter terminal, allows large negative voltages to be applied to Q_2 base. In some circuits, the reverse bias at the base of Q_1 may be excessive, and in this case a diode should be connected in series with Q_1 emitter.

Capacitor C_2 in Figure 8-1 is a speed-up capacitor to improve the transistor turn-*on* and turn-*off* times. The function of the speed-up capacitor is discussed in detail in Sec. 4-4.

8-2 DESIGN OF A COLLECTOR-COUPLED MONOSTABLE MULTIVIBRATOR

The design of a monostable multivibrator usually begins with specifications of the output pulse width, the supply voltage, the load, and perhaps, the transistors to be employed. As with other circuits, it might be possible to connect the load directly into the circuit as either R_{L1} or R_{L2}. More frequently, the load is coupled to Q_2 collector, and R_{L2} is selected to be much smaller than the load resistance. Alternatively, I_{C2} may be selected to be much larger than the maximum output load current.

I_{C2} and R_{L2} must be chosen so that transistor Q_2 is in saturation, that is, $(I_{C2}R_{L2}) \simeq V_{CC}$. Base current I_{B2} is calculated as $I_{C2}/h_{FE(min)}$, the minimum value of h_{FE} ensuring that I_{B2} is large enough to drive Q_2 to saturation. The base resistance R_{B2} is then calculated as $(V_{CC} - V_{B2})/I_{B2}$. R_{L1} usually is made equal to R_{L2}, and so $I_{C1} \simeq I_{C2}$.

Resistors R_1 and R_2 provide *on* or *off* bias to Q_1 base. For a stable bias voltage V_{B1}, the current I_2 that flows through R_1 and R_2 should be much larger than the base current to Q_1. When Q_1 is *on*, I_{B1} flows through R_1. If I_{B1} is not much smaller than I_2, variations in I_{B1} may upset the bias voltage at Q_1 base. To achieve the condition $I_2 \gg I_{B1}$, R_1 and R_2 should be selected as small as possible. However, R_1 and R_2 also constitute a load on resistance R_{L2}; thus to avoid overloading R_{L2}, R_1 and R_2 should be chosen as large as possible. These contradictory requirements are met by applying the rule of thumb that $I_2 \simeq I_{C2}/10$. Since $I_{C1} \simeq I_{C2}$, I_{B1} becomes $I_{C2}/h_{FE(min)}$. Then, I_2 is $(h_{FE(min)}/10) \times I_{B1}$. When a design is worked through, it will be seen that making $I_2 \simeq I_{C2}/10$ also results in R_1 and R_2 each being 5 to 10 times R_{L2}.

The output pulse width for a monostable circuit is dictated by the time taken for C_1 to discharge from its initial voltage level to approximately zero volt. Therefore, C_1 is calculated from Equation (2-2).

$$e_c = E - (E - E_o)\epsilon^{-t/CR} \qquad [\text{Equation (2-2)}]$$

For the collector-coupled monostable circuit,

$$e_c \simeq 0, \qquad E = V_{CC}, \qquad \text{and}$$
$$E_o = -(V_{CC} - V_{B2})$$

Note that C_1 is charged initially positive on the left-hand side and negative on the right-hand side. When Q_1 is *on* and Q_2 is *off*, C_1 tends to charge negative on the left-hand side, and positive on the right-hand side. Thus, the initial voltage E_o of C_1 must be taken as negative, and the charging voltage E as positive.

The initial value of Q_2 base voltage, that is, when Q_2 is *on*, is

$$V_{B2} = V_{BE} + V_{D1}$$

and

$$E_o = -(V_{CC} - V_{BE} - V_{D1})$$
$$t = \text{Specified PW}$$
$$C = C_1$$
$$R = R_B, \text{ which is}$$

the resistance through which C_1 is charged when Q_1 is *on* and Q_2 is *off*.

The speed-up capacitor C_2 is determined by a method similar to that employed for the inverter circuit and the Schmitt trigger circuit.

EXAMPLE 8-1

A collector-coupled monostable multivibrator is to operate from a ± 9 V supply. Transistor collector currents are to be 2 mA, and the transistors

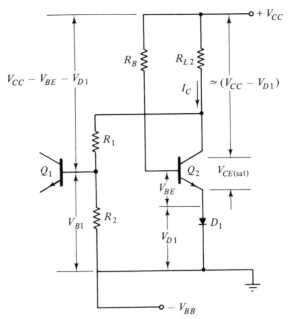

(a) Q_2 *on* in saturation
Q_1 *off*

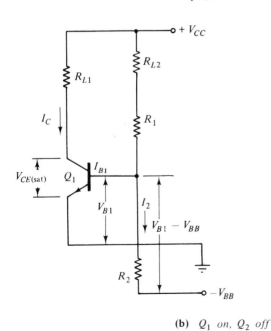

(b) Q_1 *on,* Q_2 *off*

FIGURE 8-4. Monostable multivibrator circuit (a) for Q_2 on and (b) Q_1 on.

used have $h_{FE(min)} = 50$. Neglecting the output pulse width, design a suitable circuit.

solution

The circuit is as shown in Figure 8-1. For Q_2 *on* and saturated [Figure 8-4(a)],

$$R_{L2} \simeq \frac{V_{CC} - V_{D1} - V_{CE(sat)}}{I_C}$$

$$= \frac{9\ V - 0.7\ V - 0.2\ V}{2\ mA}$$

$$= 4.05\ k\Omega \qquad \text{(use } 4.7\ k\Omega \text{ standard value)}$$

$$I_{B2(min)} = \frac{I_C}{h_{FE(min)}}$$

$$= \frac{2\ mA}{50} = 40\ \mu A$$

$$R_B = \frac{V_{CC} - V_{BE} - V_{D1}}{I_{B2}}$$

$$= \frac{9\ V - 0.7\ V - 0.7\ V}{40\ \mu A}$$

$$= 190\ k\Omega \qquad \text{(use } 180\ k\Omega \text{ standard value)}$$

For Q_1 *on* and saturated [Figure 8-4(b)],

$$R_{L1} = R_{L2} = 4.7\ k\Omega$$

To make $I_2 > I_{B1}$, let

$$I_2 \simeq \frac{I_C}{10}$$

$$= \frac{2\ mA}{10} = 200\ \mu A$$

$$V_{B1} = V_{BE} = 0.7\ V \qquad \text{(when } Q_1 \text{ is } on)$$

$$V_{R2} = V_{B1} - V_{BB}$$

$$= 0.7\ V - (-9\ V) = 9.7\ V$$

$$R_2 = \frac{V_{R2}}{I_2}$$

$$= \frac{9.7 \text{ V}}{200 \ \mu\text{A}}$$

$$= 48.5 \text{ k}\Omega \qquad (\text{use } 47 \text{ k}\Omega \text{ standard value})$$

$$I_{B1} + I_2 \simeq 200 \ \mu\text{A} + 40 \ \mu\text{A}$$

$$= 240 \ \mu\text{A}$$

$$R_{L2} + R_1 = \frac{V_{CC} - V_{B1}}{I_{B1} + I_2}$$

$$= \frac{9 \text{ V} - 0.7 \text{ V}}{240 \ \mu\text{A}}$$

$$= 34.6 \text{ k}\Omega$$

$$R_1 = (R_{L2} + R_1) - R_{L2}$$

$$= 34.6 \text{ k}\Omega - 4.7 \text{ k}\Omega$$

$$= 29.9 \text{ k}\Omega \qquad (\text{use } 27 \text{ k}\Omega \text{ standard value})$$

The circuit design is now complete (ignoring PW). V_{B1} should be calculated when Q_2 is *on* to determine that Q_1 is *off* at this time, and to check that the reverse bias is not excessive on Q_1 base-emitter junction.

When Q_2 is *on*,

$$V_{B1} = V_{C2} - V_{R1}$$

$$V_{C2} = V_{D1} + V_{CE(\text{sat})}$$

$$\simeq 0.7 \text{ V} + 0.2 \text{ V}$$

$$= 0.9 \text{ V}$$

$$V_{R1} = \frac{R_1}{R_1 + R_2}(V_{C2} - V_{BB})$$

$$= \frac{27 \text{ k}\Omega}{27 \text{ k}\Omega + 47 \text{ k}\Omega}[0.9 \text{ V} - (-9 \text{ V})]$$

$$\simeq 3.6 \text{ V}$$

$$\therefore \qquad V_{B1} = 0.9 \text{ V} - 3.6 \text{ V}$$

$$= -2.7 \text{ V}$$

This value of V_{B1} is sufficient to ensure that Q_1 is biased *off* when Q_2 is *on*. Also -2.7 V is less than the typical limit of -5 V for a reverse-biased base-emitter junction.

EXAMPLE 8-2

For the circuit designed in Example 8-1, select a suitable capacitor to give an output pulse of 250 μs.

solution

$$e_c = E - (E - E_o)\epsilon^{-t/CR} \qquad \text{[Equation (2-2)]}$$

$$(E - E_o)\epsilon^{-t/CR} = E - e_c$$

$$\epsilon^{-t/CR} = \frac{E - e_c}{E - E_o}$$

$$\epsilon^{t/CR} = \frac{E - E_o}{E - e_c}$$

$$\frac{t}{CR} = \ln \frac{E - E_o}{E - e_c}$$

$$C = \frac{t}{R \ln \left(\dfrac{E - E_o}{E - e_c} \right)}$$

$$e_c = 0 \text{ V}$$

$$E = V_{CC} = 9 \text{ V}$$

$$E_o = -(V_{CC} - V_{BE} - V_{D1})$$

$$= -(9 \text{ V} - 0.7 \text{ V} - 0.7 \text{ V}) = -7.6 \text{ V}$$

$$t = 250 \text{ } \mu\text{s}$$

$$R = R_B = 180 \text{ k}\Omega$$

$$\therefore \quad C_1 = \frac{250 \text{ } \mu\text{s}}{180 \text{ k}\Omega \ln \left[\dfrac{9 \text{ V} - (-7.6 \text{ V})}{9 \text{ V} - 0 \text{ V}} \right]}$$

$$= 2.3 \times 10^{-9}$$

$$= 0.0023 \text{ } \mu\text{F} \qquad \text{(use 0.0025 } \mu\text{F standard capacitor value)}$$

8-3   TRIGGERING THE MONOSTABLE MULTIVIBRATOR

Monostable multivibrator triggering can be effected either by switching *off* the normally *on* transistor, or by turning *on* the normally *off* transistor. Figure 8-5(a) shows a positive-going spike capacitor coupled to the base of normally *off* transistor Q_1. This raises Q_1 base above its grounded emitter, thus switching it *on*. Q_1 switch-*on* then causes Q_2 to switch *off*. The input spike "sees" resistances R_1 and R_2 in parallel as a load, as well as the

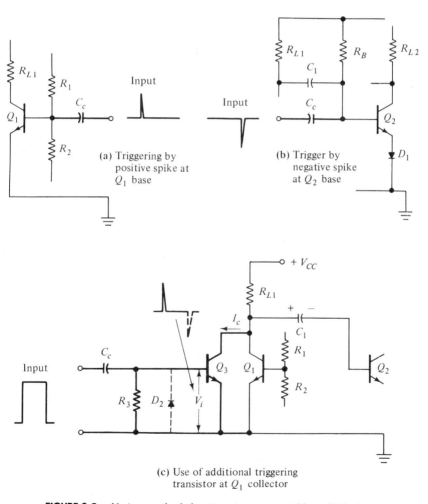

(a) Triggering by positive spike at Q_1 base

(b) Trigger by negative spike at Q_2 base

(c) Use of additional triggering transistor at Q_1 collector

FIGURE 8-5.   Various methods for triggering a monostable multivibrator.

transistor input resistance. Therefore, the spike has to supply current through R_1 and R_2 as well as to supply base current to Q_1. To ensure that Q_1 switches *on* and Q_2 switches *off*, the input current must be supplied for a time t equal to the turn-on time for Q_1 added to the turn-off time for Q_2.

The arrangement in Figure 8-5(b) provides for Q_2 (the normally *on* transistor) to be switched *off*. In this case, the negative-going spike pulls Q_2 base below ground for the transistor turn-*off* time. During this brief time, C_1 behaves as a short circuit, so that the load "seen" by the input spike is $R_B \| R_{L1}$. This is greater than the load "seen" by the positive-going spike in Figure 8-5(a). Therefore, triggering by a negative-going spike at Q_{2B} requires a larger input current than triggering by a positive-going spike at Q_{1B}.

Perhaps the most effective monostable triggering circuit is that shown in Figure 8-5(c), in which an additional transistor Q_3 is employed. Q_3 normally is biased *off* by means of resistor R_3 shorting its base and emitter terminals together. Coupling capacitor C_c and resistor R_3 operate as a differentiating circuit (see Chapter 2), so that the pulse input is differentiated, as illustrated in the figure. Only the positive-going spike will turn *on* Q_3. In the event that the negative-going spike is too large for Q_3 base-emitter, diode D_2 may be used to clip it off. When Q_3 switches *on*, its collector current causes a voltage drop across R_{L1} and the charge on capacitor C_1 causes Q_2 to be biased *off*. Thus Q_3 switch-*on* has the same effect as Q_1 switch-*on*. To correctly trigger the circuit of Figure 8-5(c) the input spike must hold Q_3 *on* for the turn-*off* of Q_2.

The design procedure for the triggering circuit of Figure 8-5(c) is similar to the capacitor-coupled inverter design in Example 5-5, except that instead of the input pulse width, t is made equal to the turn-off time for Q_2. This is also the procedure followed for selecting C_c in the circuit of Figure 8-5(a). Design of the circuit of Figure 8-5(b) is similar to the design given in Example 5-4.

8-4 EMITTER-COUPLED MONOSTABLE MULTIVIBRATOR

In the emitter-coupled monostable multivibrator (circuit in Figure 8-6), a resistance R_E connects both transistor emitter terminals to ground. Also, instead of R_2 being connected to a negative supply voltage, it now is connected to ground. The negative supply voltage is no longer required, and it is seen that one advantage of the emitter-coupled circuit is that it operates from a single supply voltage. Another advantage of this circuit is that the presence of R_E makes it easy to maintain the transistors

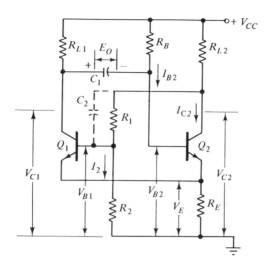

FIGURE 8-6.   Emitter-coupled monostable multivibrator.

unsaturated. Thus, the transistors can be made to switch faster than in the case of the collector-coupled multivibrator.

Reference to Figure 8-6 shows that transistor Q_2 is normally *on*, and that it is supplied with base current via R_B. At this time, there is a voltage drop V_E across resistor R_E, as shown in the figure. Also, the voltage drop across R_{L2} makes Q_2 collector voltage something less than the supply voltage level. Q_1 base is biased from Q_2 collector via potential divider R_1 and R_2; their ratio is such that with Q_2 *on*, V_{B1} is less than V_E. Therefore, Q_1 base voltage is below its emitter voltage, and Q_1 is biased *off*. With Q_1 *off*, its collector voltage equals the supply voltage. The initial voltage across C_1 at this time is $V_{CC} - V_{B2}$.

When Q_1 is triggered *on*, its collector voltage drops, and the charge on C_1 causes the base voltage of Q_2 to drop. When Q_2 begins to turn *off*, its collector voltage starts to rise, thus raising the base voltage of Q_1. With Q_1 *on*, the new level of V_E is $V_{B1} - V_{BE1}$, and because the base of Q_2 is pushed below this level (by the charge on C_1) Q_2 is biased *off*. Then, Q_2 remains *off* until C_1 has discharged enough to allow V_{B2} to rise above V_E.

Speed-up capacitor C_2 may be employed, as for the collector-coupled circuit. If the supply voltage is kept low, the reverse base-emitter voltage for Q_2 may not be large enough to require a diode in series with the emitter terminal. Triggering methods for the emitter-coupled circuit are exactly the same as those for the collector-coupled multivibrator.

The design procedure for an emitter-coupled monostable multivibrator is similar to that for the collector-coupled circuit. When the circuit is

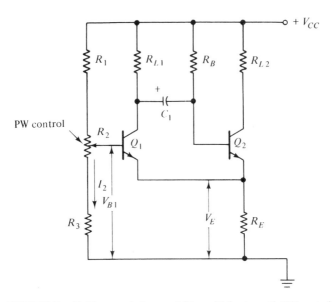

FIGURE 8-7.   Emitter-coupled monostable multivibrator with PW control.

designed for nonsaturated operation, a minimum V_{CE} level must be selected, and $h_{FE(max)}$ must be used in the calculations.

A convenient arrangement for adjusting the output pulse width of an emitter-coupled monostable multivibrator is shown in Figure 8-7. V_{B1} is adjusted by means of potentiometer R_2. The circuit is designed so that the maximum level of V_{B1} is less than the normal level of V_{B2}. When Q_1 is triggered *on*, the voltage drop at Q_1 collector and the charge on C_1 cause V_{B2} to be pushed below V_{B1}. Q_2 remains *off* until C_1 discharge allows V_{B2} to rise above V_{B1} again. The time for this to occur depends upon the actual voltage level of V_{B1}. This time is also the output pulse width. Thus, control of V_{B1} provides pulse width control.

8-5 THE IC OPERATIONAL AMPLIFIER AS A MONOSTABLE MULTIVIBRATOR

An IC operational amplifier connected to function as a monostable multivibrator is shown in Figure 8-8(a). The inverting input terminal is grounded via resistance R_3, and the noninverting terminal is biased above ground by resistances R_1 and R_2. Since the noninverting terminal has a positive input, the output is saturated near the V_{CC} level. In Figure 8-8(b), it is seen that capacitor C_1 is normally charged positive on the right-hand side and negative on the left-hand side.

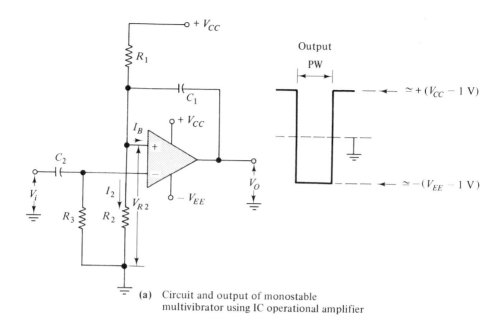

(a) Circuit and output of monostable
multivibrator using IC operational amplifier

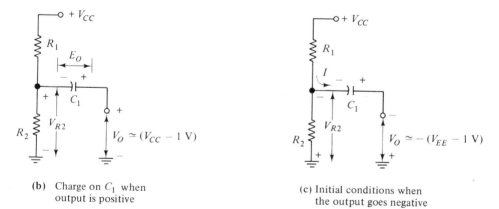

(b) Charge on C_1 when
output is positive

(c) Initial conditions when
the output goes negative

FIGURE 8-8. Monostable multivibrator using IC operational amplifier.

When a large enough positive-going input is coupled to the inverting terminal via C_2, the inverting terminal voltage is raised above the level of the noninverting terminal. The output then switches rapidly to approximately $-(V_{EE}-1 \text{ V})$, see Figure 8-8(c). This pushes the noninverting terminal down to $-(V_{EE}-1 \text{ V})-E_o$, thus ensuring that the output remains negative until C_1 discharges. C_1 begins to discharge via R_1 and R_2 as soon as the output goes negative. Eventually C_1 will charge positive on the

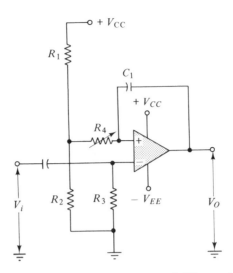

FIGURE 8-9.   IC monostable circuit with PW control.

left-hand side and negative on the right-hand side. When the voltage on the left-hand side of C_1 rises above the voltage level at the inverting terminal, the noninverting terminal again has a positive input. Now, V_o rapidly returns to approximately $(V_{CC} - 1$ V$)$, and the circuit has returned to its original condition.

The output voltage of the circuit moved from its normal level of $(V_{CC} - 1$ V$)$ to a negative level $-(V_{EE} - 1$ V$)$, and eventually returned to $(V_{CC} - 1$ V$)$. Thus a negative output pulse is generated when the circuit is triggered. The output pulse width depends upon C_1, the values of R_1 and R_2, and the bias level at the inverting terminal.

Figure 8-9 shows a modification to the operational amplifier monostable circuit to facilitate pulse width control. If R_1 and R_2 are made much smaller than R_4, then R_4 is effectively the charge and discharge resistance for C_1. When R_4 is adjustable, as illustrated, the output pulse width can be controlled.

EXAMPLE 8-3

Design a monostable multivibrator using a 741 operational amplifier with $V_{CC} = \pm 15$ V. The circuit is to be triggered by a 1.5 V input spike, and the output pulse width is to be 200 μs.

solution

To use a triggering input of 1.5 V, let

$$V_{R2} = 1 \text{ V} \qquad [\text{See Figure 8-8(a).}]$$

Make $I_2 \gg I_B$. From the 741 data sheet in Appendix 1-11,

$$I_{B(max)} = 500 \text{ nA}$$

Let

$$I_2 = 100 \times I_{B(max)}$$
$$= 100 \times 500 \text{ nA} = 50 \text{ } \mu A$$
$$R_2 = \frac{V_{R2}}{I_2} = \frac{1 \text{ V}}{50 \text{ } \mu A}$$
$$= 20 \text{ k}\Omega \qquad (\text{use } 18 \text{ k}\Omega \text{ standard value})$$

I_2 becomes

$$I_2 = \frac{1 \text{ V}}{18 \text{ k}\Omega}$$
$$\simeq 56 \text{ } \mu A$$
$$V_{R1} = V_{CC} - V_{R2} = 15 \text{ V} - 1 \text{ V}$$
$$= 14 \text{ V}$$
$$R_1 = \frac{V_{R1}}{I_2}$$
$$= \frac{14 \text{ V}}{56 \text{ } \mu A} = 250 \text{ k}\Omega \qquad (\text{use } 270 \text{ k}\Omega \text{ standard value})$$
$$R_3 = R_1 \| R_2 = 18 \text{ k}\Omega \| 270 \text{ k}\Omega$$
$$\simeq 16.9 \text{ k}\Omega \qquad (\text{use } 18 \text{ k}\Omega \text{ standard value})$$

When V_o is positive, the initial charge on C_1 is

$$E_o = V_{R2} - V_o$$
$$\simeq V_{R2} - (V_{CC} - 1 \text{ V})$$
$$\simeq 1 \text{ V} - 15 \text{ V} + 1 \text{ V}$$
$$\simeq -13 \text{ V}$$

When V_o is negative, the final charge on C_1 at switchover is

$$e_c \simeq +(V_{EE}-1\text{ V})$$

$$= +14\text{ V}$$

$$\text{Charging voltage} = E = V_{R2}-(-V_o)$$

$$= 1\text{ V}+14\text{ V}$$

$$= 15\text{ V}$$

$$\text{Charging resistance} = R_1 \| R_2$$

$$= 18\text{ k}\Omega \| 270\text{ k}\Omega$$

$$\simeq 16.9\text{ k}\Omega$$

$$C_1 = \frac{t}{R \ln\left(\dfrac{E-E_o}{E-e_c}\right)}$$

$$= \frac{200\ \mu\text{s}}{16.9\text{ k}\Omega \ln\left[\dfrac{15\text{ V}-(-13\text{ V})}{15\text{ V}-14\text{ V}}\right]}$$

$$= 3.55 \times 10^{-9}$$

$$= 0.00355\ \mu\text{F} \qquad \text{(use 0.0036 } \mu\text{F standard value)}$$

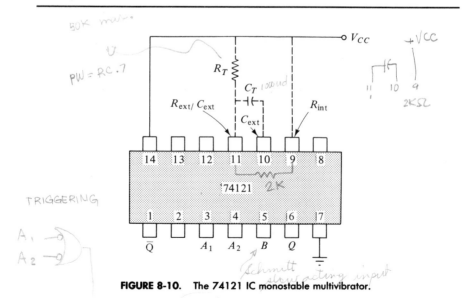

FIGURE 8-10. The 74121 IC monostable multivibrator.

8-6 IC MONOSTABLE MULTIVIBRATOR

Monostable multivibrators are available as integrated circuit components. The 74121 shown in Figure 8-10 is typical of such components. Operating from a 5 V supply, the unit provides two complementary outputs (Q and $\overline{Q}$), and has three input terminals. Two of the inputs (A_1 and A_2) typically require -1.4 V to effect triggering. The other input triggers the circuit when a typical level of $+1.55$ V is applied. The output pulse width is around 30 to 35 ns when no external components are employed and R_{int} (terminal 9) is connected directly to V_{CC}. With R_{int} left unconnected and an external timing capacitor and resistor (R_T and C_T in Figure 8-10), the output pulse width is

$$PW \simeq 0.7\, C_T R_T \qquad (8\text{-}1)$$

Pulse widths of up to 28 s are possible.

8-7 ASTABLE MULTIVIBRATOR

The *astable multivibrator* has no stable state. Instead, the circuit oscillates between the states, (Q_1 *on*, Q_2 *off*) and (Q_2 *on*, Q_1 *off*). The output at the collector of each transistor is a square wave; therefore, the circuit is applied as a *square wave generator*.

Consider the circuit of a *collector-coupled astable multivibrator* shown in Figure 8-11(a). Each transistor has a bias resistance R_B and each is capacitor-coupled to the collector of the other transistor. This is similar to the arrangement of the normally *on* transistor in a monostable multivibrator. Consequently, each transistor in an astable circuit functions in the same way as the normally-*on* transistor in a monostable circuit. When Q_1 is *on* and Q_2 is *off*, capacitor C_1 is charged to ($V_{CC} - V_{BE1}$), positive on the right-hand side. For Q_2 *on* and Q_1 *off*, C_2 is charged to ($V_{CC} - V_{BE2}$), positive on the left-hand side.

Refer to the circuit waveforms in Figure 8-11(b); it is seen that prior to time t_1, transistor Q_1 is *on* and its collector voltage is $V_{CE(sat)}$. Also, Q_2 is *off*, and its collector voltage is V_{CC}. Thus, capacitor C_1 is charged to ($V_{CC} - V_{BE1}$). At t_1, the base voltage of transistor Q_2 rises above ground, causing Q_2 to switch *on*. The collector current I_{C2} now causes Q_2 collector voltage to fall to $V_{CE(sat)}$. Since C_1 will not discharge instantaneously, the base voltage of Q_1 becomes

$$V_{B1} = V_{C2} - (\text{Charge on } C_1)$$
$$= V_{CE(sat)} - (V_{CC} - V_{BE1})$$
$$\simeq -V_{CC}$$

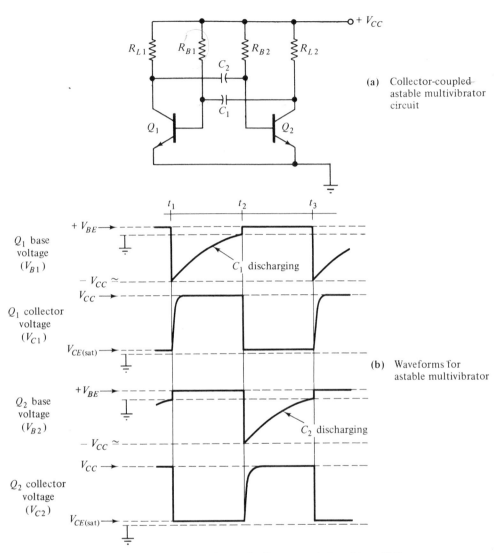

(a) Collector-coupled astable multivibrator circuit

(b) Waveforms for astable multivibrator

FIGURE 8-11. Circuit and waveforms of collector-coupled astable multivibrator.

With its emitter grounded, and its base at $-V_{CC}$, transistor Q_1 is biased *off*. Therefore, at time t_1, the collector voltage of Q_1 raises to V_{CC}. The rise of V_{C1} is not instantaneous, because capacitor C_2 charged via R_{L1} as Q_1 switches *off*.

Between times t_1 and t_2, the base voltage of Q_2 remains at V_{BE}, and Q_2 remains biased *on*. During this time, however, C_1 discharges via resistance R_{B1}. Therefore, the voltage at Q_1 base rises from $-V_{CC}$ toward

V_{CC}. When Q_1 base rises above ground, the transistor begins to switch *on*. The falling collector voltage of Q_1 is coupled to Q_2 base via capacitor C_2, thus causing Q_2 to switch *off*. As Q_2 turns *off* its collector voltage rises, and C_1 is recharged via R_{L2} and Q_1 base. This pumps a large current into the base of Q_1, making it switch *on* very fast. Consequently, the collector voltage of Q_1 falls very rapidly at switch-*on*. The switch-over process is reversed when C_2 discharges sufficiently to allow Q_2 base to rise above ground.

The output pulse width from either transistor is equal to the time during which the transistor is *off*. This is the time taken by the capacitor to discharge approximately from V_{CC} to zero volts.

$$t = CR \ln\left(\frac{E - E_o}{E - e_c}\right) \qquad [\text{Equation (2-8)}]$$

In this equation,

$$t = PW, \qquad C = C_1 = C_2, \qquad R = R_{B1} = R_{B2}$$

and E, is equal to the supply voltage, V_{CC}; E_o, the initial capacitor charge, is equal to $-V_{CC}$. (Note this is taken as negative, because the capacitor would eventually charge with reversed polarity to approximately $+V_{CC}$ if transistor switch-over did not occur.) The final capacitor charge at switch over is $e_c = 0$ V.
Equation 2-8 becomes:

$$PW = CR \ln\left[\frac{V_{CC} - (-V_{CC})}{V_{CC} - 0}\right]$$

$$= CR \ln\left[\frac{2V_{CC}}{V_{CC}}\right]$$

$$= CR \ln 2$$

$$\simeq 0.69\ CR$$

For

$$C = 0.1\ \mu F \qquad \text{and} \qquad R_B = 100\ k\Omega,$$

$$PW = 0.69 \times 0.1\ \mu F \times 100\ k\Omega$$

$$= 6.9\ ms$$

and the output frequency is

$$f = \frac{1}{2\ PW} = \frac{1}{2 \times 6.9\ ms}$$

$$\simeq 72.5\ Hz$$

EXAMPLE 8-4 _____

Design an astable multivibrator to generate a 1 kHz square wave. The supply voltage is 5 V, and the load current is to be 20 μA.

solution

The circuit is as shown in Figure 8-11(a). Make $I_C \gg$(load current):

$$I_C = 100 \text{ (load current)}$$
$$= 100 \times 20 \ \mu A = 2 \text{ mA}$$

Use 2N3904 transistors which, from the data sheet in Appendix 1-4, have a $h_{FE(min)}$ value of 70.

$$R_L \simeq V_{CC}/I_C = \frac{5 \text{ V}}{2 \text{ mA}}$$
$$= 2.5 \text{ k}\Omega \qquad \text{(use 2.7 k}\Omega \text{ standard value)}$$

$$I_{B(min)} = \frac{I_C}{h_{FE(min)}}$$
$$= \frac{2 \text{ mA}}{70} \simeq 28.6 \ \mu A$$

$$R_B = \frac{V_{CC} - V_{BE}}{I_B}$$
$$= \frac{5 \text{ V} - 0.7 \text{ V}}{28.6 \ \mu A}$$
$$= 150 \text{ k}\Omega \qquad \text{(standard value)}$$

$$PW = \frac{1}{2f}$$
$$= \frac{1}{2 \times 1 \text{ kHz}} = 0.5 \text{ ms}$$

and

$$PW = 0.69 \ C_1 R_B$$
$$C_1 = \frac{PW}{0.69 \ R_B} = \frac{0.5 \text{ ms}}{0.69 \times 150 \text{ k}\Omega}$$
$$= 4.8 \times 10^{-9}$$
$$= 0.0048 \ \mu F \qquad \begin{array}{l} \text{(use 0.005 } \mu F \\ \text{standard capacitor value)} \end{array}$$

⇒ free running, don't have to trigger or synch

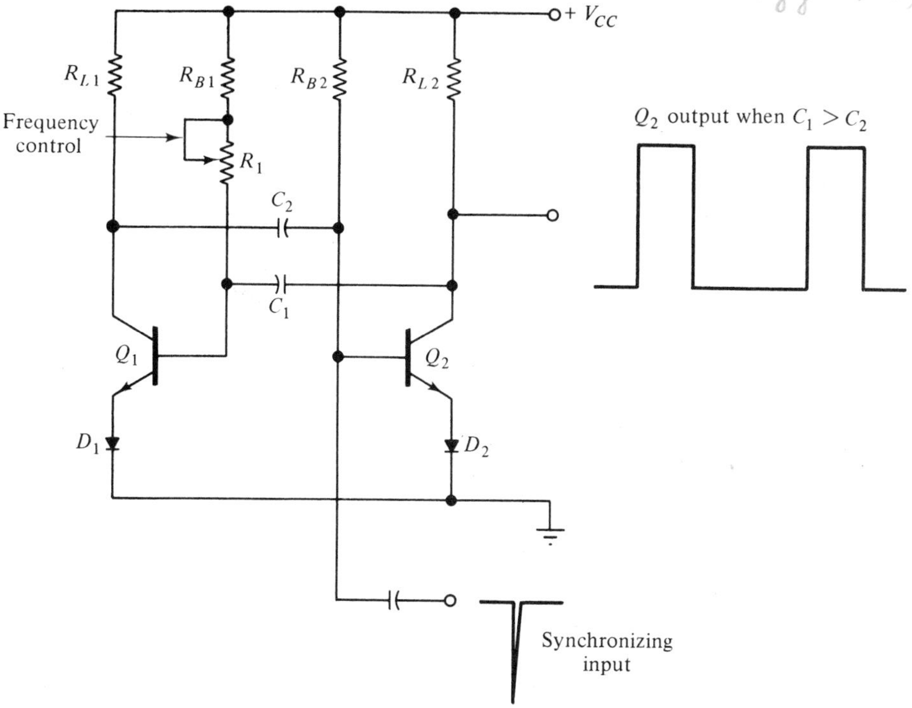

FIGURE 8-12. Astable multivibrator with frequency control, synchronizing input, and diodes for transistor protection.

The circuit of Figure 8-12 shows several possible modifications to the simple astable multivibrator circuit of Figure 8-11(a). Each transistor has its base biased to approximately $-V_{CC}$ when *off*. Consequently, if V_{CC} is greater than the maximum base-emitter reverse voltage, the transistors may be destroyed. Inclusion of diodes D_1 and D_2 (Figure 8-12) affords protection for the transistor base-emitter junctions, as explained in Sec. 8-1.

If C_1 and C_2 are not equal capacitors, one transistor will remain *off* for a longer time than the other one. In this case, the output is no longer a square wave, as the transistor with the largest capacitor at its base remains *off* longest. The output frequency of the circuit may be made adjustable by including a variable resistor R_1 in series with one of the base bias resistors. In Figure 8-12, R_1 controls the rate of discharge of capacitor C_1; thus R_1 can be used to adjust the *off*-time of Q_1.

Occasionally the frequency of an astable multivibrator has to be synchronized to some external frequency. Figure 8-12 shows a negative-going synchronizing spike input capacitor-coupled to the base of Q_2. When the spike input is applied, Q_2 is switched *off* and C_2 is recharged to its maximum voltage. Q_2 then remains *off* for its normal pulse width.

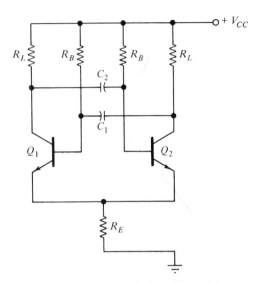

FIGURE 8-13.   Emitter-coupled astable multivibrator.

One problem with the collector-coupled astable circuit is that it may not always start oscillating when the supply voltage is switched *on*. Because of the circuit symmetry, it can happen that both transistors switch *on* and remain *on*. Oscillation can be started by shorting one of the transistor bases to its emitter terminal for a brief instant. However, this usually is not practical. The emitter-coupled astable multivibrator circuit shown in Figure 8-13 solves the problem. In this circuit, when one transistor begins to conduct, the other transistor has its emitter voltage raised and its base voltage reduced. Thus, it is almost impossible for the two to remain *on* at one time.

8-8   IC OPERATIONAL AMPLIFIER AS AN ASTABLE MULTIVIBRATOR

An inverting Schmitt trigger circuit using an IC operational amplifier can be easily converted into an astable multivibrator with the addition of a resistor and capacitor. In the circuit shown Figure 8-14, R_2, R_3; and the operational amplifier constitute a Schmitt trigger circuit. The waveforms in the illustration help to explain the circuit operation. When the output is high, current flows through R_1, charging C_1 positively until it reaches the UTP of the Schmitt. The Schmitt output then goes negative, and current commences to flow out of C_1 via R_1 until its voltage reaches the Schmitt

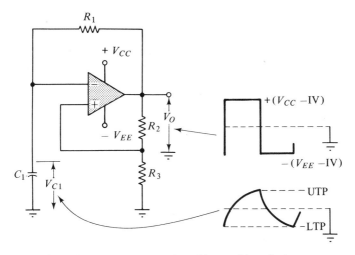

FIGURE 8-14.   IC operational amplifier astable multivibrator.

LTP. The Schmitt output then goes positive once again, and the cycle recommences.

Design of this circuit merely involves the Schmitt circuit design, then selection of suitable R and C values for the desired charge and discharge time and trigger voltage levels.

EXAMPLE 8-5 _____

Using a 741 operational amplifier, design an astable multivibrator to an output frequency of 300 Hz and an output amplitude of ± 11 V.

solution

$$V_o \simeq \pm (V_{CC} - 1 \text{ V})$$
$$V_{CC} \simeq \pm (V_o + 1 \text{ V})$$
$$\simeq \pm (11 \text{ V} + 1 \text{ V}) = \pm 12 \text{ V}$$

Let the Schmitt trigger voltage $= \dfrac{V_o}{2} = \pm 5.5$ V

$$V_{R3} = \pm 5.5 \text{ V}$$

Let $I_2 = 100 \times I_{B(\text{max})} = 100 \times 500$ nA
$$= 50 \ \mu A$$

$$R_3 = \frac{V_{R3}}{I_2} = \frac{5.5 \text{ V}}{50 \text{ }\mu\text{A}}$$

$$= 110 \text{ k}\Omega \qquad \text{(use 100 k}\Omega \text{ standard value)}$$

$$V_{R2} = V_{R3}$$

$$R_2 = R_3 = 100 \text{ k}\Omega$$

Let $I_{R1(min)} = 100 \times I_{B(max)} = 50 \text{ }\mu\text{A}$

$$V_{R1(min)} = V_o - \text{trigger voltage}$$

$$= 11 \text{ V} - 5.5 \text{ V} = 5.5 \text{ V}$$

$$R_1 = \frac{5.5 \text{ V}}{50 \text{ }\mu\text{A}} \simeq 100 \text{ k}\Omega$$

$$T = \frac{1}{300 \text{ Hz}} = 3.3 \text{ ms}$$

C_1 charging time $t = \frac{1}{2} T = 1.67$ ms

From Eq. 2-9, $C = \dfrac{t}{R_1 \ln \left[\dfrac{E - E_o}{E - e_c} \right]}$

$$= \frac{1.67 \text{ ms}}{100 \text{ k}\Omega \ln \left[\dfrac{11 \text{ V} - (-5.5 \text{ V})}{11 \text{ V} - 5.5 \text{ V}} \right]}$$

$$= 0.015 \text{ }\mu\text{F}$$

8-9 THE 555 IC TIMER

The 555 integrated circuit *timer* can be applied to a myriad of timing applications: monostable multivibrators, astable multivibrators, ramp generators, sequential timers, etc. Although at first glance the functional block diagram (Figure 8-15) may look a little complex, it is really quite easily understood by anyone familiar with flip-flops and comparators. Calculation of external component values for various applications is also very simple.

Referring to the function block diagram, the 555 timer is seen to consist of: a *potential dividing network* R_1, R_2, and R_3; two *voltage comparators*; a *set-reset flip-flop*; an *output stage*, and two transistors. The 555 data sheet in Appendix 1-15 indicates that the circuit functions

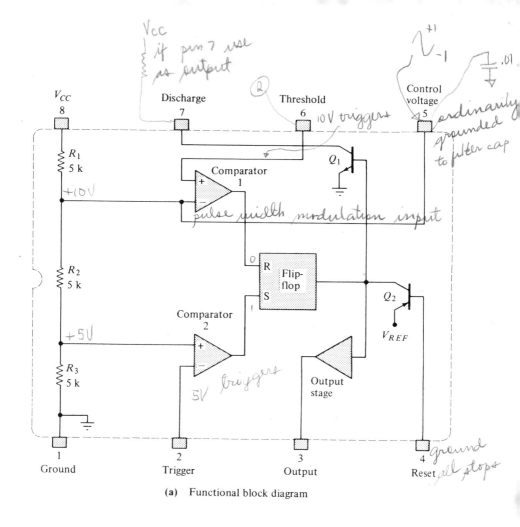

Vcc if pin 7 use as output

+1 −1 .01

V_CC 8

Discharge 7

② **Threshold** 6 10 V triggers

Control voltage 5 ordinarily grounded to filter cap

R_1 5 k

Comparator 1

Q_1

+10V

pulse width modulation input

R_2 5 k

0 R **Flip-flop** S

Q_2

Comparator 2

+5V

+

V_{REF}

R_3 5 k

5V triggers

Output stage

1 **Ground**

2 **Trigger**

3 **Output**

4 **Reset** ground all stops

(a) Functional block diagram

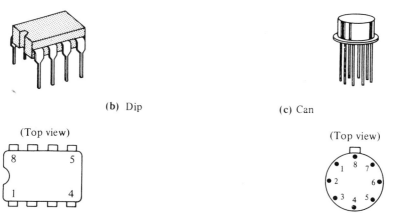

(b) Dip

(c) Can

(Top view)

(Top view)

8 5

1 4

1 8 7
2 6
3 4 5

FIGURE 8-15. 555 timer.

satisfactorily with supply voltages (V_{CC}) ranging from 5 V to 18 V. The set-reset flip-flop is explained in Chapter 9. For now note that its output goes *low* when a positive input is applied to the *set* terminal, and that the output goes *high* when a positive input appears at the *reset* terminal.

The potential divider provides a bias voltage to the inverting input terminal of comparator 1, and a different bias voltage level to the noninverting terminal of comparator 2. Access to the other inputs of the comparators is available via terminals 2 and 6, identified as *trigger* and *threshold*, respectively.

The comparator output levels control the flip-flop, and the flip-flop output is fed to the output stage and to the base of *npn* transistor Q_1. When the flip-flop output is high, Q_1 is biased *on*. In this condition the transistor could *discharge* a capacitor connected to terminal 7, for example. Q_1 is *off* when the flip-flop output is low.

The *output stage* provides a low output resistance and also inverts the output level of the flip-flop. The voltage at terminal 3 is low when the flip-flop output is high, and when the flip-flop output is low terminal 3 voltage is at a high level. The output stage can *sink* or *source* (at output terminal 3) a maximum current of 200 mA (see the 555 data sheet in Appendix 1-15).

Transistor Q_2 is a *pnp* device with its emitter connected to an internal reference voltage V_{REF}, which is always less than V_{CC}. If *reset* terminal 4 is connected to V_{CC} the base-emitter junction of Q_2 is reverse-biased, causing the transistor to remain *off*. When terminal 4 is pulled below V_{REF} (i.e., towards ground level), Q_2 switches *on*. This turns Q_1 *on*, causes the output at terminal 3 to go to ground level, and resets the flip-flop to its high output state.

The complete functioning of the timer circuit is best understood by considering a typical application, such as the monostable multivibrator in the next section.

8-10   555 AS A MONOSTABLE MULTIVIBRATOR

A basic 555 monostable circuit is shown in Figure 8-16. The supply voltage is connected across terminal 8 ($+ V_{CC}$) and terminal 1 (ground). Terminal 2 (*trigger*) is directly connected to a (negative-going) trigger pulse source. C_A is a capacitor which charges from V_{CC} via resistor R_A when *npn* transistor Q_1 (see Figure 8-15) is *off*. Terminal 4 is connected directly to V_{CC} to ensure that *pnp* transistor Q_2 (see Figure 8-15) remains *off* at all times Terminal 5 is left open-circuited, and the output is taken from terminal 3.

Operation of the 555 monostable circuit is explained below in point form. Refer to Figure 8-15 and Figure 8-16.

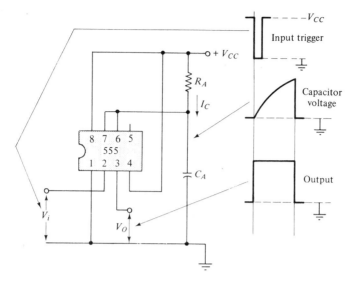

FIGURE 8-16.   555 monostable multivibrator.

Initial state

- Terminal 2 *high* because trigger source level is *high*.
- Comparator 2 output is *low* because terminal 2 is *high* (inverting input), and the voltage at the noninverting input of the comparator is V_{R3},

where $$V_{R3} = V_{CC} \times \frac{R_3}{R_1 + R_2 + R_3} = \tfrac{1}{3}V_{CC}$$

- Comparator 1 output is *low* because terminal 6 is *low* (noninverting input) and the inverting input of the comparator is at $V_{(R_2 + R_3)}$,

where $$V_{(R_2 + R_3)} = V_{CC} \times \frac{R_2 + R_3}{R_1 + R_2 + R_3} = \tfrac{2}{3}V_{CC}$$

- Flip-flop output is *high*, in its *reset* condition.
- Q_1 is *on* because the flip-flop output is *high*.
- Capacitor C_1 is in its discharged state because Q_1 is *on*.
- Terminals 6 and 7 are at a *low* voltage level because the capacitor is in its discharged state.
- Output voltage at terminal 3 is *low* because the flip-flop output is *high*.

Triggered state

- The trigger input causes terminal 2 to go below V_{R3}; i.e., the inverting input to comparator 2 is driven below the voltage level at the noninverting input.

- Comparator 2 output goes *high* because of the trigger input.
- The flip-flop is driven into its *set* condition (output level *low*) by the *high* output of comparator 2.
- Q_1 is switched *off* by the flip-flop output going *low*.
- Terminal 3 *output* goes *high* because of the flip-flop output going *low*.
- With Q_1 *off*, C_A (connected to terminals 6 and 7) commences to charge exponentially via R_A.
- When the trigger input at terminal 2 goes *high* once again comparator 2 output goes *low*. The flip-flop remains in its *set* condition.

Final state

- Comparator 1 output remains *low* until the capacitor voltage (connected to terminal 6) becomes equal to $V_{(R_2+R_3)} = \frac{2}{3}V_{CC}$. Then comparator 1 output goes *high*.
- The flip-flop is driven into its *reset* condition by the *high* from comparator 1, and its output is *high* once again.
- Q_1 is switched *on* by the *high* output from the flip-flop.
- C_A is discharged by Q_1, and the voltage level at terminals 6 and 7 falls.
- The output voltage at terminal 3 goes to a *low* level because the flip-flop output is *high*.
- Comparator 1 output goes *low* once again as terminal 6 voltage falls below $V_{(R_2+R_3)}$. The flip-flop remains in its *reset* condition.
- The final state of the monostable multivibrator is the same as its initial state. The circuit is now ready for triggering once again.

The 555 monostable circuit gives an output pulse each time it is triggered. The output pulse width depends upon the values of R_A and C_A, and upon the internal voltage levels of the 555 circuit.

8-11 DESIGN OF A 555 MONOSTABLE CIRCUIT

Design of the monostable circuit in Fig. 8-16 involves nothing more than selection of R_A and C_A. The supply voltage (V_{CC}) can be anything from 4.5 V to 18 V, (see Appendix 1-15). No matter what the value of V_{CC}, $V_{R3} = \frac{1}{3}V_{CC}$ and $V_{(R_2+R_3)} = \frac{2}{3}V_{CC}$, as already shown above. Also, as already seen, when the circuit is triggered, C_A charges up to $\frac{2}{3}V_{CC}$, and then the circuit returns to its initial state. The time (t) required for C_A to charge through $\frac{2}{3}V_{CC}$ determines the output pulse width. This time can be readily calaculated from Equation (2-9),

$$t = CR \ln \left[\frac{E - E_o}{E - e_c} \right]$$

For the circuit in Figure 8-16:

$$C = C_A, \qquad R = R_A, \qquad E = V_{CC}, \qquad E_o = 0$$

and e_c is the final capacitor voltage, $e_c = \frac{2}{3} V_{CC}$. Applying the quantities to Eq. (2-9) gives

$$t = 1.1 C_A R_A \tag{8-2}$$

C_A should normally be chosen as small as possible to ensure that Q_1 (see Figure 8-15) has no difficulty in discharging it rapidly. However, C_A should not be so small that is is affected by stray capacitance. If C_A is to be as small as possible, then the charging current must also be as small as possible. The minimum level of charging current occurs when the capacitance voltage is at its maximum level, i.e. when $e_c = \frac{2}{3} V_{CC}$. At this instant the voltage across R_A is

$$V_{RA} = V_{CC} - \frac{2}{3} V_{CC} = \frac{1}{3} V_{CC}$$

and the capacitor charging current is

$$I_{C(min)} = \frac{1}{3} V_{CC} / R_A$$

or

$$R_A = \frac{V_{CC}}{3 I_{C(min)}} \tag{8-3}$$

$I_{C(min)}$ should be chosen much greater than the threshold current I_{th} which flows into terminal 6. This is to ensure that I_{th} does not divert a significant amount of I_C away from the capacitor.

The design procedure now becomes:

1. Note that value of I_{th} from the 555 specification sheet.
2. Select $I_{C(min)} \gg I_{th}$.
3. Calculate R_A, using Eq. (8-3).
4. Calculate C_A, using Eq. (8-2).

EXAMPLE 8-6

Design a 555 monostable circuit to have an 1 ms output pulse width. The supply voltage is to be $V_{CC} = 15$ V.

solution

From the 555 data sheet Appendix 1-15, maximum $I_{th} = 0.25 \ \mu A$.

$$I_{C(min)} \gg I_{th}$$

Let $I_{C(min)} = 100 \times I_{th}$
$\qquad = 100 \times 0.25 \ \mu A = 25 \ \mu A$
Equation (8-3) is

$$R_A = \frac{V_{CC}}{3 \ I_{C(min)}} = \frac{15 \ V}{3 \times 25 \ \mu A}$$

$$= 200 \ k\Omega \qquad \text{(use 220 k}\Omega \text{ standard value)}$$

From Equation (8-2), $C_A = \dfrac{t}{1.1 \ R_A} = \dfrac{1 \ ms}{1.1 \times 220 \ k\Omega}$

$$\simeq 4000 \ pF \qquad \text{(use 0.039 } \mu F \text{ standard value)}$$

8-12 MODIFICATIONS TO THE BASIC 555 MONOSTABLE CIRCUIT

Figure 8-17 shows a monostable circuit with the trigger input *capacitor-coupled* to terminal 2. Terminal 2 is connected to V_{CC} via resistor R_1. This is to ensure that the inverting input (*trigger*) terminal of comparator 2 remains above the noninverting input voltage until the trigger input is applied. The input waveform is differentiated by C_1 and R_1, and diode D_1 clips off the unwanted positive spike (see the waveforms in Figure 8-17).

R_1 in Figure 8-17 performs exactly the same function as R_B in the *normally-off capacitor-coupled inverter* discussed in Sec. 5-2; i.e., it holds a transistor (inside the comparator) in the *off* condition. As explained in Sec. 5-2, a 22-kΩ resistor is a reasonable maximum value to use for R_1.

The minimum value of C_1 is determined by considering the trigger input current (specified in Appendix 1-15 as $I_T = 0.5 \ \mu A$ typical). The current that flows through R_1 when the trigger voltage is present must also be considered. C_1 is then calculated in a similar way to that used for the capacitor -coupled inverter circuit (see Example 5-4), except that t is made

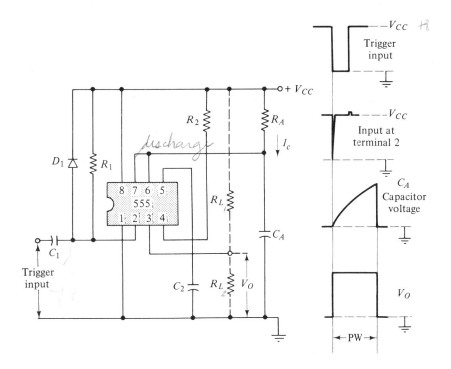

FIGURE 8-17.   555 monostable with capacitor-coupled trigger.

equal to the 555 *output rise time* instead of the input pulse width. According to Appendix 1-15, the output rise time is 100 ns.

Capacitor C_2 (usually 0.01 μF) prevents pickup of unwanted noise signals at the inverting input of comparator 1 when there is no other external connection to the *control voltage* terminal (terminal 5). C_2 also functions as a by-pass capacitor to maintain the dc voltage constant across resistors R_2 and R_3 (inside the device, see Fig. 8-15) at high trigger frequencies.

As illustrated in Figure 8-17, the load (R_L) can be connected from *output* terminal 3 to either $+V_{CC}$ or ground level. This has no effect on the voltage level at the 555 output terminal, but it obviously affects the current direction through the load.

A resistor (R_2) may be connected between the *reset* terminal (terminal 4) and V_{CC}. This allows a reset voltage to be capacitor-coupled to terminal 4, in order to reset the circuit back to its initial state, i.e., prior to the end of the normal pulse width. A suitable maximum value for R_2 is 22 kΩ (see Sec. 5-2).

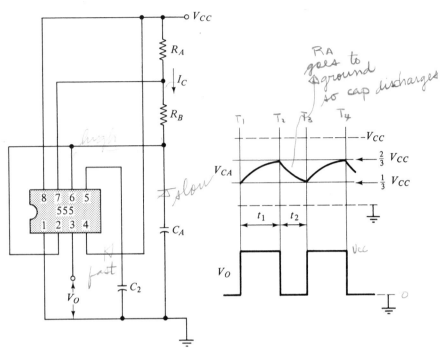

(a) Basic astable circuit

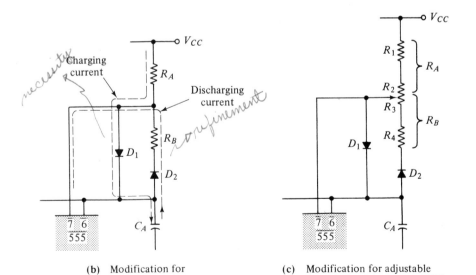

(b) Modification for 50% duty cycle

(c) Modification for adjustable duty cycle with constant PRF

FIGURE 8-18. 555 astable multivibrator.

8-13   555 ASTABLE MULTIVIBRATOR

The monostable circuit is converted into an astable circuit simply by connecting the *trigger* terminal (terminal 2) directly to the *threshold* terminal (terminal 6) [(see Figure 8-18(a)]. The charging resistor is replaced by two resistors, R_A and R_B, and the *discharge* terminal (terminal 7) is connected to the junction of the two.

When the capacitor voltage (connected to terminal 6 and terminal 2) goes below $\frac{1}{3}V_{CC}$, the inverting input of comparator 2 is below the level of the noninverting input (which is a $V_{R3} = \frac{1}{3}V_{CC}$). Comparator 2 output goes *high*, and triggers the flip-flop into its *set* condition, in which its output is a a *low* level. Q_1 is now *off*, and C_A charges via R_A and R_B.

C_A continues to charge until it reaches $\frac{2}{3}V_{CC}$, at which point the noninverting input of comparator 1 (connected to C_A via terminal 6) is raised above the level of the inverting input [at $V_{(R3+R3)} = \frac{2}{3}V_{CC}$]. The output of comparator 1 now goes *high*, triggering the flip-flop into its *reset* state (output *high*) and causing Q_1 to switch *on*. Capacitor C_A is now discharged by Q_1 via resistor R_B. Discharge of C_A continues until its voltage falls below $\frac{1}{3}V_{CC}$. At this point the output of comparator 2 goes *high*, triggering the flip-flop to its *low* output state and switching Q_1 *off* once again. The cycle has now recommenced, and it continues repetitively.

Design of a 555 astable multivibrator involves only the calculation of R_A, R_B, and C_A. Capacitor C_2 is usually 0.01 μF, as already explained. Since C_A charges via $(R_A + R_B)$ from $\frac{1}{3}V_{CC}$ to $\frac{2}{3}V_{CC}$, the initial capacitor voltage is $E_o = \frac{1}{3}V_{CC}$, and the final level is $e_c = \frac{2}{3}V_{CC}$. Also, the charging voltage is $E = V_{CC}$. Substituting the values into Equation (2-9) gives

$$t_1 = 0.693 C_A (R_A + R_B) \qquad (8\text{-}4)$$

Similarly, for the discharge period: $E_o = \frac{2}{3}V_{CC}$, $e_c = \frac{1}{3}V_{CC}$, and $E = 0$. Using these quantities, Equation (2-9) yields

$$t_2 = 0.693 C_A R_B \qquad (8\text{-}5)$$

EXAMPLE 8-7

Design a 555 astable multivibrator to give a pulse output with PRF = 2 kHz and duty cycle = 66%. Use $V_{CC} = 18$ V.

solution

$$t_1 + t_2 = \frac{1}{PRF} = \frac{1}{2 \text{ khz}}$$
$$= 500 \ \mu s$$
$$t_1 = (\text{duty cycle}) \times (t_1 + t_2)$$
$$= \frac{66}{100} \times 500 \ \mu s$$
$$= 330 \ \mu s$$
$$t_2 = 500 \ \mu s - 330 \ \mu s = 170 \ \mu s$$
$$I_{C(min)} \gg I_{th}$$
$$I_{C(min)} \gg 0.25 \ \mu A \quad \text{(from data sheet)}$$
Let $I_{C(min)} = 1$ mA

From Equation (8-3),

$$R_A + R_B = \frac{V_{CC}}{3 \ I_{C(min)}} = \frac{18 \text{ V}}{3 \times 1 \text{ mA}}$$
$$= 6 \text{ k}\Omega$$

From Equation (8-4),

$$C_A = \frac{t_1}{0.693(R_A + R_B)} = \frac{330 \ \mu s}{0.693 \times 6 \text{ k}\Omega}$$
$$\simeq 0.08 \ \mu F \quad \text{(use 0.082 } \mu F \text{ standard value)}$$

From Equation (8-5),

$$R_B = \frac{t_2}{0.693 C_A} = \frac{170 \ \mu s}{0.693 \times 0.08 \ \mu F}$$
$$= 3.07 \text{ k}\Omega \quad \text{(use 2.7 k}\Omega \text{ standard value)}$$
$$R_A = (R_A + R_B) - R_B$$
$$= 6 \text{ k}\Omega - 2.7 \text{ k}\Omega$$
$$= 3.3 \text{ k}\Omega \quad \text{(standard value)}$$

EXAMPLE 8-8 _____

Analyze the circuit designed in Example 8-7 to determine the actual PRF and duty cycle.

solution

Equation (8-4),

$$t_1 = 0.693C_A(R_A + R_B)$$
$$= 0.693 \times 0.082 \ \mu F \times (3.3 \ k\Omega + 2.7 \ k\Omega)$$
$$= 341 \ \mu s$$

Equation (8-5),

$$t_2 = 0.693C_A R_B$$
$$= 0.693 \times 0.082 \ \mu F \times 2.7 \ k\Omega$$
$$= 153 \ \mu s$$
$$t_1 + t_2 = 341 \ \mu s + 153 \ \mu s$$
$$= 494 \ \mu s$$
$$PRF = 1/(t_1 + t_2) = 1/494 \ \mu s$$
$$= 2.02 \ kHz$$
$$\text{Duty cycle} = \frac{t_1}{t_1 + t_2} \times 100\% = \frac{341 \ \mu s}{153 \ \mu s + 341 \ \mu s} \times 100\%$$
$$= 69\%$$

8-14   MODIFICATIONS TO THE BASIC ASTABLE CIRCUIT

When Equations (8-4) and (8-5) are applied to the circuit of Figure 8-18(a) it is seen that t_1 must always be greater than t_2 (i.e., the duty cycle of the pulse output is greater than 50%)/ This is because Equation (8-4) involves $(R_A + R_B)$, whereas Equation (8-5) involves only R_A. Clearly, for $t_1 = t_2$, $(R_A + R_B)$ must be equal to R_B, and this is impossible.

Figure 8-18(b) shows a modification which allows the duty cycle to be made 50% or less. During the charging cycle, diode D_2 is reverse-biased and D_1 is forward-biased. C_A is charged from V_{CC} via R_A and D_1. When Q_1 (in Figure 8-15) is *on*, C_A is discharged via D_2 and R_B. Equation (8-4) now becomes

$$t_1 = 0.693 C_A R_A$$

while Equation (8-5) remains

$$t_2 = 0.693 C_A R_B$$

Now, with $R_A = R_B$, t_1 and t_2 are equal, and the duty cycle is 50%. When R_A is less than R_B, t_1 is less than t_2, and the duty cycle becomes less than 50%.

The further modification in Figure 8-18(c) allows the duty cycle to be adjusted without altering the PRF of the circuit. The duty cycle depends upon the relative values of R_A and R_B. Since the variable resistor $(R_2 + R_3)$ can be adjusted to increase R_A while decreasing R_B, and *vice versa*, the duty cycle is variable.

The pulse repetition frequency is the reciprocal of $t_1 + t_2$:

$$PRF = \frac{1}{t_1 + t_2}$$

and
$$t_1 + t_2 = 0.693 C_A R_A + 0.693 C_A R_B$$
$$= 0.693 C_A (R_A + R_B)$$
$$= 0.693 C_A [(R_1 + R_2) + (R_3 + R_4)]$$

or
$$PRF = 1 / [0.693 C_A (R_1 + R_2 + R_3 + R_4)]$$

Clearly, the PRF remains constant no matter what the distribution of the variable resistance $(R_2 + R_3)$ between R_A and R_B.

A common requirement is for a square wave generator with a frequency control that does not affect the duty cycle of the square wave output. This could be very simply achieved by making C_A a variable capacitor. A wide range of output frequencies could be produced by including a multiposition switch to select different fixed capacitor values. A variable capacitor could also be permanently connected for continuous frequency adjustment between switched ranges [see Figure 8-19(a)]. This arrangement could also be employed with the basic astable circuit in Figure 8-18(a), or with either of the modifications shown in Figures 8-18(b) and (c).

Another method of constructing a variable frequency square wave generator with an almost constant duty cycle uses the basic astable circuit

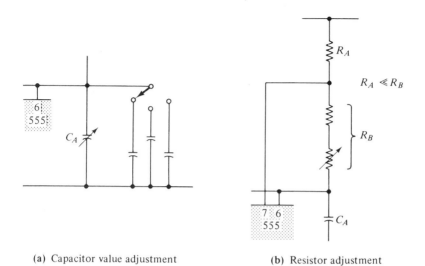

(a) Capacitor value adjustment   (b) Resistor adjustment

FIGURE 8-19. Astable circuit modifications to make a variable frequency square wave generator.

in Figure 8-18(a). R_B is replaced with a variable resistor in series with a fixed value resistor [see Figure 8-19(b)]. R_A is made very much smaller than the minimum value of R_B. The output frequency is adjusted by altering R_B. Because ($R_A + R_B$) is always a little larger than R_B, the duty cycle is always a little greater than 50%, but is not significantly changed when the frequency is adjusted. When calculating component values, R_A should be not less than about 1 kΩ. This is to avoid overloading discharge transistor Q_1 (see Figure 8-15).

In Figure 8-20 yet another method of constructing a 50% duty cycle astable multivibrator is shown. In this case capacitor C_A is charged via R_A and R_B from the low impedance output terminal 3. As in Figure 8-18, terminals 2 and 6 detect the low and high limits of capacitor voltage. Discharge terminal 7 is left unconnected.

When the output (at terminal 3) is high, C_A charges positively until its voltage arrives at 2/3 V_{CC}, (detected at terminal 6). The output then switches to a low level, and C_A is discharged through the same two resistors (R_A and R_B) until it falls to 1/3 V_{CC}, (detected at terminal 2). Once again the output switches to a high level and the cycle repeats.

The one problem with this circuit is that V_o at terminal 3 does not go all the way up to V_{CC}, instead it is usually about 1 V below the level of the supply. Equation 8-4, which should be applicable here is not quite correct, because C_A is being charged to 2/3 V_{CC} and discharged to 1/3 V_{CC}, but is being charged from a voltage of $V_o \simeq V_{CC} - 1$ V. The use of resistor R_C

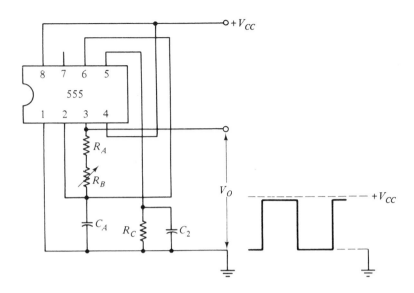

FIGURE 8-20.   555 square wave generator.

connected at terminal 5 can correct the error. Referring to Figure 8-15, it is seen that a resistor from terminal 5 to ground shunts R_2 and R_3. Selection of a suitable value for R_C can make $V_{R3} = 1/3$ V_o, and $(V_{R2} + V_{R3}) = 2/3$ V_o. The output from the circuit of Figure 8-20 now has a 50% duty cycle, and adjustment of R_B affords output frequency control without affecting the duty cycle. If a diode D_1 is connected in series with R_A, and another series resistor R_D and a diode D_2 (with opposite polarity to D_1) are connected in parallel with R_A and D_1, the charge and discharge paths can have different time durations.

8-15   OTHER 555 TIMER APPLICATIONS

Figures 8-21 and 8-22 illustrate just two more of the many applications of the 555 IC timer. In Figure 8-21 three monostable multivibrators are cascade-connected to make a *sequential timer*. An input pulse (V_i) triggers monostable 1 *on*. This produces a positive output pulse at terminal 3 with a pulse width PW 1. When monostable 1 goes *off*, the negative-going trailing edge of its output pulse triggers monostable 2, which then produces output PW 2. Similarly, at the end of PW 2 monostable 3 is triggered, producing PW 3.

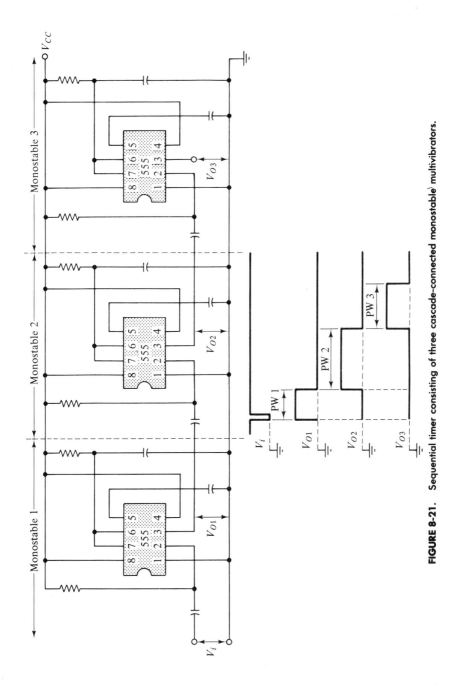

FIGURE 8-21. Sequential timer consisting of three cascade-connected monostable multivibrators.

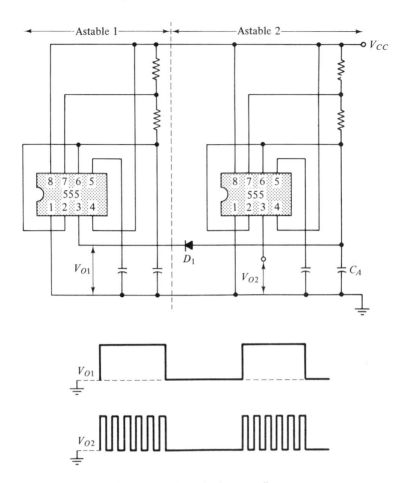

FIGURE 8-22.   Pulsed tone oscillator.

The *pulsed tone oscillator* circuit in Figure 8-22 generates a repetitive succession of groups of high-frequency square waves. The output of low-frequency astable 1 controls astable 2. When output terminal 3 on astable 1 is low, it holds C_A in astable 2 in an uncharged state via the interconnecting diode D_1. When the output of astable 1 goes high, D_1 is reverse-biased. This allows C_A in astable 2 to charge, and thus astable 2 oscillates for the duration of the output pulse from astable 1.

A linear ramp generator can easily be constructed by modifying the monostable circuit in Figure 8-17. Resistor R_A is replaced by the kind of constant current circuit (Q_2, R_1, R_2, and R_3) in Figure 7-3. This circuit has the effect of linearizing the capacitor output voltage. The same kind of modification can be made to the astable circuit in Figure 8-18(a) to create a free-running ramp generator.

8-16 CMOS TIMER CIRCUIT

The ICM7555 manufactured by Intersil is typical of timers using CMOS technology. The 7555 can be substituted in all situations for a bipolar 555 IC. However, the 7555 draws only 80 μA typically from the supply, whereas the 555 requires 10 mA. The 7555 can also operate from a supply voltage as low as 2 V, compared to 4.5 V minimum for the 555. One other advantage of the CMOS 7555 is that its output voltage (at terminal 3) exhibits very low offset, i.e., the output effectively swings from ground to $+V_{CC}$. This means, for example, that the square wave generator circuit of Figure 8-20 using a 7555 timer could operate satisfactorily without resistor R_C.

REVIEW QUESTIONS AND PROBLEMS

8-1 Sketch the circuit of a collector-coupled monostable multivibrator. Sketch the waveforms and explain the operation of the circuit. Also explain the function of each component.

8-2 A collector-coupled monostable multivibrator is to operate from a ± 12 V supply. The transistor collector currents are to be 3 mA, and the transistors have $h_{FE(min)} = 70$. Neglecting the output pulse width, design a suitable circuit.

8-3 Select suitable capacitors for the circuit in Problems 8-2 to give an output pulse of 330 μs.

8-4 The monostable multivibrator designed for Problems 8-2 and 8-3 is to be triggered at 10 μs intervals between output pulses. Calculate the maximum size of the speed-up capacitor that may be employed.

8-5 Sketch and explain the various methods of triggering a monostable multivibrator. For the multivibrator in problem 8-2, design a triggering system using an additional transistor. The triggering input is a 3 V, 10 μs pulse with a source resistance of 3.3 kΩ.

8-6 Sketch the circuit of an emitter-coupled monostable multivibrator. Carefully explain how the circuit operates. Discuss the relative advantages and disadvantages of emitter-coupled and collector-coupled monostable multivibrators. Show how the emitter-coupled circuit may be modified to provide pulse width control.

8-7   Sketch the circuit of a monostable multivibrator employing an IC operational amplifier. Explain how the circuit operates; also show how the output pulse width may be controlled.

8-8   Design a monostable multivibrator using a 741 operational amplifier with $V_{CC} = \pm 9$ V. The circuit is to be triggered by a 0.5 V input spike, and the output pulse width is to be 300 μs.

8-9   A 74121 IC monostable multivibrator is to have an output pulse width of 2 μs. Select suitable external components, and show how they should be connected to the circuit.

8-10   Sketch the circuit of a collector-coupled astable multivibrator. Also sketch the waveforms of collector and base voltages. Carefully explain how the circuit operates.

8-11   Derive an expression for the output pulse width of an astable multivibrator. Design an astable multivibrator to have 5 kHz output square wave. The available supply is 9 V, and the load current is 50 μA.

8-12   Sketch the circuit of an astable multivibrator in which the output frequency may be adjusted. Also show how the multivibrator frequency may be synchronized with an external frequency.

8-13   Sketch the circuit of an emitter-coupled astable multivibrator. Explain its operation, and discuss its advantages compared to a collector-coupled circuit.

8-14   Sketch the circuit of an IC operational amplifier employed as an astable multivibrator. Briefly explain how the circuit functions.

8-15   Using a 741 IC operational amplifier, design an astable multivibrator to produce a square wave output with an amplitude of approximately ± 9 V and frequency of approximately 500 Hz.

8-16   Sketch the functional block diagram of a 555 IC timer. Briefly explain the function of each component.

8-17   Sketch the circuit and waveforms generated for a 555 monostable multivibrator. Referring to the monostable circuit and the 555 functional block diagram, explain how the 555 operates as a monostable circuit.

8-18   Using a supply of 18 V, design a 555 monostable circuit to produce a 0.5 ms output pulse.

8-19   Show how a basic 555 monostable circuit should be modified to: (a) use capacitor-coupled triggering, (b) prevent unwanted signals being picked up at the control voltage terminal, (c) permit a reset signal to be coupled to the reset terminal. Sketch the various waveforms and briefly explain.

8-20 Sketch the circuit of a 555 timer employed as an astable multivibrator. Show the capacitor and output waveforms and explain how the circuit functions.

8-21 Show how a basic 555 astable circuit may be modified: (a) to produce a 50% duty cycle, (b) to provide an adjustable duty cycle with a constant PRF, (c) to create a variable frequency square wave generator. Briefly explain in each case.

8-22 Design a 555 astable multivibrator to generate an output with PRF = 5 kHz and a duty cycle of 75%. Use V_{CC} = 15 V.

8-23 Analyze the circuit designed in Problem 8-22 to determine the actual PW and duty cycle.

8-24 Sketch circuit diagrams to show how 555 timers may be used to construct: (a) a sequential timer, (b) a pulsed tone oscillator. Explain how each circuit operates.

8-25 Design the 555 square wave generator illustrated in Fig. 8-20. Output amplitude is to be $\simeq +10$ V, and frequency is to be adjustable over the range 1 kHz to 10 kHz. Assume that V_o is approximately 1 V less than V_{CC}.

8-26 Compare the 7555 CMOS timer circuit to the 555 bipolar device. Design a square wave generator using the 7555 IC to have V_o = 5 V, and f = (500 Hz to 15 kHz).

CHAPTER 9

Bistable
Multivibrators

INTRODUCTION

The *bistable multivibrator*, also known as a *flip-flop*, is a switching circuit with two stable states. The circuit can be triggered from one state to the other by applying an input voltage via a suitable triggering circuit. The triggering input may be applied to the collectors, bases, or emitters of the transistors. Bistable multivibrators can be either collector-coupled or emitter-coupled circuits. They are also available in integrated circuit form. Because of its application in computing systems, the bistable multivibrator is the most important of all multivibrator circuits.

9-1 THE COLLECTOR-COUPLED BISTABLE MULTIVIBRATOR

The collector-coupled bistable multivibrator circuit shown in Figure 9-1(a) has two stable states. Either Q_1 is *on* and Q_2 is *off*; or Q_2 is *on* and Q_1 is biased *off*. The circuit is completely symmetrical. Load resistors R_{L1} and R_{L2} are equal, and potential dividers (R_1, R_2) and (R'_1, R'_2) form similar bias networks at the transistor bases. Each transistor is biased from the

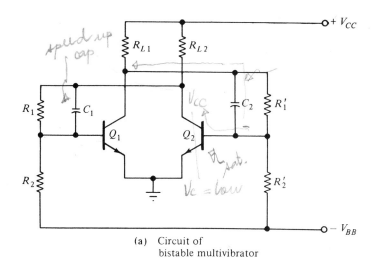

(a)   Circuit of
bistable multivibrator

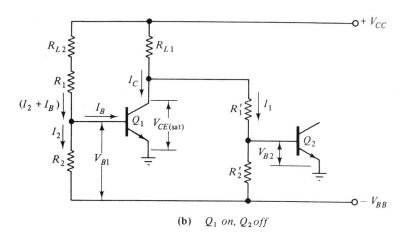

(b) Q_1 on, Q_2 off

FIGURE 9-1.   Collector-coupled bistable multivibrator circuit, and circuit
condition when Q_1 is on and Q_2 is off.

collector of the other device. When either transistor is *on*, the other
transistor is biased *off*.

Consider the condition of the circuit when Q_1 is *on* and Q_2 is *off*. With
Q_2 *off*, there is no collector current flowing through R_{L2}. Therefore, as
shown in Figure 9-1(b), R_{L2}, R_1, and R_2 can be treated as a potential
divider biasing Q_1 base from V_{CC} and $-V_{BB}$. With Q_1 *on* in saturation, its
collector voltage is $V_{CE(\text{sat})}$, and R_1' and R_2' bias V_{B2} below ground level.
Since the emitters of the transistors are grounded, Q_2 is *off*. The circuit can

remain in this condition (Q_1 *on*, Q_2 *off*) indefinitely. When Q_1 is triggered *off*, Q_2 switches *on*, and remains *on* with its base biased via R_{L1}, R_1', and R_2'. At this time, the base of Q_1 is biased negatively from Q_2 collector. Thus, Q_1 remains *off* and Q_2 remains *on* indefinitely. The output voltage at each collector is approximately V_{CC}.

Capacitors C_1 and C_2 operate as speed-up capacitors to improve the switching speed of the transistors. However, in the bistable circuit, C_1 and C_2 are also termed *commutating* or *memory capacitors*.

Consider the conditions when Q_1 is *on* and Q_2 is *off*. Capacitor C_1 is charged to the voltage across R_1, and C_2 is charged to the voltage across R_1'. As will be seen when design of a bistable circuit is considered, the voltage across R_1 (at the base of the *on* transistor) is several volts greater than across R_1' (at the base of the *off* transistor). Therefore, when Q_1 is *on*, C_1 is charged to a voltage greater than the voltage on C_2. Now, consider what occurs when the *on* transistor is triggered *off* for a brief instant. With both transistors *off*, both collector voltages are approximately at the level of V_{CC}. Also, each base voltage becomes $V_B \simeq V_{CC} -$ (charge on the capacitor at the transistor base). Since C_2 has a smaller charge than C_1, V_{B2} is greater than V_{B1}. One transistor must begin to switch *on*, and the one with the highest base voltage switches *on* first. Thus Q_2 (the formerly *off* transistor) switches *on* before Q_1 and, in doing so, it biases Q_1 *off*. Once switchover occurs, C_2 becomes charged to a greater voltage than C_1.

It is seen that the charge on the capacitors enables them to "remember" which transistor was *on* and which was *off*, and facilitates the circuit changeover from one state to another.

9-2 DESIGN OF A COLLECTOR-COUPLED BISTABLE MULTIVIBRATOR

Design of a bistable multivibrator commences with a specification of supply voltage and load resistance. Alternatively, the output current may be specified, or I_C may simply be specified as a level much larger than the output current. As in the case of the monostable and Schmitt circuits, the bias resistances R_1 and R_2 must be chosen small enough to provide a stable bias level, and large enough so that they do not overload R_L. The rule of thumb that (bias current) $I_2 \simeq \frac{1}{10} I_C$ again can be applied, and the circuit design procedure is then fairly simple. When the value of R_L is calculated, the next larger standard resistance should be selected. This will ensure sufficient voltage drop across R_L to have the transistor in saturation. The bias resistances should be selected as the next standard resistance size that is smaller than that calculated. This will provide slightly more base current than required for saturation.

The voltage on the commutating capacitors must not change significantly during the turn-*off* time of the transistors. If the capacitors are allowed to discharge by 10% of the difference between maximum and minimum capacitor voltages, Equation (2-10) may be applied.

$$t = 0.1 \ CR \qquad\qquad [\text{Equation (2-10)}]$$

therefore,

$$C = \frac{t_{(\text{off})}}{0.1 \ R}$$

In this case, $C = C_1 = C_2$ and $t_{(\text{off})}$ is the turn-*off* time for the transistors; R is the resistance "seen" looking into the terminals of R_1 or R'_1. With one transistor *on*, the minimum value of R approximates $R_1 \| R_2$. This gives

$$C_1 = C_2 = \frac{t_{(\text{off})}}{0.1(R_1 \| R_2)} \qquad\qquad (9\text{-}1)$$

As for other switching circuits, the presence of capacitors limits the maximum frequency at which the bistable circuit may be triggered. To determine the maximum triggering frequency, the *recovery time* for the capacitors must be calculated. This is the time for the capacitors to discharge from maximum voltage to minimum voltage, or *vice versa*. The maximum triggering frequency is then calculated as 1/(recovery time). Using Equation (2-9), the recovery time is

$$t_{re} = 2.3 \ CR \qquad\qquad [\text{Equation (2-9)}]$$

where again $R = (R_1 \| R_2)$ and maximum triggering frequency is

$$f_{\text{max}} = \frac{1}{t_{re}} = \frac{1}{2.3 \ C(R_1 \| R_2)} \qquad\qquad (9\text{-}2)$$

EXAMPLE 9-1

Design a collector-coupled bistable multivibrator to operate from a ±5 V supply. Use 2N3904 transistors, with $I_C = 2$ mA.

solution

Refer to Figure 9-1(b):

$$V_{CE(sat)} = 0.2 \text{ V (typically)}$$

$$R_L \simeq \frac{V_{CC} - V_{CE(sat)}}{I_C} \text{ (i.e., neglecting } I_1)$$

$$= \frac{5 \text{ V} - 0.2 \text{ V}}{2 \text{ mA}}$$

$$= 2.4 \text{ k}\Omega \text{ (use 2.7 k}\Omega \text{ standard value)}$$

From the 2N3904 data sheet in Appendix 1-4, $h_{FE(min)} = 70$, so

$$I_{B(min)} = \frac{I_C}{h_{FE(min)}}$$

$$= \frac{2 \text{ mA}}{70} = 28.6 \text{ } \mu\text{A}$$

With Q_1 *on*,

$$V_{R2} = V_{BE1} - V_{BB}$$

$$= 0.7 \text{ V} - (-5 \text{ V}) = 5.7 \text{ V}$$

$$I_2 \simeq \frac{1}{10} I_C = \frac{2 \text{ mA}}{10} = 200 \text{ } \mu\text{A}$$

$$R_2 = \frac{V_{R2}}{I_2} = \frac{5.7 \text{ V}}{200 \text{ } \mu\text{A}}$$

$$= 28.5 \text{ k}\Omega \text{ (use 27 k}\Omega \text{ standard value)}$$

Now I_2 becomes

$$I_2 = \frac{5.7 \text{ V}}{27 \text{ k}\Omega}$$

$$= 211 \text{ } \mu\text{A}$$

$$R_{L2} + R_1 = \frac{V_{CC} - V_{BE}}{I_2 + I_B}$$

$$= \frac{5 \text{ V} - 0.7 \text{ V}}{211 \text{ } \mu\text{A} + 28.6 \text{ } \mu\text{A}}$$

$$= 17.9 \text{ k}\Omega$$

$$R_1 = (R_{L2} + R_1) - R_{L2}$$

$$= 17.9 \text{ k}\Omega - 2.7 \text{ k}\Omega$$

$$= 15.2 \text{ k}\Omega \text{ (use 15 k}\Omega \text{ standard value)}$$

Analysis.

$$V_{C(on)} = V_{CE(sat)} = 0.2 \text{ V}$$

$$V_{C(off)} = V_{CC} - V_{RL2} \text{ (for } Q_2 \text{ off)}$$

$$\text{Voltage across } (R_{L2} + R_1) = V_{CC} - V_{BE1}$$

$$= 5 \text{ V} - 0.7 \text{ V} = 4.3 \text{ V}$$

$$V_{RL2} = (V_{CC} - V_{BE1}) \times \frac{R_{L2}}{R_1 + R_{L2}}$$

$$= 4.3 \text{ V} \times \frac{2.7 \text{ k}\Omega}{15 \text{ k}\Omega + 2.7 \text{ k}\Omega} = 0.66 \text{ V}$$

$$V_{C(off)} = 5 \text{ V} - 0.66 \text{ V} = 4.34 \text{ V}$$

$$V_{B(off)} = V_{CE(sat)} - V_{R'1} \text{ (for } Q_2 \text{ off)}$$

$$\text{Voltage across } (R_1' + R_2') = V_{CE(sat)} - V_{BB}$$

$$= 0.2 \text{ V} - (-5 \text{ V})$$

$$= 5.2 \text{ V}$$

$$V_{R'1} = (V_{CE(sat)} - V_{BB}) \times \frac{R_1'}{R_1' + R_2'}$$

$$= 5.2 \text{ V} \times \frac{15 \text{ k}\Omega}{15 \text{ k}\Omega + 27 \text{ k}\Omega} = 1.86 \text{ V}$$

$$V_{B(off)} = 0.2 \text{ V} - 1.86 \text{ V} = -1.66 \text{ V}$$

Figure 9-2(a) shows the bistable circuit as designed in Example 9-1, with resistor values and voltage levels indicated.

In Figure 9-2(b) a bistable circuit using 2N3906 *pnp* transistors is shown. The circuit design is exactly the same as for the *npn* circuit, except that all voltage polarities and current directions are reversed.

9-3 EMITTER-COUPLED BISTABLE MULTIVIBRATOR

The emitter-coupled bistable multivibrator circuit (Figure 9-3) is the same as the collector-coupled circuit, except that an emitter resistor R_E has been added. Load resistors R_{L1} and R_{L2} are equal, as are capacitors C_1 and C_2, and potential dividers (R_1, R_2) and (R_1', R_2').

The presence of R_E allows the circuit to be operated from a single polarity supply. The emitter resistor also limits the collector current of the

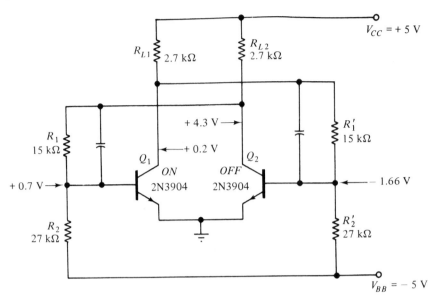

(a) Bistable circuit using *NPN* transistors

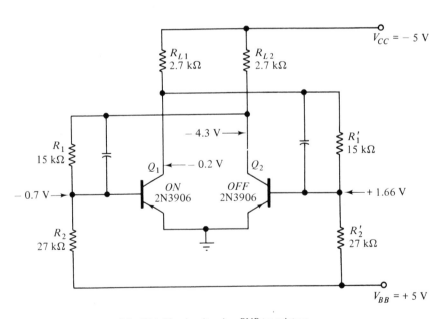

(b) Bistable circuit using *PNP* transistors

FIGURE 9-2. Bistable multivibrator circuit designed in example 9-1, (a) using *npn* transistors, (b) using *pnp* transistors.

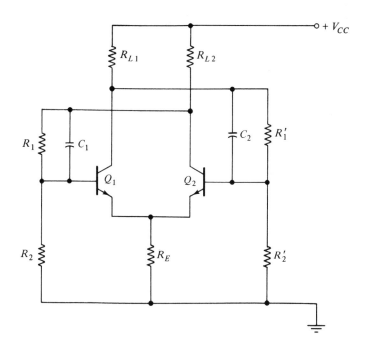

FIGURE 9-3.   Emitter-coupled bistable circuit.

on transistor to any desired level, so that the transistor may be saturated or unsaturated.

In designing a nonsaturated emitter-coupled bistable circuit, the voltage drop across R_E must be made several times the base-emitter voltage of the device. This is necessary to maintain reasonably stable bias conditions. A minimum V_{CE} of about 3 V should be designed into the circuit in order to avoid device saturation. A good rule of thumb is to divide $(V_{CC} - V_{CE})$ equally between V_{RL} and V_{RE}. Also to avoid saturation, the *maximum* transistor h_{FE} must be employed in the design calculation. When R_L is calculated, the next *smaller* standard value should be selected, again to avoid saturation. R_1 and R_2 must also be carefully chosen with nonsaturation in mind. As with other multivibrator circuits, bias current I_2 should be made approximately $\frac{1}{10} I_C$. This is to ensure that R_1 and R_2 are small enough to provide a stable bias voltage, but not so small that they will overload R_L.

9-4 COLLECTOR TRIGGERING

Bistable multivibrator triggering circuits normally are designed to turn *off* the *on* transistor. The triggering may be *asymmetrical* or *symmetrical*. In *asymmetrical triggering*, two trigger inputs are employed, one to *set* the circuit in one particular state, and the other to *reset* to the opposite state. This process is sometimes referred to as *set-reset triggering*. Symmetrical triggering uses only one trigger input, and the state of the circuit is changed each time a trigger pulse is applied.

Refer to the asymmetrical collector trigger circuit shown in Figure 9-4, and assume that Q_1 is *on* and Q_2 is *off*. With Q_2 *off*, the voltage at its collector is approximately V_{CC}. The negative-going step input, coupled via C_3, forward-biases D_1 and pulls its cathode down by ΔV. At this time, D_2 is reverse-biased by the negative-going input and has no function to perform. The anode of D_1 is pulled down by $\Delta V - V_{D1}$; that is, by ΔV minus the diode forward voltage drop. Thus, Q_2 collector voltage is changed from approximately V_{CC} to $[V_{CC} - (\Delta V - V_{D1})]$. See the waveforms in Figure 9-4. Capacitor C_1, which does not discharge instantaneously, acts initially like a battery. Consequently, the voltage change at Q_2 collector also appears at Q_1 base. Q_1 base voltage initially is V_{BE} (with Q_1 *on*) and it falls by $\Delta V - V_{D1}$. Thus the input voltage at C_3 causes the base of Q_1 to be pushed below the level of its emitter voltage.

When the step input is applied, C_3 immediately starts to charge via R_{L2} (the polarity is shown in Figure 9-4). Both collector and base voltages rise from their minimum levels, as shown. To ensure that Q_1 switches *off*, its base voltage must remain below the emitter voltage level for the

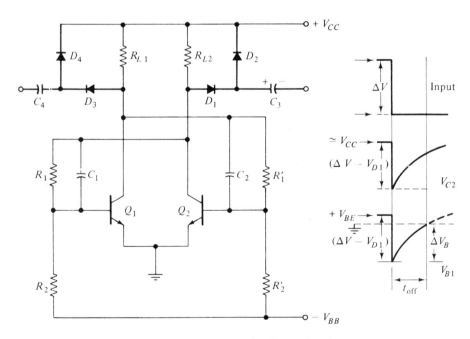

FIGURE 9-4. Asymmetrical collector triggering.

transistor turn-*off* time $t_{(off)}$. This may be achieved by use of a large value coupling capacitor for C_3. However, it is best to choose C_3 as small as possible. The smallest suitable capacitor is one that will allow V_B to rise to the level of the emitter voltage during the transistor turn-*off* time. The waveforms of Figure 9-4 show that V_{B1} rises to ground level during $t_{(off)}$.

When the triggering input becomes positive, returning to its normal dc level, D_1 is reverse-biased and the state of the bistable circuit is unaffected. The charge on C_3 which resulted from the negative-going input remains until the trigger input becomes positive. D_2 is then forward-biased by the capacitor charge, and C_3 is rapidly discharged via D_2. The triggering circuit now is ready to receive another negative-going input. However, with Q_1 already *off*, the state of the circuit will not be altered by a triggering input via C_3. Instead, Q_2 must be triggered *off* by an input applied to C_4.

Symmetrical collector triggering is shown in Figure 9-5. With Q_1 *on* and Q_2 *off*, I_C flows through R_{L1} causing V_{C1} to be approximately zero volts. Also, V_{C2} is approximately V_{CC}. The amplitude of the negative-going trigger input does not exceed the voltage drop across R_L (*i.e.*, less than V_{CC}), so that diode D_3 does not become forward-biased. However, diode D_1 is forward-biased by the negative-going input, as has been explained. Thus, Q_1 base is pushed down exactly as discussed for asymmetrical triggering and Q_1 is turned *off*. With Q_1 *off*, V_{C1} rises to approximately V_{CC} and V_{C2} drops to near zero. The next negative-going input forward-biases

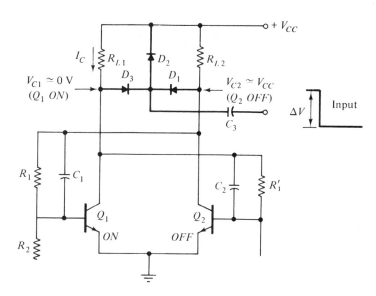

FIGURE 9-5.   Symmetrical collector triggering.

D_3 and causes Q_2 base to be pushed below its emitter level. Hence, Q_2 switches *off* and the circuit returns to its original state.

It is seen that the circuit changes state each time a negative-going trigger voltage is applied. D_2 functions as before, becoming forward-biased and discharging C_3 each time the input returns to its upper level. A resistor could function in place of D_2, but it would load the trigger signal and would take a relatively long time to discharge C_3.

The design of collector triggering circuits mainly involves determination of the smallest suitable coupling capacitor. The allowable change in voltage at the base of the transistor to be switched *off* dictates the voltage through which the coupling capacitor may be charged.

Refer to Figures 9-4 and 9-5 again. Note that when the trigger voltage pulls the collector of Q_2 down to near ground level, capacitor C_1 begins to discharge via $R_1 \| R_2$. As already explained in Sec. 9-2, the commutating capacitor voltage should not be allowed to discharge by more than 10% of the difference between maximum and minimum capacitor voltages. Equation (9-1) can be applied to calculate C_1 and C_2.

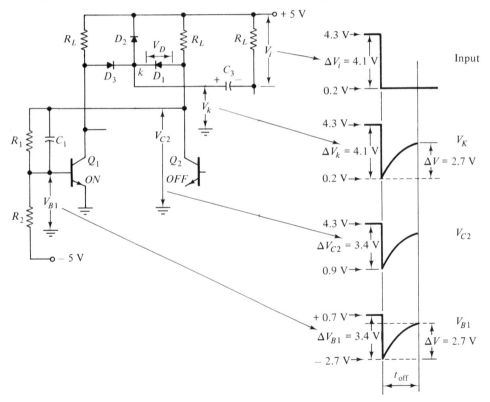

FIGURE 9-6. Triggering voltage waveforms for symmetrical collector triggering design.

EXAMPLE 9-2

The saturated collector-coupled flip-flop designed in Example 9-1 is to be triggered by the collector output of a previous similar stage. Design a suitable symmetrical collector triggering circuit.

solution

The waveforms of the triggering voltage as it appears at various points in the circuit are shown in Figure 9-6. From Example 9-1, the collector voltage of the flip-flop changes from 4.3 V to 0.2 V. This change is used as an input triggering voltage (see Figure 9-6).

$$V_i = 4.3 \text{ V} - 0.2 \text{ V} = 4.1 \text{ V}$$

At the diode cathodes,

$$\Delta V_K \simeq 4.1 \text{ V}$$
$$\Delta V_{C2} = \Delta V_K - V_{D1}$$
$$= 4.1 \text{ V} - 0.7 \text{ V} = 3.4 \text{ V}$$
$$\Delta V_{B1} = \Delta V_{C2} = 3.4 \text{ V}$$

To keep $V_{B1} < V_E$ during $t_{(off)}$:

$$\Delta V = \Delta V_{B1} - V_{BE}$$
$$= 3.4 \text{ V} - 0.7 \text{ V} = 2.7 \text{ V}$$

$\therefore C_3$ can charge by 2.7 V during $t_{(off)}$.
For C_3:

$$\text{Initial voltage} = E_0 \simeq 0 \text{ V}$$
$$\text{Final voltage} = e_c \simeq \Delta V = 2.7 \text{ V}$$
$$\text{Charging voltage} = E = \Delta V_i - V_{D1} = 3.4 \text{ V}$$
$$\text{Charging resistance} \simeq R_L = 2.7 \text{ k}\Omega$$
$$\text{Turn-}off \text{ time (for 2N3904)} = t_{(off)} = 250 \text{ ns}$$

From Equation (2-8):

$$C_3 = \frac{t}{R_L \ln\left(\dfrac{E - E_0}{E - e_c}\right)}$$

$$= \frac{250 \text{ ns}}{2.7 \text{ k}\Omega \times \ln\left(\dfrac{3.4 \text{ V} - 0}{3.4 \text{ V} - 2.7 \text{ V}}\right)}$$

$$\simeq 59 \text{ pF (use 62 pF standard value)}$$

The diodes required for the triggering are low-current devices capable of surviving a peak inverse voltage greater than V_{CC}. 1N914 diodes are more than adequate (see the data sheet in Appendix 1-1).

EXAMPLE 9-3 _____

Determine the value of suitable commutating capacitors for the flip-flop designated in Example 9-1 when collector triggering is employed. Also calculate the maximum triggering frequency for the circuit.

solution

By Equation (9-1),

$$C_1 = C_2 = \frac{t_{(off)}}{0.1(R_1 \| R_2)}$$

$$= \frac{250 \text{ ns}}{0.1(15 \text{ k}\Omega \| 27 \text{ k}\Omega)} = 259 \text{ pF (use 270 pF standard value)}$$

By Equation (9-2),

$$f_{(max)} = \frac{1}{2.3\, C(R_1 \| R_2)}$$

$$= \frac{1}{2.3 \times 270 \text{ pF}(15 \text{ k}\Omega \| 27 \text{ k}\Omega)}$$

$$= 167 \text{ kHz}$$

9-5 BASE TRIGGERING

Base triggering circuits are subdivided into *asymmetrical base triggering*, *symmetrical base triggering* and *collector-steered base triggering* circuits. The first two of these are shown schematically in Figure 9-7. In Figure 9-7(a), a negative-going input coupled via C_3 forward-biases D_1 and pulls Q_2 base below its emitter voltage level. C_3 charges via R_{L1} (C_2 by-passes R_1') and allows the base voltage to rise to ground over a time period equal to the

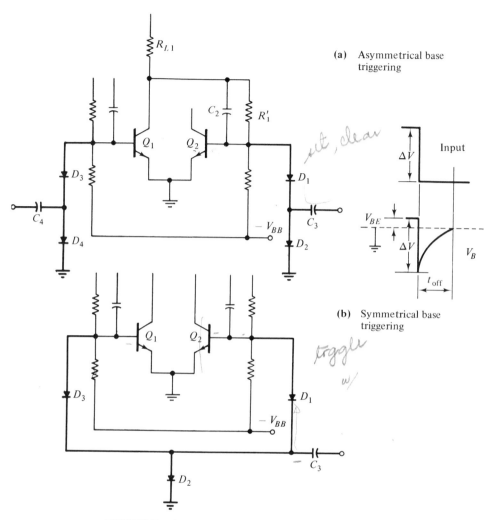

FIGURE 9-7.   Asymmetrical and symmetrical base triggering.

transistor turn-*off* time. C_3 is discharged via D_2 when the input becomes positive, to return to its normal dc level.

In Figure 9-7(a), the negative-going input to C_3 can only switch Q_2 *off*. To turn Q_1 *off*, a negative-going input must be coupled via C_4 and D_3. Positive-going inputs at either terminal reverse-bias D_1 or D_3 and have no effect on the bistable circuit.

For the *symmetrical base triggering circuit* in Figure 9-7(b), D_2 is common and the input is applied to the common cathode terminal of D_3 and D_1. If Q_1 is *off*, its base voltage is biased negatively with respect to

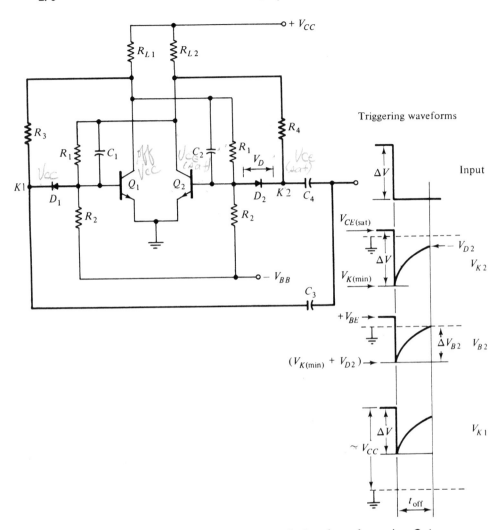

FIGURE 9-8. Collector-steered base triggering circuit and waveforms when Q_2 is initially *on*.

ground. Thus, if the input signal is kept small enough, D_3 does not become forward-baised. Q_2 base is at V_{BE} with respect to ground (with Q_2 *on*), so that the negative-going input forward-biases D_1 and pulls the transistor base below its emitter level. When Q_2 is *off* and Q_1 is *on*, D_3 is forward-biased by the triggering input, and D_1 remains reverse-biased. D_2 becomes forward-biased only when the input becomes positive. At this time, C_3 is discharged.

The amplitude of the trigger input to the symmetrical base triggering circuit is very critical. The input must be greater than 0.7 V in order to switch *off* the *on* transistor. If the base voltage of the *off* transistor is, say, -1.5 V, then the input voltage should be less than 1.5 V to avoid affecting the *off* transistor. The *collector-steered base triggering circuit* overcomes the problem of critical triggering amplitude.

In the *collector-steered base triggering circuit*, shown in Figure 9-8, diode D_2 has its anode connected to Q_2 base and its cathode connected via *steering resistance R_4* to Q_2 collector. Similarly, the anode of D_1 is connected to Q_1 base, and the cathode is connected via R_3 to Q_1 collector. The common triggering signal is applied to D_1 and D_2 cathodes via separate capacitors C_3 and C_4, respectively.

Consider the circuit conditions for Figure 9-8 when Q_1 is *off* and Q_2 is *on*. With Q_1 *off*, the voltage at its collector is approximately V_{CC}. Therefore, the cathode of D_1 is approximately at V_{CC}. In this case, triggering inputs with amplitude less than V_{CC} will *not* forward-bias D_1. Since Q_2 is *on*, its collector voltage is $V_{CE(sat)}$ above ground level; consequently, the cathode of D_2 is at $V_{CE(sat)}$. A negative-going trigger input with an amplitude of a few volts will forward-bias D_2 and pull the base of Q_2 below ground. Thus, to cause the circuit to correctly change state, the trigger voltage amplitude can have a value anywhere between about 2V and V_{CC}. The triggering voltage at each of the bases and at the diode cathodes are illustrated by the waveforms in Figure 9-8.

For collector-steered base triggering (Figure 9-8), R_3 and R_4 should be selected much larger than R_L. This is to ensure that the steering resistors do not constitute a significant load on R_{L1} and R_{L2}. When Q_2 base voltage is pulled down by the input trigger voltage, commutating capacitor C_2 commences to charge via R_L. Since Q_2 is to be switched *off*, C_2 voltage can be allowed to increase slightly. However, the charge on C_1 (at the base of the transistor that is to switch *on*) should not change significantly. Once again Equation (9-1) applies.

9-6 THE *T* FLIP-FLOP

When used as a logic element the bistable multivibrator is usually termed a *flip-flop*. Flip-flops are available as integrated circuits in a wide variety of forms and combinations. The simplest of all is the *T flip-flop*, or *toggle* flip-flop.

The *T* flip-flop is a bistable multivibrator symmetrically triggered by an input to one terminal. Any one of the symmetrically triggered bistable circuits already studied can be employed as a *T* flip-flop (e.g., the circuit in Fig 9-5). The logic symbol for a *T* flip-flop is shown in Figure 9-9(a). The Q and $\overline{Q}$ terminals are the circuit output points. These would be the collector terminals of transistors Q_1 and Q_2 for the circuit in Figure 9-5.

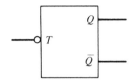

(a)   Logic symbol for T flip-flop
triggered by a negative-going edge.

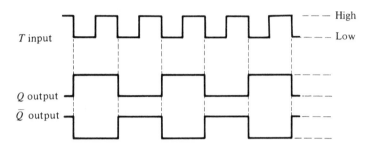

(b)   Timing diagram

FIGURE 9-9.   Logic symbol and timing diagram for *T* flip-flop triggered by a
negative-going edge.

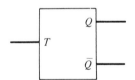

(a)   Logic symbol for T flip-flop triggered
by a positive-going edge.

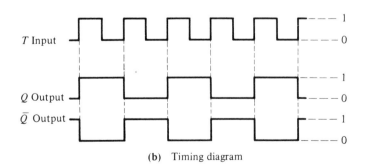

(b)   Timing diagram

FIGURE 9-10.   Logic symbol and timing diagram for *T* flip-flop triggered by a
positive-going edge.

The T terminal is the input triggering (or toggling) terminal for the flip-flop. Note that the small circle at the toggle input terminal on the logic symbol indicates that a negative-going signal is required to trigger this flip-flop (again refer to Fig. 9-5).

Figure 9-9(b) shows the timing diagram for the flip-flop. Each time the T input goes from a *high* to a *low* level, the Q and $\overline{Q}$ outputs are seen to change state. The circuit is said to be triggered by a negative-going edge.

In Figure 9-10 the logic symbol and timing diagram are shown for a flip-flop which is triggered by a positive-going edge. In this case the Q and $\overline{Q}$ outputs change state at the instant that the T input changes from *low* to *high*. The bistable circuit in Figure 9-5 could be converted into a positive-going edge triggered flip-flop by connecting a normally-*off* inverter (see Section 5-2) at the triggering input.

Another way of referring to *high* and *low* levels is to term the *high* level *logic 1* and the *low* level *logic 0*. Thus, in Figure 9-10(b) the Q output is seen to go from *0* to *1* and $\overline{Q}$ output shifts from *1* to *0* at the first (from the left) T input change from *0* to *1*.

9-7   THE *SC* FLIP-FLOP

The *set-clear* or *SC* flip-flop which has the logic symbol illustrated in Figure 9-11(a) could have a circuit similar to the asymmetrically triggered flip-flop in Figure 9-4. Previous consideration of that circuit showed that it can be triggered into one particular state (i.e., it can be *set*) when a trigger voltage is applied to one input terminal. Also, it can be triggered (or *cleared*) into the opposite state by a voltage applied to the other input terminal. Another name for the *SC* flip-flop is the *RS* flip-flop or *reset-set* flip-flop.

Referring to Figure 9-11, note the small circles at the S and C inputs on the logic symbol, once again indicating that this circuit triggers on a negative-going edge. Following the timing diagram waveforms from left to right:

At t_1, Q and $\overline{Q}$ change to *set* state ($Q=1, \overline{Q}=0$) when S goes from *1* to *0*.

At t_2, Q and $\overline{Q}$ do not change state when S goes from *0* to *1*.

At t_3, Q and $\overline{Q}$ do not change state when S goes from *1* to *0* because the flip-flop is already in the *set* condition.

At t_4, Q and $\overline{Q}$ change to the *cleared* state when C goes from *1* to *0*.

At t_5, Q and $\overline{Q}$ do not change state when S goes from *0* to *1*.

At t_6, Q and $\overline{Q}$ do not change state when C goes from *0* to *1*.

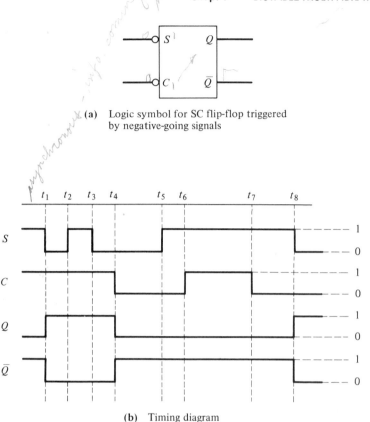

(a) Logic symbol for SC flip-flop triggered
     by negative-going signals

(b) Timing diagram

FIGURE 9-11.   *SC flip-flop logic symbol and timing diagram.*

At t_7, Q and $\bar{Q}$ do not change state when C goes from 1 to 0, because the
flip-flop is already in the *cleared* condition.

At t_8, Q and $\bar{Q}$ go into the *set* condition when S goes from 1 to 0.

The circuit of a positive-going edge triggered SC flip-flop is shown in
Figure 9-12. In this circuit the set-clear function is performed by transistors
Q_3 and Q_4. Terminals Q, $\bar{Q}$, S, and C correspond with the logic symbol
terminals in Figure 9-13(a). When both inputs are at 0, triggering transistors Q_3 and Q_4 are *off* and the flip-flop could be in either state. When a 1
input is applied to terminal S, transistor Q_3 is biased *on* into saturation.
This causes Q_2 collector to be pulled down, and Q_1 base is biased below
ground via R_1 and R_2. Consequently, Q_1 switches *off* and Q_2 switches *on*
into saturation. The output at terminal Q (from Q_1 collector) is now a
positive voltage, i.e., logic 1. The output at terminal $\bar{Q}$ (from Q_2 collector)

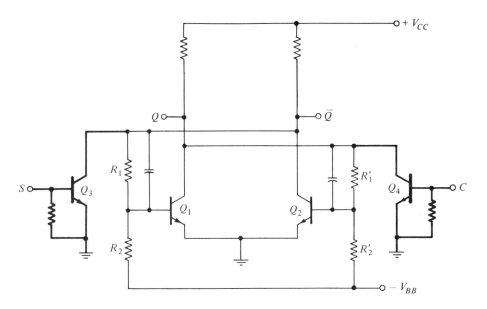

FIGURE 9-12.   Circuit of SC flip-flop.

is near zero volts, or logic *0*. The flip-flop is now in *set* condition, and this was obtained by applying a *1* to the *set* terminal, terminal *S*.

If a *0* input is applied to terminal *S*, the circuit remains *set*. Also further *1* inputs to terminal *S* do not affect the *set* condition of the flip-flop. A *1* input to terminal *C* (the *clear* terminal), switches Q_4 on. This pulls the collector of Q_1 down, biasing Q_2 *off* (via R_1' and R_2') and switching Q_1 *on*. The output is now *0* at terminal *Q* and *1* at terminal $\overline{Q}$, and the circuit is said to be *cleared*. To return to the *set* condition, *1* must be applied to the *S* input while the *C* input is at *0*.

In Figure 9-13(a) there are no small circles at *S* and *C* inputs, showing that the logic diagram is that of a flip-flop requiring positive-going inputs for triggering (as for the circuit in Fig. 9-12). The timing diagram for this circuit is similar to that in Figure 9-11(b), except that the *S* and *C* waveforms are inverted to show that the outputs change state only when the input levels change from 0 to 1.

Figure 9-13(b) shows a *truth table* for the flip-flop of Figure 9-12 and Figure 9-13(a). This is an alternative to the timing diagram as a method of describing the effect of various input combinations. When *S* and *C* are both *0*, there is no change in the flip-flop outputs. The outputs could be in either *set* or *clear* state. When a *1* is applied to *S*, and *C* remains *0*, the outputs go the *set* state with *Q=1* and *$\overline{Q}$=0*. When a *1* is applied to *C*,

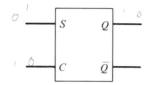

(a)   Logic symbol for *SC* flip-flop triggered
by positive-going inputs.

S	C	Q	$\overline{Q}$
0	0	No change	No change
1	0	1	0
0	1	0	1
1	1	Unpredictable	

(b)   Truth table

FIGURE 9-13.   Logic symbol and truth table for *SC* flip-flop triggered by
positive-going inputs.

and *S* remains at *0*, the circuit is *cleared* to *Q = 0* and $\overline{Q} = 1$. When a *1* is
applied to both *S* and *C* at once, both outputs may tend to go to *0*, but the
final state of the flip-flop is unpredictable.

9-8   THE CLOCKED *SC* FLIP-FLOP

A *clocked SC flip-flop* (also known as a *reset-set-toggle* or *RST* flip-flop) is
one that combines the triggering facilities of a toggle flip-flop with the
set-clear arrangement of a *SC* flip-flop. *Clock* is the logic term for an
accurate trigger frequency source. In Figure 9-14(a) the toggle input is
identified as *CLK* for a clock.

Referring to the timing diagram in Figure 9-14(b), it is seen that the
outputs change state only when the clock input goes from *0* to *1*. With
the outputs in the *clear* condition ($Q = 0, \overline{Q} = 1$), and with *S = 1* and *C = 0*,
the output changes to the *set* condition at the first positive-going edge of
the clock. The circuit then remains in *set* condition even when *S* goes to *0*.
When *C* has a *1* input applied (and *S* is at *0*), the outputs return to the
clear condition as soon as the clock goes from *0* to *1* again. With both *S*

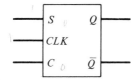

(a)   Logic symbol for clocked SC flip-flop

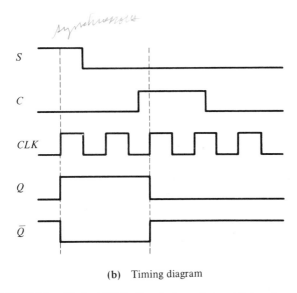

(b)   Timing diagram

FIGURE 9-14.   Clocked *SC* flip-flop; logic symbol and timing diagram.

and *C* inputs at *0*, the clock has no further effect on the outputs. If both *S* and *C* inputs had *1* applied, the flip-flop final state would be unpredictable.

9-9   THE *D* FLIP-FLOP

The *D flip-flop* is similar to the clocked *SC* flip-flop except that it has a single control terminal [see Figure 9-15(a)]. As the timing diagram in Figure 9-15(b) illustrates, the *Q* output of the flip-flop changes to the same state as the *D* input level. When the *D* input is *1*, *Q* becomes *1*. When the *D* input is *0*, *Q* changes to *0*. The output changes occur only when the clock input changes from *0* to *1*, and, of course, the $\overline{Q}$ output always assumes the opposite state of the *Q* output.

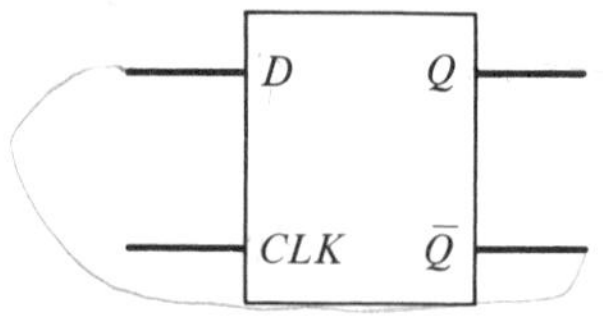

(a) Logic symbol for D flip-flop

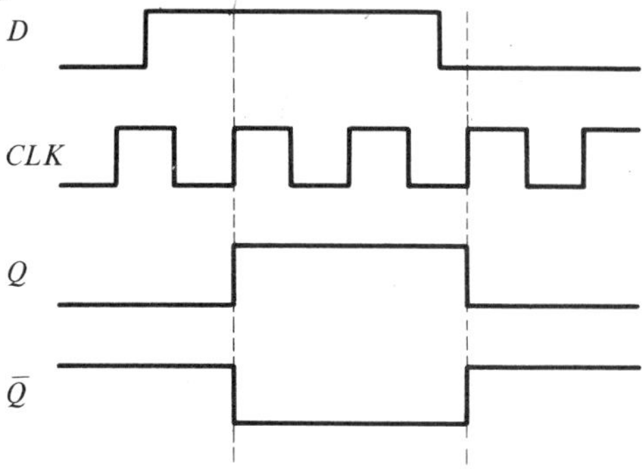

Q follows D upon ↑ clock
↑ triggered
clock when D what info.

(b) Timing diagram

FIGURE 9-15. Logic symbol and timing diagram for *D* flip-flop.

latch — output won't change when you don't clock it anymore

9-10 THE *JK* FLIP-FLOP

The *JK flip-flop* is identical to the clocked *SC* flip-flop with one exception. The exception occurs when both *J* and *K* inputs are at *1*. In this condition the flip-flop toggles each time an input is applied to the clock. Figure 9-16 shows the logic symbol and timing diagram.

Appendix 1-16 shows the data sheet for a 7476 *JK* flip-flop (and others). Note that as well as the input and output terminals already discussed, the logic symbol on the data sheet shows *preset* and *clear* terminals. These are inputs which operate independently of the *J*, *K*, and *clock* terminals, and which allow the flip-flop to be set in any desired state prior to a clock signal's being applied. Note also that only the *preset* and *clear* terminals have small circles. Thus, these two are affected by negative-going signals, while all other inputs are triggered by positive-going inputs.

The required supply voltage for this circuit is nominally 5 V, but may be as low as 4.5 V or as high as 5.5 V (see V_{CC} on the data sheet). The *high-level input voltage* (V_{IH}) is specified as a minimum of 2 V for correct

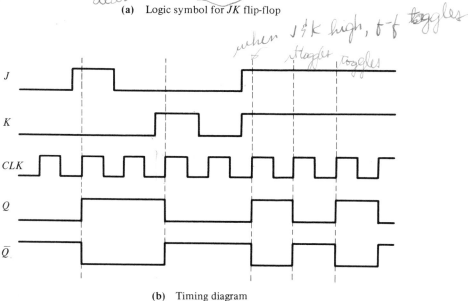

special purpose set — *clear input*

low priority clear — *clear input*

(a)   Logic symbol for JK flip-flop

when J & K high, f-f toggles

toggles   toggles

(b)   Timing diagram

FIGURE 9-16.   Logic symbol and timing diagram for *JK* flip-flop.

circuit operation. It can be anything greater than 2 V up to the same level as V_{CC}, but it should not be less than 2 V. The *low-level input voltage* (V_{IL}) should be less than 0.8 V. V_{IL} and V_{IH} specifications are essentially statements that the voltage representing logic *1* (V_{IH}) must not be less than 2 V, and that representing logic *0* (V_{IL}) must not be greater than 0.8 V.

The *high-level output voltage* (V_{OH}) is nominally 3.4 V but could be higher, according to the data sheet. It could also be as low as 2.4 V. The *low-level output voltage* (V_{OL}) is nominally 0.2 V, but could be as large as 0.4 V.

The *high-level input current* (I_{IH}) is stated as a maximum of either 40 μA or 80 μA, depending upon which of the input terminals is involved. This is essentially a leakage current, with the (internal) input transistors in an *off* condition. When the input voltage is *low*, the *low-level input current* (I_{IL}) is specified as -1.6 mA maximum for all but *clear* and *preset* inputs. For the *clear* and *preset* terminals, $I_{IL} = -3.2$ mA maximum. The negative sign here indicates that the currents flow *out* of the input terminals (conventional current direction from + to −).

The *low-level output current* (I_{OL}) is specified as a maximum of 16 mA. The absence of a negative sign on this quantity means that the current flows *into* the output terminal. Another way of stating this is to say that the output terminal can *sink* 16 mA. Of course, there could be less than 16 mA flowing into the output terminal when the output voltage is *low*, but there cannot be more than 16 mA. Considering the fact that I_{IL} is -1.6 mA for the *J* and *K* inputs of this type of flip-flop, the number of (*J* and/or *K*) inputs for other flip-flops that may be connected to one output terminal is 16 mA/1.6 mA = 10.

REVIEW QUESTIONS AND PROBLEMS

9-1 Sketch the circuit of a collector-coupled bistable multivibrator employed *npn* transistors. Explain how the circuit operates.

9-2 Repeat Problem 9-1 for a circuit using *pnp* transistors.

9-3 Repeat Problem 9-1 for an emitter-coupled circuit.

9-4 Design a collector-coupled bistable multivibrator to operate from a ± 6 V supply. Use 2N3904 transistors and make $I_C \simeq 1$ mA.

9-5 Design a collector-coupled bistable multivibrator using 2N3906 transistors. The supply voltage is ± 9 V, and the collector current is to be approximately 2 mA.

9-6 Sketch the circuit for asymmetrical collector triggering. Show the triggering waveforms, and explain how the circuit functions.

9-7 Repeat Problem 9-6 for symmetrical collector triggering.

9-8 The bistable multivibrator designed for Problem 9-4 is to use symmetrical collector triggering. The trigger input is to be the collector output of a previous similar stage. Design a suitable triggering circuit.

9-9 Calculate the value of the commutating capacitors for the flip-flop and triggering circuits designed for Problems 9-4 and 9-8. Also, calculate the maximum triggering frequency that should be employed.

9-10 Sketch the circuits for asymmetrical base triggering and symmetrical base triggering. Explain how each circuit operates, and discuss the major disadvantages of symmetrical base triggering.

9-11 Sketch a collector-steered base triggering circuit. Show the triggering waveforms, and explain how the circuit operates.

9-12 Design a collector-steered base triggering circuit for the flip-flop designed for Problem 9-4. The triggering input is to be the collector output from a previous similar stage.

9-13 Determine the value of suitable commutating capacitors for the flip-flop and triggering circuits designed for Problems 9-4 and 9-12. Also calculate the maximum triggering frequency for the circuits.

9-14 Sketch the logic symbol for a T flip-flop triggered by a positive-going input. Also sketch the timing diagrams, and briefly explain.

9-15 Repeat Problem 9-14 for an SC flip-flop. Also sketch a basic flip-flop circuit and show how two additional transistors can be employed to perform the *set* and *clear* functions.

9-16 Sketch logic symbols for T and SC flip-flops which can be triggered by negative-going inputs. Also draw the timing diagram and write a truth table for each circuit.

9-17 Repeat Problem 9-14 for a clocked SC flip-flop.

9-18 Sketch the logic symbol and timing diagram for a D flip-flop, and explain its operation.

9-19 Sketch the logic and timing diagram for a JK flip-flop. Explain how the JK flip-flop differs from the SC and D flip-flops.

9-20 Referring to the data sheet for the 7476 JK flip-flop, discuss the various quantities listed.

CHAPTER 10

Logic Gates

The two basic logic gates are the **and** gate and the **or** gate. The **and** gate produces a change in output voltage only when appropriate input voltages are applied to all of its input terminals. With the **or** gate, an output is generated when any one of its input terminals receives a signal. **Nand** and **nor** gates function in a slightly different way from **and** and **or** gates, respectively. Each type of gate has its own logic symbol.

Logic gates can be constructed by use of diodes and resistors **(diode logic)**, by use of resistors and transistors **(resistor transistor logic)**, or by using combinations of transistors **(transistor transistor logic)**. Other logic families are named according to the circuit configurations, and all are available as integrated circuits. The performances of the various IC logic families are expressed in terms of switching speed, power dissipation, noise immunity, and fan-out.

10-1 DIODE *AND* GATE

The circuit of a diode *AND* gate with three input terminals is shown in Figure 10-1(a). If one or more of the input terminals (*i.e.*, diode cathodes) are grounded, then the diode (or diodes) are forward-biased. Consequently, the output voltage V_O is equal to the diode forward voltage drop V_D. Suppose an input of 5 V is applied to terminal A, while terminals B and C are grounded. Diode D_1 is reverse-biased while D_2 and D_3 remain

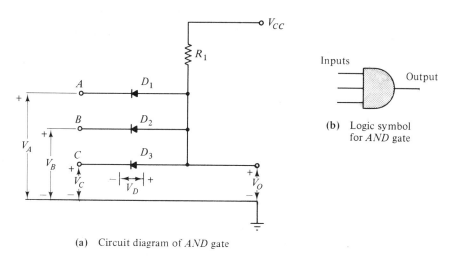

(a)   Circuit diagram of AND gate

(b)   Logic symbol for AND gate

FIGURE 10-1.   Circuit of a three-input diode *AND* gate and logic symbol for the *AND* gate.

forward-biased, and $V_O = V_D$. If levels of 5 V are applied to all three inputs, no current flows through R_1, and $V_O = V_{CC} = 5$ V. Thus, a *high* output voltage is obtained from the *AND* gate only when high input voltages are present at input *A*, *AND* at input *B*, *AND* at input *C*. Hence the name *AND* gate.

An *AND* gate may have as few as two or a great many input terminals. In all cases an output is obtained only when the correct input voltage levels are provided at every input terminal.

For all logic gates, the level of input and output voltages usually are described as either *high* or *low*. Depending upon the particular gate circuit, a *high* level might be between 3 V and 6 V, while a *low* voltage level might be less than 1 V. The high level is usually designated *1*, and the low level is designated *0*. The logic symbol for the *AND* gate is shown in Figure 10-1(b). When one or more of the input levels is *0*, the gate output is also *0*. When all three input levels are *1*, the output also goes to *1*.

EXAMPLE 10-1

An *AND* gate has three input terminals which are connected to the collectors of saturated transistors. The transistors can each take an *additional* collector current of 0.5 mA. Design a suitable circuit and determine the low and high output levels from the gate. Use $V_{CC} = 5$ V.

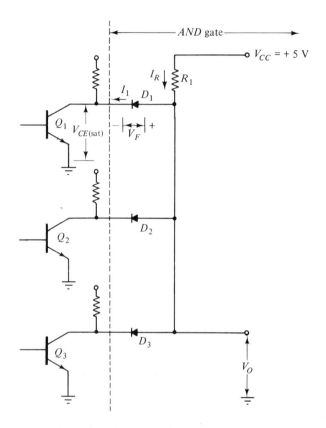

FIGURE 10-2. Diode *AND* gate with inputs controlled by transistors.

solution

The circuit is as shown in Figure 10-2. In the figure, maximum additional collector current I_1 flows through Q_1 when Q_1 is *on* and Q_2 and Q_3 are *off*.

$$I_R = I_{1(max)}$$

$$V_{CC} = (I_R R_1) + V_D + V_{CE(sat)}$$

$$R_1 = \frac{V_{CC} - V_D - V_{CE(sat)}}{I_R}$$

$$= \frac{5\,V - 0.7\,V - 0.2\,V}{0.5\,mA}$$

$$= 8.2\,k\Omega$$

This is the minimum value for R_1 to limit the transistor additional collector current to 0.5 mA. R_1 could be made larger than 8.2 kΩ, in which case the current would be smaller.

Output voltages are

$$V_{O(low)} = V_{CE(sat)} + V_D$$

$$= 0.2 \text{ V} + 0.7 \text{ V} = 0.9 \text{ V}$$

$$V_{O(high)} = V_{CC} = 5 \text{ V}$$

10-2   DIODE *OR* GATE

A three-input diode *OR* gate and its logic symbol are shown in Figure 10-3. It is obvious from the gate circuit that the output is zero when all three inputs are at ground level. If a 5 V input is applied to terminal A, D_1 is forward-biased, and V_O becomes ($5 \text{ V} - V_D$). If terminals B and C are grounded at this time, diodes D_2 and D_3 are reverse-biased. Instead of terminal A, the positive input might be applied to terminal B or C to obtain a positive output voltage. A *high* output voltage is obtained from an *OR* gate, when a high input is applied to terminal A, *OR* to terminal B, *OR* to terminal C. Hence the name *OR* gate.

As in the case of the *AND* gate, an *OR* gate may have only two or a great many input terminals.

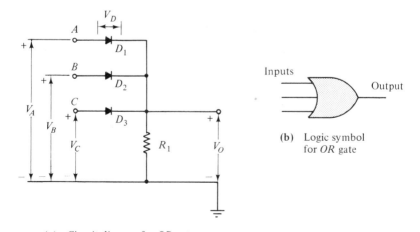

(a)   Circuit diagram for *OR* gate

(b)   Logic symbol for *OR* gate

FIGURE 10-3.   Circuit of a three-input diode *OR* gate and logic symbol for the *OR* gate.

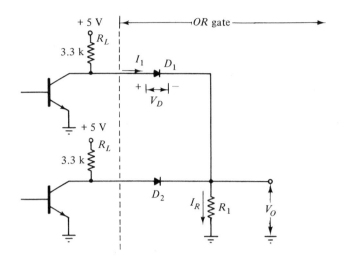

FIGURE 10-4. Diode *OR* gate with inputs controlled by transistors.

EXAMPLE 10-2

An *OR* gate has two input terminals, each of which is supplied from flip-flops having $R_L = 3.3$ kΩ. The supply voltage to the flip-flops is $V_{CC} = 5$V. The gate output voltage is to be at least 3.5 V. Design a suitable circuit.

solution

Refer to the circuit shown in Figure 10-4. When one input is *high*.

$$V_O = V_{CC} - (I_1 R_L) - V_D$$

and

$$I_R = I_1 = \frac{V_{CC} - V_D - V_O}{R_L}$$

$$I_R = \frac{5 \text{ V} - 0.7 \text{ V} - 3.5 \text{ V}}{3.3 \text{ k}\Omega} = 0.24 \text{ mA}$$

Also,

$$R_1 = \frac{V_O}{I_R}$$

$$= \frac{3.5 \text{ V}}{0.24 \text{ mA}}$$

$$\simeq 14.6 \text{ k}\Omega$$

This is a minimum value for R_1 to maintain the output voltage at a minimum of 3.5 V. R_1 could be made larger, in which case V_O would be larger.

10-3   POSITIVE LOGIC AND NEGATIVE LOGIC

The diode *AND* and *OR* gates already discussed both provide positive outputs when positive input signals are applied. This is referred to as *positive logic*. When negative input and output voltages are involved, the operation is termed *negative logic*.

Consider the situation with the diode *AND* gate shown in Figure 10-5. In this case, there is no supply voltage and the load resistance is grounded. If a negative input voltage V_A is now applied to terminal A, diode D_1 is

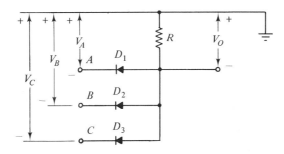

FIGURE 10-5.   Negative logic *OR* gate.

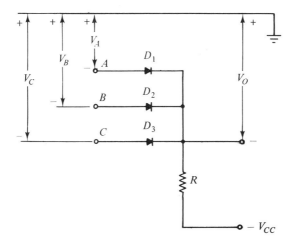

FIGURE 10-6.   Negative logic *AND* gate.

forward-biased, and the output is pulled negatively. Therefore, a negative output is obtained when a negative input is applied to terminal A, OR terminal B, OR terminal C. Thus, the positive logic AND gate can also function as a negative logic OR gate.

The OR gate of Figure 10-3 is reproduced in Figure 10-6 with a negative supply connected to R and the input terminals grounded. While any of the inputs remain at ground level, one diode is forward-biased and the output remains $V_O = -V_D$. A (large) negative output voltage is obtained only when negative inputs are applied to terminal A, AND to terminal B, AND to terminal C. Thus, the positive logic OR gate can also be used as a negative logic AND gate.

10-4   DTL $NAND$ GATE

As already explained, a positive logic diode AND gate has a low voltage output when one or more of its inputs are low, and a high output when all inputs are high. If a transistor inverter is connected at the output of the AND gate, the inverter output is high when one or more of the AND inputs are low, and low when all AND gate inputs are high. Used in this fashion, the inverter is termed a NOT gate. The combination of the NOT gate and the AND gate is then referred to as a $NOT\text{-}AND$ gate, or a $NAND$ gate.

Figure 10-7(a) shows an integrated circuit DTL (diode transistor logic) $NAND$ gate composed of a diode AND gate and an inverter. R_1, D_1, D_2, and D_3 constitute the AND gate. The inverter is formed by transistor Q_1 with load resistor R_L and bias resistor R_B. When all input terminals are at ground level, the voltage at point X is the voltage drop across the input diodes (i.e., $V_X = V_D$). If diodes D_4 and D_5 were not present, V_D would be sufficient to forward-bias the base-emitter junction of Q_1. The negative supply $-V_{BB}$ keeps diodes D_4 and D_5 forward-biased, so that when the inputs are at 0 V the transistor base voltage is

$$V_B = V_X - (V_{D4} + V_{D5})$$
$$= V_D - V_{D4} - V_{D5}$$

For silicon devices,

$$V_B \simeq 0.7\text{ V} - 0.7\text{ V} - 0.7\text{ V}$$
$$= -0.7\text{ V}$$

Therefore, when any one input to the $NAND$ gate is at 0 V, Q_1 is biased *off*, and the output voltage is V_{CC}.

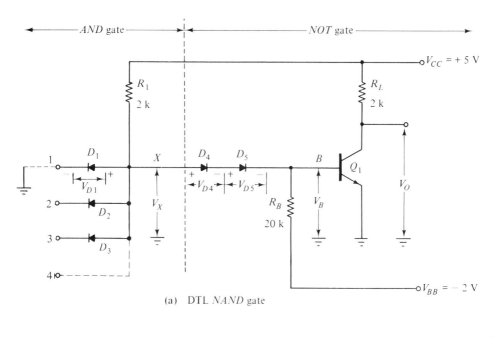

(a) DTL *NAND* gate

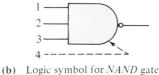

(b) Logic symbol for *NAND* gate

FIGURE 10-7.   Diode transistor *NAND* gate and logic symbol.

Suppose all inputs to the *NAND* gate are made sufficiently positive to reverse-bias D_1, D_2, and D_3. Now V_B depends upon the values of R_1 and R_2, and upon the levels of V_{CC} and $-V_{BB}$. If these quantities are all correctly selected, V_B is positive at this time, Q_1 is driven into saturation, and the output voltage goes to $V_{CE(sat)}$. When any one input to the *NAND* gate is at logic 0, the gate output is at *1*. When input *A AND* input *B AND* input *C* are at *1*, the output of the *NAND* gate is level *0*. The logic symbol employed for a *NAND* gate is shown in Figure 10-7(b). The symbol is simply that of an *AND* gate with a small circle at the output to indicate that the output voltage is inverted.

Input terminal *4* in Figure 10-7(a) is a direct connection to the diode anodes. Thus it provides for the connection of additional diodes to increase the number of input terminals. With this facility the gate is said to be *expandable*. Terminal *4* is shown on the logic symbol as a direct input [Figure 10-7(b)].

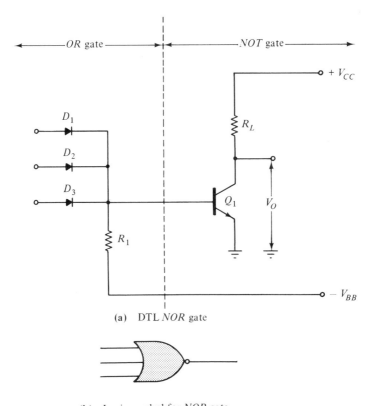

(a)   DTL *NOR* gate

(b)   Logic symbol for *NOR* gate

FIGURE 10-8.   Diode transistor *NOR* gate and logic symbol.

10-5   DTL *NOR* GATE

A transistor inverter (or *NOT* gate) connected at the output of a positive logic *OR* gate generates a negative-going output when any one of the inputs is positive. The circuit is termed a *NOT-OR* gate or *NOR* gate. A diode transistor *NOR* gate is shown in Figure 10-8(a), with its logic symbol shown in Figure 10-8(b). As in the case of the *NAND* gate, a small circle is employed to denote the polarity inversion at the output.

10-6   LOGIC GATE PERFORMANCE FACTORS

10-6.1   Propagation Delay Time

The switching speed of a logic gate is defined in terms of a *propagation delay time*. This is the time required for the gate to switch from its *low* output state to its *high* output state, or *vice versa*. The quantity varies with the transistor collector current and output load conditions. The switching time is also dependent upon the circuit configuration; for example, the transistors may have to be switched out of saturation, or they may be unsaturated. In DTL, the transistors are saturated. Typical propagation delay time for integrated circuit DTL is 25 to 30 ns. Some other types of logic circuits switch from one state to another in a time of 2 ns or less.

The method of measuring the propagation delay time is illustrated on the data sheet for the MC306.MC307 ECL gates in Appendix 1-18. (ECL is discussed in Section 10-10). Refer to the propagation delay illustration for the OR function in Appendix 1-18. The input (e_{in}) is shown as a solid line, and the output (e_{out}) is shown broken. It is seen that the time t_{d1} is measured as the time between input falling to its 50% amplitude level and output going to its 50% level. Similarly, t_{d2} is the time between input rising to 50% and output arriving at 50%. For the NOR function, t_{d1} and t_{d2} are measured in a similar way, except (as illustrated) e_{out} goes high when e_{in} goes low, and vice versa.

The average propagation delay time is frequently stated on data sheets, and this is simply

$$t_{d(ave)} = \frac{t_{d1} + t_{d2}}{2}$$

10-6.2   Noise Immunity

The minimum input voltage at which a logic gate switches is termed the *threshold voltage*. Consider the DTL gate in Figure 10-7(a). For Q_1 to be switched *on*, D_4 and D_5 must be forward-biased. For this to occur, the minimum input voltage at the cathode of D_3 is

$$V_i = V_{D4} + V_{D5} + V_{BE1} - V_{D1}$$
$$= 0.7 \text{ V} + 0.7 \text{ V} + 0.7 \text{ V} - 0.7 \text{ V} = 1.4 \text{ V}$$

Therefore, the threshold voltage for a DTL gate is approximately 1.4 V.

This (1.4 V) can also be termed the *minimum high input* level. A higher voltage than this (usually no larger than V_{CC}) can also be considered a *high* input, or *logic 1* input. There is also a *maximum low input* level, or upper limit for a *logic 0* input. The difference between these two is termed the *noise margin* of the circuit. *Noise spikes* with amplitudes on the same order as the noise margin could produce unwanted triggering of the gate. Thus the noise margin of a circuit gives a good indication of how susceptible the circuit may be to noise (i.e., in comparison to other circuits). However, the *noise immunity* of a logic gate does not depend solely on the noise margin.

Where a circuit has a low input impedance, noise spikes are potentially divided, and therefore are less likely to cause switching. Similarly, a gate that is driven from a low impedance source may be more immune to noise voltages. Also, a gate that switches slowly is less sensitive to fast spikes than one that has a very short propagation delay time. It is seen that many factors are involved in the *ac noise immunity* for given circuit. Instead of attempting to rate the noise immunity of a circuit in terms of voltage levels, switching time, and impedance, noise immunity is usually described as *poor, fair, good,* or *excellent.*

10-6.3 Fan-In and Fan-Out

Since logic gates usually are connected in complex combinations, the output of each circuit must be capable of driving the inputs of many other similar circuits. The number of similar gates that any one gate can drive is limited. This limit is termed the *fan-out* of the gate. If I_L is the total output current that a gate can handle, and I_i is the drive current for each input (or *unit load*), then *fan-out* $= I_L/I_i$.

The *fan-in* of a gate is the number of inputs that can be connected to a gate. The circuit in Figure 10-8 has three input terminals; therefore, it has a *fan-in* of 3. *Expandable* IC gates have facilities for connections of

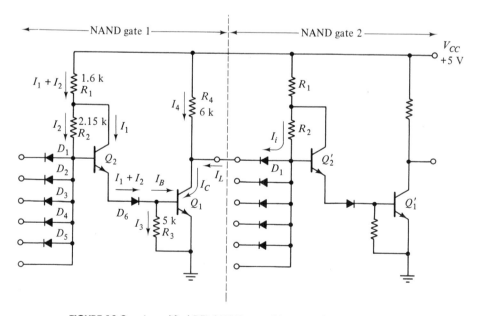

FIGURE 10-9. A modified DTL *NAND* gate driving another *NAND* gate.

additional input. The stated *fan-in* for the gate is then the total number of inputs that may be connected.

Figure 10-9 shows one DTL NAND gate driving another similar gate. The circuits are those of the Motorola MC930 IC *modified* DTL. In the modified version, resistor R_B is 5 kΩ instead of 20 kΩ, and it simply ties the transistor base terminal to the emitter, i.e. to ground level. The separate -2 V supply is now not required. Diode D_4 in Figure 10-7 is replaced by emitter follower Q_2 in Figure 10-9. The emitter follower provides base current to Q_1, and allows R_1 to be replaced by a larger resistance $(R_1 + R_2)$, thus reducing the maximum drive current to each input. The base-emitter voltage of Q_2 substitutes for the voltage drop across diode D_4 in Figure 10-7. The advantages of modified DTL are: Operation from a single supply voltage, greater output current, lower power dissipation, and lower input current. Note that the load current I_L actually flows *into* the collector of transistor Q_1, and *out of* the input terminal of the gate being controlled. In this situation transistor Q_1 is sometimes referred to as a *current sink*. From a knowledge of the supply voltage, resistor values, and transistor $h_{FE(min)}$, the *fan-out* of the circuit can be calculated.

EXAMPLE 10-3

Assuming that Q_1 and Q_2 have $h_{FE(min)} = 20$, determine the fan-out for the DTL *NAND* gates shown in Figure 10-9.

solution

$$I_3 = \frac{V_{BE1}}{R_3} = \frac{0.7 \text{ V}}{5 \text{ k}\Omega}$$

$$= 140 \text{ } \mu\text{A}$$

$$V_{B2} = V_{BE2} + V_{D6} + V_{BE1}$$

$$= 2.1 \text{ V}$$

$$V_{R1} + V_{R2} = V_{CC} - V_{B2}$$

$$= 5 \text{ V} - 2.1 \text{ V}$$

$$= 2.9 \text{ V}$$

and

$$V_{R1} + V_{R2} = (I_2 + I_1)R_1 + I_2 R_2$$

$$I_1 = h_{FE}I_2 = 20I_2$$

$$2.9 \text{ V} = (I_2 + 20I_2)R_1 + I_2 R_2$$

$$= I_2 [21R_1 + R_2]$$

$$I_2 = \frac{2.9\ V}{21\ R_1 + R_2} = \frac{2.9\ V}{(21 \times 1.6\ k\Omega) + 2.15\ k\Omega}$$

$$= 81\ \mu A$$

$$I_1 = 20 \times I_2 = 20 \times 81\ \mu A$$

$$= 1.62\ mA$$

$$I_B = (I_1 + I_2) - I_3$$

$$= (1.62\ mA + 81\ \mu A) - 140\ \mu A$$

$$\simeq 1.56\ mA$$

$$I_{C1} = h_{FE} I_B = 20 \times 1.56\ mA$$

$$= 31.2\ mA$$

$$I_4 = \frac{V_{CC} - V_{CE(sat)}}{R_4} = \frac{5\ V - 0.2\ V}{6\ k\Omega}$$

$$= 0.8\ mA$$

maximum output current

$$I_L = I_{C1} - I_4$$

$$= 31.2\ mA - 0.8\ mA$$

$$= 30.4\ mA$$

unit load $= I_i$

with Q_2' and Q_1' off, $I_i = \dfrac{V_{CC} - V_{D1} - V_{CE1(sat)}}{R_1 + R_2}$

$$= \frac{5\ V - 0.7\ V - 0.2\ V}{2.15\ k\Omega + 1.6\ k\Omega} = 1.09\ mA$$

$$\frac{I_L}{I_i} = \frac{30.4\ mA}{1.09\ mA}$$

$$= 27.9$$

fan-out $= 27$

It is obvious from Example 10-3 that a gate can operate with a load less than the fan-out. When the fan-out is exceeded, the gate may not operate correctly. Even when the fan-out is not exceeded, each additional load connected to a gate output increases the capacitance at the transistor collector and hence increases the gate switching time. For high-speed

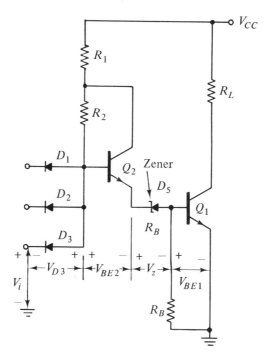

FIGURE 10-10.   High threshold logic (HTL) gate.

operation IC manufacturers usually recommend a maximum *loading factor* which is less than the dc fan-out capability of the circuit. The recommended loading factor for MC930 *NAND* gates is 8.

10-7   HIGH THRESHOLD LOGIC (HTL)

To improve upon the noise immunity of DTL *high threshold logic* (HTL) was developed. In the HTL circuit in Figure 10-10, D_5 is a Zener diode with $V_z \simeq 6.8$ V. For Q_1 *on*, the Zener diode must be in breakdown and Q_2 must be *on*. Now, the minimum input voltage for Q_1 and Q_2 *on* is

$$V_i = V_{BE2} + V_z + V_{BE1} - V_{D3}$$
$$= 0.7 \text{ V} + 6.8 \text{ V} + 0.7 \text{ V} - 0.7 \text{ V}$$
$$= 7.5 \text{ V}$$

The threshold voltage for the HTL gate is approximately 7.5 V. It is obvious that the HTL gate is more immune than the DTL circuit to noise spikes. Note also that because of the presence of the Zener diode the HTL circuit uses $V_{CC} = 15$ V, compared to $V_{CC} = 5$ V for DTL gates.

10-8 RESISTOR TRANSISTOR LOGIC (RTL)

The *resistor transistor logic* circuit (RTL) shown in Figure 10-11 has a *high* output voltage when all three inputs are at ground level. All three transistors are biased *off* so that no collector current flows, and $V_0 \simeq V_{CC}$. A collector current flows and the output drops to a *low* level, if a positive input voltage is applied to terminal *A OR* terminal *B OR* terminal C. Thus the circuit is that of a *NOR* gate.

The minimum input voltage that will begin to switch any one transistor *on* in the circuit of Figure 10-11 is the normal transistor V_{BE} of approximately 0.7 V. Although this voltage cannot drive the output to $V_{CE(\text{sat})}$, it could affect the input current to other gates [*i.e.*, those being driven from the output]. Therefore, an RTL gate may be susceptible to noise voltages around 0.7 V.

Typical integrated circuit RTL uses a supply of 3 V, has a fan-out of 5, gate power dissipation of approximately 20 mW, and propagation delay time of 12 ns.

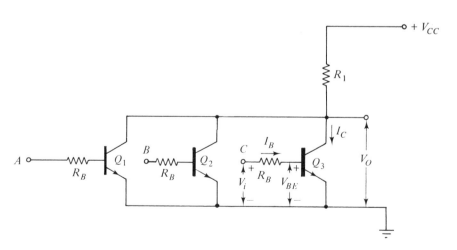

FIGURE 10-11. Circuit of RTL *NOR* gate.

10-9 TRANSISTOR TRANSISTOR LOGIC (TTL)

10-9.1 Standard TTL

In *transistor transistor logic* (TTL or T^2L), the input signals are applied directly to transistor terminals.

Consider the basic TTL circuit shown in Figure 10-12(a). The output transistor Q_2 is controlled by the voltage at the collector terminal of transistor Q_1. When the input terminal (*i.e.*, Q_1 emitter) is grounded, sufficient base current I_B flows to keep Q_1 in saturation. The collector voltage of Q_1 is $V_{CE(sat)}$ above ground. Typically, $V_{CE(sat)}$ is 0.2 V, which is not high enough to bias Q_2 on. Therefore, when the input voltage is low, Q_2 is *off* and the output level is *high*.

If a positive voltage is applied to the input terminal, Q_1 remains in saturation (I_B is still large enough) and Q_1 collector voltage goes to $V_i +$ $V_{CE(sat)}$. Depending upon the actual level of input voltage, sufficient base current can be supplied to Q_2 to drive it into saturation. Figure 10-12(b) shows Q_1 replaced by diodes representing the base-emitter and collector-base junctions. The arrangement is similar to that of a DTL circuit. It is seen that the input voltage could easily be made large enough to reverse-bias the base-emitter junction. When this occurs the collector-base junction remains forward-biased, and base current flows to saturate the output transistor.

Figure 10-12(c) shows a basic three-input TTL circuit. Q_1 is seen to be a transistor with three emitter terminals. This is fabricated easily in integrated circuit form. The three emitters are the input terminals to the gate. For Q_1 collector to rise above $V_{CE(sat)}$, input *A AND* input *B AND* input *C* must be *high* positive levels. Because of this, and because the output voltage level goes from *high* to *low*, the circuit is a *NAND* gate.

When each input terminal of a TTL gate is *high*, only a very low (emitter-base leakage) input current flows. When the input voltage is low, a (emitter) current flows *out* of the input terminal. The *high-level input current* (I_{IH}) is listed on the data sheet as 40 μA (see Appendix 1-17). The *low-level input current* (I_{IL}) is stated as -1.6 mA. The minus sign indicates that current flows out of the input terminal. This (-1.6 mA) is the *unit load* for this type of logic circuitry.

When the output terminal of a TTL gate is *low*, it can *sink* the low-level input current I_{IL} from the input terminals of several TTL gates. The maximum *low-level output current* (I_{OL}) is specified on the data sheet as 16 mA. Therefore, the maximum fan-out is

$$\frac{I_{OL}}{I_{IL}} = \frac{16 \text{ mA}}{1.6 \text{ mA}} = 10$$

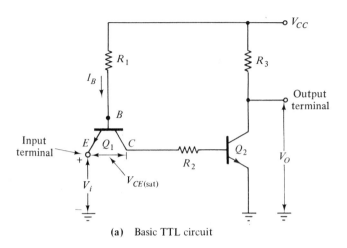

(a) Basic TTL circuit

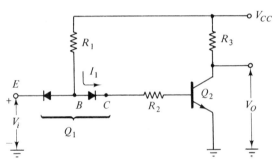

(b) Q_1 replaced with
diode equivalent of
its junctions

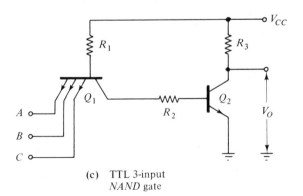

(c) TTL 3-input
NAND gate

FIGURE 10-12. TTL gate circuits.

Standard TTL is usually referred to as *74* or *7400* series, although the individual circuits may have a *54* number instead of a *74* number. Circuits with a *54* number can operate over a temperature range of $-55°C$ to $+125°C$, and the supply voltage limits are 4.5 V to 5.5 V. For *74* number circuits the temperature range is 0°C to 70°C, and the supply voltage must be between 4.75 V and 5.25 V.

The circuit of a three-input *7400 series* integrated circuit TTL *NAND* gate is shown in Figure 10-13. The diodes connected from ground to each input terminal become forward-biased only when the input voltage goes negative. Their function is to limit the amplitude of negative spikes appearing at the gate inputs. The arrangement of the output transistors (Q_2, Q_3, and Q_4) is referred to as *totem pole*. They function as follows: When Q_2 is *off*, R_3 biases Q_4 *off*, and R_2 biases Q_3 *on*. Thus Q_3 provides *active pull-up* (or low output impedance) when the gate output voltage is *high*. When Q_2 is *on* in saturation, base current supplied to Q_4 drives Q_4 into saturation. Consequently, the output voltage is pulled down, and Q_4 offers a low output impedance when the gate output is in its *low* state. At this time, Q_3 is biased *off* by the voltage drop across R_2. This is assisted by the presence of diode D_4.

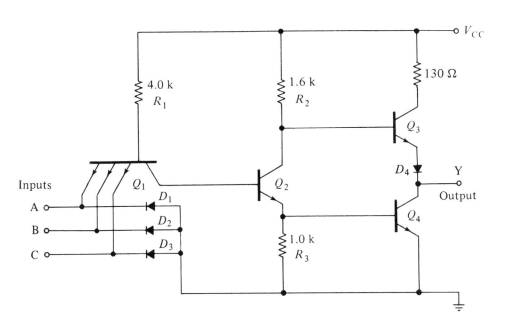

FIGURE 10-13.   54/74 series integrated-circuit three-input TTL *NAND* gate.

The multi-emitter input transistor used with TTL is normally in saturation, (i.e. its collector-base junction is forward biased), even when all the input voltages are *high*. An exception to this occurs at the instant that one input terminal goes *low*. With Q_2 still *on*, Q_1 collector-base junction becomes reverse biased, and a large current flows *from* Q_2 base into Q_1 collector. The effect of this is to cause Q_2 to switch *off* very rapidly. Q_2 also switches *on* very rapidly when Q_1 inputs go high. This is because Q_1 collector-base junction remains forward biased during the *on* switching time for Q_2. These two effects make TTL one of the fastest of all integrated circuit logic types.

Standard (54/74 series) TTL has a typical propagation delay time of 10 ns. One gate circuit dissipates approximately 10 mW, and has a fan-out of 10. Because the gate input terminals are transistor emitters, they have a low input impedance; consequently, TTL is said to have *good* noise immunity.

10-9.2   Other TTL Logic Families

As well as the standard 54/74 series, the TTL data sheet in Appendix 1-17 lists four other types of TTL gates:

High-speed TTL (54H/74H)

The circuit speed is increased by reducing the values of the resistors, and by including an additional emitter-follower transistor to drive one of the output transistors (see the H00–H30 circuit diagram on the data sheet). Because of the reduced resistor values, the supply current is approximately double that for standard TTL, resulting in an average per-gate power dissipation of 22.5 mW. The typical propagation delay time for 54H/74H TTL is 6 ns.

Low-power TTL (54L/74L)

In this case the resistor values are increased above those normally employed in standard TTL (see the L00–L30 circuit and table of component values on the data sheet). The result is lower supply currents and an average power dissipation per gate of only 1 mW. Reduced switching speed is another consequence of the increased resistance values. The typical propagation delay time is 35 ns for low-power TTL. The major applications of this logic series are found in portable battery-operated equipment, where supply currents must be minimized.

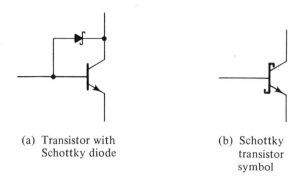

(a) Transistor with   (b) Schottky
    Schottky diode   transistor
                                 symbol

FIGURE 10-14. Construction and symbol for a Schottky transistor.

Schottky TTL (54S/74S) *switches fast*

This logic family employs *Schottky transistors* to further increase the circuit
switching speed. A *Schottky transistor* is a bipolar transistor with a
Schottky diode connected between its collector and base terminals, as
illustrated in Figure 10-14(a). A *Schottky diode* has a junction of silicon
and metal. Like other diodes it is a one-way device, but its major characteristic is that it switches very fast. The presence of the Schottky diode
prevents the transistor from going into saturation, and consequently the
transistor switching speed is minimized.

The Schottky transistor circuit symbol is illustrated in Figure 10-14(b),
and a S00–S132 Schottky TTL *NAND* gate circuit is shown on the data
sheet in Appendix 1-17.

The typical propagation delay time for Schottky TTL is 3 ns, and the
average power dissipation per gate is around 20 mW. An improved version
known as *advanced Schottky TTL* (54/74 AS) boasts 1.5 ns typical gate
delay time at 20 mW per gate power dissipation. Obviously, this type of
logic circuit should be used where high speed is the most important
consideration.

Low-power Schottky TTL (54LS/74LS)

In this family, as in the 54L/74L family, the resistor values are increased
in order to minimize power dissipation. However, because Schottky transistors are used, the typical propagation delay time is relatively small at 9
ns. Power dissipation per gate is around 2 mW. *Advanced low power
Schottky TTL* has a typical gate delay of 4 ns with 1 mW per gate power
dissipation.

Tri-state TTL (TSL)

A *tri-state* (or *three-state*) *TTL* logic gate has a *control input* as well as the
usual input and output terminals. Figure 10-15 shows the circuit arrange-

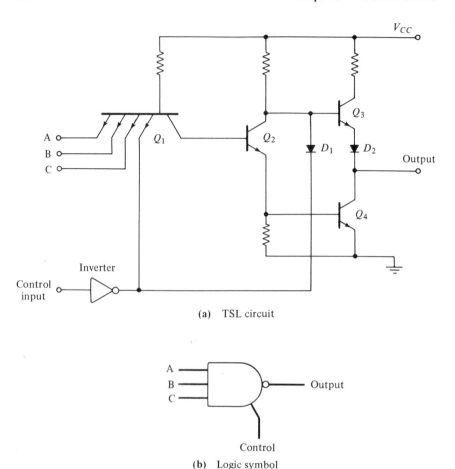

(a)   TSL circuit

(b)   Logic symbol

FIGURE 10-15.   Tri-state logic (TSL) circuit and logic symbol.

ment and logic symbol for a TSL *NAND* gate. Note that the control input
terminal goes to an inverter. The output of the inverter is connected to one
emitter on transistor Q_1, and via diode D_1 to the base of Q_3.

When the control input is *low*, the inverter output is *high*. This
reverse-biases D_1 and provides a *high* input to the connected emitter of Q_1.
In this condition the *NAND* gate functions normally; when all the gate
inputs are *high*, the output is *low*; when one or more inputs are *low*, the
output is *high*.

With a *high* input applied to the control terminal, the inverter output
goes *low*, forward-biasing D_1 and the connected emitter of Q_1. Now Q_1 is
held in a *low* state, no matter what the level of the other gate input
terminals. Thus, Q_2 and Q_4 are *off*. As well as this, the base of Q_3 is held

in a *low* state by (forward-biased) diode D_1. Consequently, Q_3 is *off*. Both output transistors Q_3 and Q_4 are *off*, and the output terminals offers a high impedance to all circuits that are connected to it. This condition is the *third state* of the TSL circuit. The output of a TSL gate may be *high* or *low* or have a high output impedance.

TSL gates are used in logic systems where the outputs of several gates are connected in parallel to a single input of another circuit. All gates are usually maintained in the high output impedance state, and are *sampled* (or switched *on* briefly) by the control signals applied in sequence. This avoids the possibility of the output of one gate short-circuiting another gate output.

Another aspect of the TSL gate is that the circuit *input* impedance also becomes high when the gate is placed in its high *output* impedance state.

then

10-10   EMITTER-COUPLED LOGIC (ECL)

One major limitation to the switching speed of logic circuits is the *storage time* of saturated transistors. The storage time is the time required to drive a transistor out of saturation, that is, to reverse the forward bias on the collector-base junction. In *emitter-coupled logic* (ECL), also termed *current mode logic*, the transistors are maintained in an unsaturated condition. This eliminates the transistor storage time, and results in logic gates which switch very fast indeed.

The schematic diagram of a typical integrated circuit ECL gate is shown in Figure 10-16. The circuit uses a negative supply $-V_{EE}$, and the positive supply terminal V_{CC} is grounded. Transistor Q_5 has its base bias voltage provided by the potential divider composed of R_5, R_6, D_1, and D_2. The diodes provide temperature compensation for changes in the V_{BE} of Q_5. Q_5 operates as an emitter follower to provide a low impedance bias to the base of transistor Q_4. With a constant bias voltage at Q_4 base, the voltage drop across emitter resistor R_2 is also maintained constant so long as the input voltages are low enough to keep transistors Q_1, Q_2, and Q_3 in the *off* state. In this circumstance, the emitter current and collector current of Q_4 are held constant and the transistor is maintained in an unsaturated condition. With Q_4 *on*, the output voltage via emitter follower Q_7 is *low*, and that via emitter follower Q_6 is *high*. When a positive voltage is applied to terminal A *OR* terminal B *OR* terminal C, the emitter voltage of Q_4 is pulled up above its base level. Consequently, Q_4 switches *off* as Q_1, Q_2, or Q_3 switches *on*. When this occurs, the voltage at the base of Q_6 falls and that at Q_7 rises.

It is seen that when the input voltages at terminals A, B, and C are *low*, the output voltage at Q_7 emitter is also *low*. Q_7 output becomes *high*

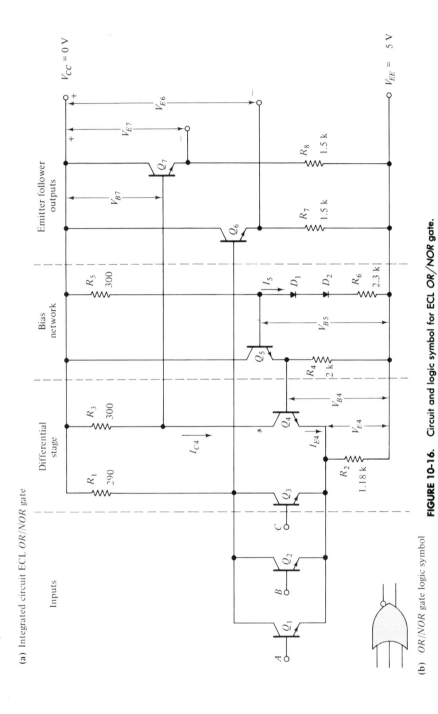

(a) Integrated circuit ECL *OR/NOR* gate

(b) *OR/NOR* gate logic symbol

FIGURE 10-16. Circuit and logic symbol for ECL *OR/NOR* gate.

when a *high* input is applied to terminal *A* OR terminal *B* OR terminal *C*. Thus, the gate functions as an *OR* gate when the output is derived from Q_7 emitter. Q_6 emitter voltage is *high* when the inputs are *low*, and *low* when terminal *A*, *B*, OR *C* inputs are *high*. Therefore, with output taken from Q_6 emitter, the circuit functions as a *NOR* gate. The *OR/NOR* logic symbol is shown in Figure 10-16(b).

EXAMPLE 10-4

The *OR/NOR* gate circuit in Figure 10-16(a) has supply voltages of -5 V and ground. Determine the output voltages when inputs *A*, *B*, and *C* are low.

solution

With inputs *A*, *B*, and *C* low,

$$I_5 = \frac{(0 - V_{EE}) - V_{D1} - V_{D2}}{R_5 + R_6}$$

$$= \frac{0 - (-5 \text{ V}) - 0.7 \text{ V} - 0.7 \text{ V}}{300 \ \Omega + 2.3 \text{ k}\Omega} \simeq 1.4 \text{ mA}$$

$$V_{B5} = (I_5 R_6) + V_{D1} + V_{D2}$$

$$= (1.4 \text{ mA} \times 2.3 \text{ k}\Omega) + 0.7 \text{ V} + 0.7 \text{ V}$$

$$\simeq 4.6 \text{ V}$$

$$V_{B4} = V_{B5} - V_{BE5}$$

$$= 4.6 \text{ V} - 0.7 \text{ V} = 3.9 \text{ V}$$

$$V_{E4} = V_{B4} - V_{BE4}$$

$$= 3.9 \text{ V} - 0.7 \text{ V}$$

$$= 3.2 \text{ V}$$

$$I_{E4} = \frac{V_{E4}}{R_2}$$

$$= \frac{3.2 \text{ V}}{1.18 \text{ k}\Omega} \simeq 2.7 \text{ mA}$$

$$I_{C4} \simeq I_{E4} = 2.7 \text{ mA}$$

$$V_{B7} \simeq V_{CC} - I_{C4} R_3 \quad \text{(neglect } I_{B7})$$

$$= 0 \text{ V} - (2.7 \text{ mA} \times 300 \ \Omega)$$

$$= -0.18 \text{ V}$$

$$V_{E7} = V_{B7} - V_{BE7}$$

$$= -0.18 \text{ V} - 0.7 \text{ V} \simeq -1.5 \text{ V}$$

This is the *low* state of the output at Q_7 emitter.

With Q_1, Q_2, and Q_3 biased *off*, only I_{B6} flows through R_1. Consider $(I_{B6} \times R_1)$ as negligible. Then,

$$V_{E6} \simeq V_{CC} - V_{BE6}$$
$$= 0 - 0.7 \text{ V}$$
$$= -0.7 \text{ V}$$

This is the *high* state of the output at Q_6 emitter.

From Example 10-4, the *high* output level for the ECL gate is -0.7 V, and the *low* output level is -1.5 V. When applied to the input of another gate, these *high* and *low* levels must be capable of switching the gate from one state to another. Consider the circuit in Figure 10-16(a), and assume that terminal C is connected to the output of another similar gate. When the *low* output level (-1.5 V) is applied to terminal C, V_{B3} is 3.5 V above V_{EE}. In Example 10-4, V_{E4} was found to be 3.2 V above V_{EE}, and this is also the voltage at the emitter of Q_3. Since, $V_{BE3} = (3.5 \text{ V} - 3.2 \text{ V}) = 0.3$ V, Q_3 base-emitter actually is forward-biased by 0.3 V. This is not sufficient to bias a silicon transistor into conduction, so Q_3 remains *off*. However, an increase of approximately 250 mV at the base of Q_3 (*e.g.*, a noise spike) could cause the transistor to at least partially switch *on*. A similar analysis of the circuit conditions when Q_3 is *on* and Q_4 is *off* shows that switching could again occur with a -250 mV spike.

The principal drawback of integrated circuit ECL compared to other IC logic families is now evident. That drawback is its sensitivity to low-level noise on the order of ± 250 mV. The high input resistance and very fast switching speed of ECL also contributes to its low noise immunity. However, the low output resistance of ECL improves the noise immunity at the input of another gate that is being driven. Another aspect of the noise sensitivity of logic gates is that most types of logic circuits generate noise spikes when transistors are switched into or out of saturation. This is not the case with ECL, because each time one transistor is switched *off* another is switched *on*. Thus the current drawn from the supply remains approximately constant.

Another disadvantage of ECL is its relatively high power dissipation, approximately 25 mW per gate. The major advantage of ECL over other types of logic undoubtedly is the very fast switching speed. Because of the nonsaturated condition of the *on* transistors, the propagation delay time can be 2 ns or less.

Appendix 1-18 shows the data sheet for MC306.MC307 3-input ECL gates manufactured by Motorola. In the schematic diagram on the data sheet, the three transistors with their bases connected to terminals *6*, *7*, and *8* correspond to Q_1, Q_2, and Q_3 and Q_3 in Figure 10-16(a). Also, the transistor with its base connected to terminal *1* corresponds to Q_4 in Figure 10-16(a). The remaining two transistors in the MC306.MC307 circuit are the emitter follower outputs. No bias network is provided in this IC gate. Instead, an external bias driver must be connected to terminal *1*.

The listed electrical characteristics of the MC306.MC307 show that the low output voltage (NOR logic 0) is -1.750 V. The high output voltage (NOR logic 1) is -0.795 V. This gives an output voltage change of 0.55 V. The shortest propagation delay time for the MC306.MC307 is listed as 5.5 ns.

10-11 MOSFET LOGIC

10-11.1 P-MOS and N-MOS

As already discussed in Secs. 4-6 and 4-7, MOSFET switches have an extremely high input resistance, very small drain to source voltage drop, and very little power dissipation. The *n*-channel *enhancement mode* MOSFET is normally *off* when its gate is at the same potentials as its substrate. When the gate is made positive with respect to the substrate, an *n-type* channel is created from drain to source, and drain current flows. Similarly, the *p*-channel device has no drain current while its gate and substrate are at the same potential. The *p-type* channel appears when the gate is made negative with respect to the substrate.

P-MOS logic gates are made up of *p*-channel MOSFET transistors. No resistors or capacitors are involved. N-MOS gates are composed only of *n*-channel MOSFETs. N-MOS circuits are very similar to P-MOS circuits, with the important exception that all voltage polarities and current directions are reversed. One other important difference between P-MOS and N-MOS is that N-MOS is the faster of the two types of logic. This is due to the fact that charge carriers in *n*-channel devices are electrons while those in *p*-channel FETs are holes, and electrons have *greater mobility* than holes (i.e., they move faster).

The circuit of a P-MOS *NAND* gate is shown in Figure 10-17. Note that Q_1 has a channel resistance (or $R_{D(on)}$ value) around 100 kΩ, while the $R_{D(on)}$ value for each of Q_2 and Q_3 is on the order of 1 kΩ. The gate of Q_1 is biased to its drain terminal. When the source terminal of Q_1 is less than V_{DD}, the gate is negative with respect to the source. This is the condition necessary to bias Q_1 *on*. Consequently Q_1 is always in the *on* condition, and its $R_{D(on)}$ acts as a load resistor for Q_2 and Q_3.

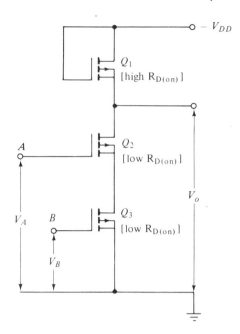

FIGURE 10-17. P-MOS *NAND* gate.

The circuit in Figure 10-17 is a negative logic gate using a supply voltage $-V_{DD}$. When input A and input B are *low* (near ground) transistors Q_3 and Q_2 are both *off*. No drain current flows and there is no voltage drop across Q_1. The output voltage at this time is a *high* (negative) level close to $-V_{DD}$. When a *high* (negative) input is applied to the gate of Q_3, Q_3 tends to switch *on*. However, with the gate of Q_2 still held near ground, Q_2 remains an open circuit and the output remains at its *high* level. When *high* inputs are applied to the gates of Q_3 and Q_2, both transistors are switched *on* and current flows through the channels of all three transistors. The total $R_{D(on)}$ of Q_2 and Q_3 adds up to about 2 kΩ, while that of Q_1 is around 100 kΩ. Therefore, the voltage drop across Q_2 and Q_3 is much smaller than that across Q_1, and the output voltage is now at a *low* level.

It is seen that the circuit performs as a negative logic *NAND* gate. When any one of the inputs is *low*, the output is a *high* negative voltage. When input A and input B are high negative levels, the output voltage is *low*. As already stated, an N-MOS *NAND* gate is exactly similar to the circuit in Figure 10-17, except that V_{DD} must be positive and the circuit functions as a positive logic gate.

An N-MOS *NOR* gate circuit is shown in Figure 10-18. Here again Q_1 is permanently biased *on*, and its $R_{D(on)}$ value is around 100 kΩ. Q_2 and Q_3 each have $R_{D(on)}$ values of about 1 kΩ. When both input levels are low, Q_2 and Q_3 are *off*. At this time the voltage drop across Q_1 is almost zero

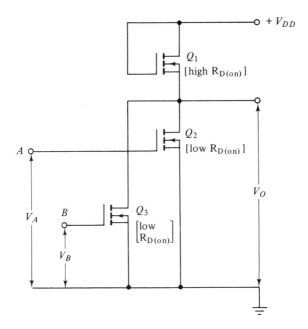

FIGURE 10-18.   N-MOS NOR gate.

and the output level is *high*, close to V_{DD}. When a *high* (positive) input is applied to terminal A or terminal B, Q_2 and Q_3 switch *on*, causing current to flow through Q_1. The voltage drop across either Q_2 or Q_3 (or both) is much smaller than that across Q_1, since the $R_{D(on)}$ of Q_1 is around $100\,\text{k}\Omega$, while $R_{D(on)}$ for Q_2 and Q_3 is approximately $1\,\text{k}\Omega$. Therefore, when a *high* input is applied to terminal A or terminal B, the output voltage goes to a *low* level.

A P-MOS *NOR* gate is exactly similar to the circuit of Figure 10-18, except that V_{DD} must be a negative quantity and the circuit functions as negative logic.

10-11.2   CMOS Logic Gates

CMOS was introduced in Sec. 4-7, and the operation of the CMOS inverter was explained in that section. Although the integrated circuit fabrication process for CMOS is more complicated than that for P-MOS or N-MOS, CMOS has the very important advantage that its power dissipation per gate is much less than that for any other logic family. (Integrated injection logic can be an exception to this, see Section 10-12). Other CMOS advantages are: operation from supply voltages as low as 1V, fan-out in excess of 50, excellent noise immunity.

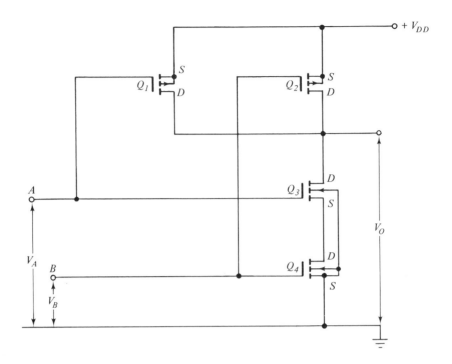

FIGURE 10-19.   CMOS *NAND* gate.

Consider the CMOS *NAND* gate shown in Figure 10-19. The parallel-connected transistors Q_1 and Q_2 are p-channel MOSFETs, and the series-connected devices Q_3 and Q_4 are n-channel MOSFETs. When input terminals A and B are grounded, the gates of Q_1 and Q_2 are negative with respect to the source terminals. Therefore, Q_1 and Q_2 are biased *on*. Also, the gates of Q_3 and Q_4 are at the same potential as the device source terminals, and consequently Q_3 and Q_4 are *off*. Depending upon the actual load current and $R_{D(\text{on})}$ values, there will be a small voltage drop along the channels of Q_1 and Q_2. Thus, the output voltage V_O is close to the level of the supply voltage V_{DD}. When both A and B are grounded, V_O is approximately equal to V_{DD}.

When a *high* positive input voltage (equal to 0.7 V_{DD} or greater) is applied to terminal B, Q_4 is biased *on* and Q_2 is biased *off*. However, with terminal A still grounded, Q_3 remains *off* and Q_1 is still *on*, and the output voltage remains at $V_O \simeq V_{DD}$. When *high* inputs are applied to terminal A and terminal B, both p-channel devices (Q_1 and Q_2) are biased *off* and both n-channel FETs (Q_3 and Q_4) are biased *on*. The output now goes to $V_O \simeq 0$ V.

The circuit of a CMOS *NOR* gate is shown in Figure 10-20. Once again, two p-channel devices (Q_1 and Q_2) and two n-channel transistors

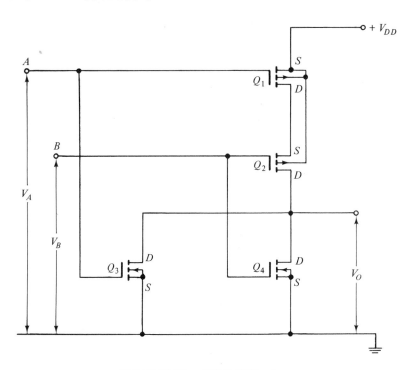

FIGURE 10-20. CMOS *NOR* gate.

(Q_3 and Q_4) are employed. When both inputs are at ground level, Q_3 and Q_4 are biased *off*, and Q_1 and Q_2 are *on*. In this condition there is about a 10 mV drop from drain to source terminals in the *p*-channel transistors, and V_O is very close to V_{DD}. When terminal *A* has a *high* positive input, Q_1 switches *off* and Q_3 switches *on*. The series combination of Q_1 and Q_2 now is open-circuited, and the output is shorted to ground via Q_3. Similarly, if terminal *A* remains grounded and terminal *B* has a *high* input applied, Q_2 switches *off* and Q_4 switches *on*. Again the output goes to ground level.

The major advantage of integrated circuit CMOS logic over all other logic systems is its extremely low power dissipation. At a minimum of 10 nW per gate, the low dissipation allows greater circuit density within a given size of IC package. The resultant low supply current demand also makes CMOS ideal for battery-operated instruments. Typical supply voltages employed for CMOS are 5 V to 10 V; however, operation with a supply of 1 V to 18 V is possible. The circuitry is immune to noise levels as high as 30% of the supply voltage. The extremely high input resistance of MOSFETs gives CMOS gates typical input resistances of 10^9 Ω, and this makes it possible to have fan-outs greater than 50. Typical propagation delay time for CMOS is 25 ns.

10-12   INTEGRATED INJECTION LOGIC (I^2L)

As already discussed, integrated circuit logic systems are compared in terms of: switching speed, power dissipation, fan-in, fan-out, and noise margin. Two other very important factors are physical size and cost of manufacture. As will be explained, individual I^2L gates require a fraction of the area of other logic types, i.e. the circuit density is much greater. Also, power dissipation per gate can be comparable with CMOS logic, very fast switching is possible, and fabrication techniques are simple and inexpensive. These improvements are due to two factors: elimination of resistors, and what is termed *merging* of transistors.

To understand the operation of I^2L consider Figure 10-21 (a) which shows a simple (normally-on) resistor-transistor inverter. Npn transistor Q_2

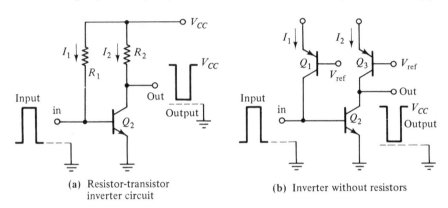

(a) Resistor-transistor
     inverter circuit

(b) Inverter without resistors

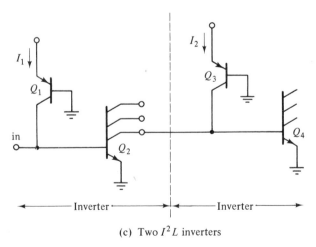

(c) Two I^2L inverters

FIGURE 10-21.   Integrated injection logic (I^2L).

is supplied with base current I_1 via resistor R_1, and the collector current I_2 causes a voltage drop across R_2 which makes the transistor saturated. When the inverter input is at ground level, I_1 is diverted away from the transistor base. Thus Q_2 is *off*, and the output is *high*. With the input high once again, Q_2 goes *on*, and the output is *low*.

In Figure 10-21 (b), R_1 and R_2 are replaced by *pnp* transistors Q_1 and Q_3 respectively. When sufficient emitter current (I_1) is supplied to Q_1, Q_1 performs exactly the same function as R_1 in Figure 10-21 (a), i.e. it supplies base current to Q_2. Similarly, with an adequate level of I_2, Q_3 passes collector current to Q_2, as does R_2 in Figure 10-21 (a). Here again, when the input is at ground level, I_1 is diverted and Q_2 goes *off*. When the input is *high*, Q_2 is *on*, and its collector voltage is *low*.

Integrated circuit resistors can easily occupy ten times the area of a transistor. Consequently, by replacing the resistors with transistors, a big reduction in gate area is achieved. Putting it another way: twenty (or more) purely transistor inverters might be fabricated on the area normally occupied by one resistor-transistor inverter. Furthermore, the large area occupied by each resistor results in much unwanted capacitance. Eliminating the resistors, reduces the capacitance and improves the switching speed of each inverter.

Now refer to Figure 10-21 (c), which shows two I^2L inverters, one of which has its output connected to the input of the other. It is seen that each inverter consists of only two transistors, and that the collector load for each is the input stage of the next inverter. Q_3 is the collector load for Q_2. Note that Q_2 (and Q_4) have several separate collectors. These are isolated from each other, so that they may be connected to the inputs of several different gates. When Q_2 is *off*, these collectors could be at different (high or low) levels depending upon the state of the other gates.

In Figure 10-21 (c) the base driver transistors Q_1 and Q_3 have their own bases grounded. Each of the pnp transistor emitters is typically 750 mV above the level of the base. When Q_2 is open-circuited, Q_3 is in saturation and its collector voltage should be about 600 mV to 700 mV above ground. This is sufficient to bias Q_4 into saturation. With Q_2 on, its saturation voltage is perhaps 50 mV to 100 mV, which pulls the base of Q_4 low enough to switch it *off*. The actual supply voltage at the emitter of each pnp transistor is typically:

$$V_{CC} = V_{BE} + V_{CE(sat)} = 750 \text{ mV} + 100 \text{ mV} = 850 \text{ mV}$$

Instead of thinking in terms of voltages, I^2L operation is more easily described by considering currents. In Figure 10-21 (c), when Q_2 is *off* I_2 flows into the base of Q_4 to drive it into saturation. With Q_2 on, I_2 is diverted through Q_2, and Q_4 is *off*.

The current supplied to the emitters of the *pnp* transistors must be regulated by the power supply. This is easily done by using one external

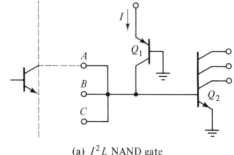

(a) I^2L NAND gate

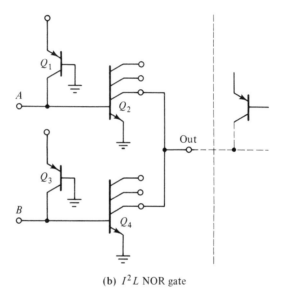

(b) I^2L NOR gate

FIGURE 10-22. I^2L *NAND* and *NOR* gates.

resistor to supply current to many transistors. The charge carriers constituting the current are said to be *injected* into the transistor emitters, hence the name *integrated injection logic*. If the supply current is kept low, the power dissipation per gate is obviously minimized. However, a disadvantage of low current is that the switching time of transistors is increased. A choice must be made between fast switching and low power dissipation.

I^2L NAND and NOR gates are shown in Figure 10-22 (a) and (b) respectively. The NAND gate is simply an inverter stage with several commoned input terminals. If all the collectors of (previous stage) transistors connected to terminal A, B, and C are open-circuited, then all

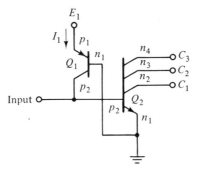

(a) p and n regions of inverter transistors

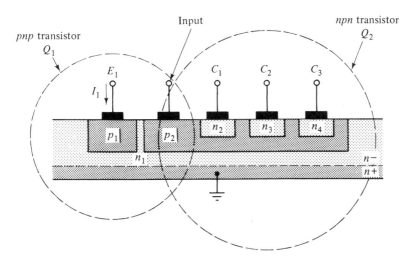

(b) Cross-section of inverter transistors

FIGURE 10-23.   **Merged transistor construction of I^2L.**

inputs can be described as *high*. The result is, that Q_2 is *on*, and its output is *low*. When any one of the transistors connected to A, B or C is *on*, the input is *low*, Q_2 is *off*, and the gate output is *high*.

The NOR gate uses two inverters with their output terminals paralleled. With both inputs *low*, Q_2 and Q_4 are *off* and the output is *high*. A *high* input to either A or B causes one of the output transistors to switch *on*, pulling the gate output level *low*.

The fabrication advantages of I^2L are illustrated in Figure 10-23. The basic inverter circuit is reproduced in Figure 10-23 (a), with its terminals identified as: *input*; emitter E_1; (output) collectors C_1, C_2 and C_3. The regions of each transistor are also identified as: p_1, n_1 and p_2 for Q_1; n_1, p_2

and n_2, n_3, n_4 for Q_2. Because the base of Q_1 and the emitter of Q_2 are both grounded, and because they are both n-type material, a single identification n_1 is employed. In fact, one single n-type region of semiconductor can be used for both. Similarly, the collector of Q_1 and the base of Q_2 are connected together and are both p-type material. Consequently, they are both identified as p_2, and a single p-type region of semiconductor material can be used for the two.

The cross-section of such an I^2L inverter is shown in Figure 10-23(b). A single bed of n-type forms region n_1. Regions p_1 and p_2 are diffused into n_1, and n_2, n_3, n_4 are each diffused into p_2. Transistor Q_1 consists of p_1, n_1 and p_2, while Q_2 is made up of n_1, p_2 and n_2, n_3, n_4. The common n_1 region is grounded via the low resistive $n+$ plane; current I_1 is injected into p_1; the common p_2; terminal is the input; while the outputs are C_1, $C_2 C_3$. Since they share common regions of n-type and p-type material, transistors Q_1 and Q_2 are said to be *merged*. This gives I^2L its other name: *merged transistor logic* (MTL). Many additional inverters can be fabricated using the same n-type bed, since they too have their n_1 regions grounded.

To fully appreciate the advantages of the I^2L merged transistor technique, it must be realized that other forms of integrated circuit logic require transistors to be isolated from each other. This involves a much more expensive fabrication process, and almost always results in unwanted junctions which must be kept in a reverse biased state. The unwanted junctions also add speed-reducing capacitances. With I^2L, the need to isolate transistors is eliminated and there are no unwanted junctions.

Integrated injection logic can operate from low or high level supply voltages. Typical switching times range from 10 ns up to 250 ns, depending upon the level of injection current, however, switching times less than 1 ns are possible. Power dissipation per gate can be anywhere from 6 nW to 70 μW, again depending upon the injection current. Input and output voltage swings are approximately 700 mV. A single output of an I^2L gate can *sink* (i.e. take in) a current of 20 mA.

10-13 COMPARISON OF MAJOR TYPES OF IC LOGIC

The major integrated circuit logic families are compared in Table 10-1. TTL, ECL, and CMOS are the most widely used logic systems today. They are the only types that should be seriously considered for any major application. However, the large amount of hardware already in the field does not just disappear when something new is developed. So a knowledge of all currently used circuitry is important to anyone studying logic circuits.

Table 10-1
Comparison of IC Logic Types

	DTL	RTL	HTL	TTL							ECL	P-MOS	CMOS	I²L
				74	74H	74L	74S	74AS	74LS	74ALS				
Propagation delay time (ns)	30	12	119	10	6	33	3	1.5	9	4	2	100	25	*10 to 250
Power dissipation per gate (mW)	15	15	50	10	22.5	1	20	20	2	1	25	1	10 nW	*6 nW to 70 µW
Noise margin (V)	1.4	0.7	7.5	0.4	0.4	0.4	0.4	0.4	0.4	0.4	0.25	2	0.3 V_{DD}	0.25
Noise immunity rating	good	poor	excellent	good	good	good	good	good	good	good	fair	excellent	excellent	fair
Fan-out	8	5	10	10	10	10	10	10	10	10	25	>50	>50	*depends upon injection current

Handwritten annotations: 25MH, 40MH, 13MH, 80MH, 80MH, 30MH, 120MH, 5MH, 5MH; "7400 ⇒ 25 MH"; "wide band longer time"

In situations where speed is important, any of the TTL families (except for 74L) may be suitable. Where very high speed is desirable, ECL or Schottky TTL (74S) must be chosen. ECL offers a fan-out of 25 but only *fair* noise immunity, compared to Schottky TTL with a fan-out of 10 but *good* noise immunity. Schottky TTL has slightly lower power dissipation at 20 mW per gate, while the P_D for ECL is 25 mW per gate.

If switching speed is not the paramount consideration, then either CMOS or low-power TTL (74L) might be appropriate. CMOS is the faster of the two and has by far the lowest power dissipation per gate. CMOS also has *excellent* noise immunity, and low-power TTL is said to have *good* noise immunity. CMOS also has a fan-out in excess of 50, while the fan-out for low-power TTL is 10. In situations where low-power dissipation and/or large fan-out is required CMOS is the only choice.

I^2L is suitable for applications where low power and high gate density are important. For medium or large scale integrated systems, I^2L could be the least expensive of all available options.

REVIEW QUESTIONS AND PROBLEMS

10-1   Sketch the circuit and logic symbol for a diode *AND* gate. Briefly explain the operation of the circuit.

10-2   Design a four-input diode *AND* gate using a 9 V supply. The gate inputs are to be controlled from the collectors of saturated transistors which can pass an additional collector current of 1 mA. Determine the *low* and *high* output levels for the gate.

10-3   Sketch the circuit and logic symbol for a diode *OR* gate. Briefly explain the operation of the circuit.

10-4   A diode *OR* gate is to have an output voltage which goes from a *low* level of 0 V to a *high* level of at least 2 V. The inputs to the *OR* gate are connected to flip-flops with $R_L = 4.7$ kΩ and $V_{CC} = 9$ V. Design a suitable circuit.

10-5   Explain *positive logic* and *negative logic*. Sketch the circuits of negative logic *AND* and *OR* gates. Compare these to positive logic circuits.

10-6   Sketch the circuit and logic symbol for a DTL *NAND* gate. Carefully explain the operation of the circuit and discuss the function of each component.

10-7   Repeat Question 10-6 for a DTL *NOR* gate.

10-8 (a) Explain *propagation delay time*, and discuss the characteristics that affect the switching speed of logic circuits. (b) Define *noise margin* and discuss the factors that affect the noise immunity of a logic circuit.

10-9 Define *fan-in* and *fan-out* and discuss the relationship of fan-in and fan-out to: gate switching speed, input current, and output current.

10-10 Calculate the fan-out for the DTL *NOR* gate shown in Figure 10-8. Take $R_L = 2$ kΩ, $R_1 = 20$ kΩ, $V_{CC} = 5$ V, and $V_{BB} = -2$ V.

10-11 Sketch the circuit of a typical HTL *NAND* gate, and carefully explain the function of every component.

10-12 Sketch the circuit of a two-input RTL *NOR* gate, and explain its operation.

10-13 Using illustrations, explain the principle of TTL. Discuss the reasons for the fast switching speed and good ac noise immunity of TTL.

10-14 Sketch the circuit of a typical integrated circuit TTL gate. Explain the function of each component.

10-15 For standard TTL logic gates define: I_{OL}, I_{IL} and I_{IH}. State typical values for each quantity and show how the fan-out may be calculated.

10-16 Explain the differences between standard TTL and: high-speed TTL, low-power TTL, Schottky TTL, and low-power Schottky TTL.

10-17 Sketch a circuit diagram to show how a tri-state TTL gate operates. Explain how the circuit functions and discuss its applications.

10-18 Sketch the circuit of a P-MOS *NAND* gate and the circuit of an N-MOS *NOR* gate. Carefully explain how each circuit functions.

10-19 Repeat Question 10-19 for a P-MOS *NOR* gate and an N-MOS *NAND* gate.

10-20 Sketch the complete circuit and logic symbol for an ECL *OR/NOR* gate. Carefully explain the operation of the circuit, and discuss the major advantages and disadvantages of ECL.

10-21 Sketch the circuits of CMOS *NAND* and *NOR* gates. Carefully explain the operation of each circuit, and list the advantages and disadvantages of CMOS logic.

10-22 Sketch the circuit of an I^2L inverter. Explain the operation of the circuit and compare it to a resistor-transistor inverter.

10-23 Sketch I^2L NAND and NOR gates and explain how they operate.

10-24 Using illustrations, explain the construction of an I^2L circuit, and discuss its advantages compared to other logic types.

10-25 Compare the various types of IC logic in terms of propagation delay time, power dissipation, noise immunity, and fan-out.

CHAPTER 11

Sampling Gates

INTRODUCTION

A **sampling gate** is a switching circuit which usually is employed to sample the amplitude of dc or low-frequency signals. Sampling gate circuits can be constructed using diodes, bipolar transistors, or FETs. For large signal voltages, diodes or bipolar transistors may be satisfactory. For very small signals, JFETs or MOSFETs produce the best results.

11-1 DIODE SAMPLING GATE

A very simple diode gate which may be applied to voltage level sampling is shown in Figure 11-1. The signal V_s to be sampled is applied to the cathode of D_1. The output voltage V_o is derived from the cathode of D_2. A pulse *control* input V_1 is applied via R_1 to the anodes of D_1 and D_2. The signal source resistance is R_S, and the load is R_L. When the control voltage is zero or negative, diodes D_1 and D_2 are reverse-biased, and $V_o \simeq 0$ V. When the control voltage becomes positive, D_1 and D_2 are forward-biased. Then,

$$V_A = V_s + I_s R_s$$

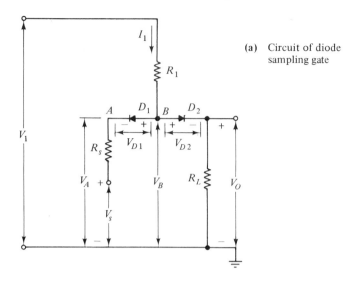

(a) Circuit of diode
     sampling gate

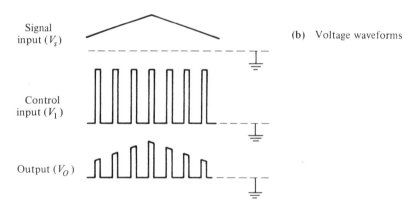

(b) Voltage waveforms

FIGURE 11-1.   Circuit and waveforms for diode sampling gate.

if

$$I_s R_s \ll V_s, \qquad V_A \simeq V_s$$

and

$$V_B = V_A + V_{D1}$$
$$\simeq V_s + V_{D1}$$
$$V_o = V_B - V_{D2}$$
$$\simeq V_s + V_{D1} - V_{D2}$$
$$\simeq V_s$$

It is seen that the signal voltage is passed to the output terminals when the control voltage pulses positively. The waveforms of input voltage, control voltage, and output voltage are illustrated in Figure 11-1(b). The circuit shown can sample only positive input signals. Reversing the diodes and the control input would permit sampling of a negative signal voltage.

The diode sampling gate has errors due to differences in the voltage drops across each diode, and due to diode leakage currents. Consequently, diode gates are applicable only where large signal amplitudes are involved and where accuracy is not important.

11-2 BIPOLAR TRANSISTOR SERIES GATE

The circuit of a bipolar transistor series sampling gate is shown in Figure 11-2. The low-frequency signal to be sampled is applied to the collector, and the output is derived from the emitter terminal. A pulse waveform at the base acts as a control, driving the transistor into saturation and cutoff. When the control voltage is positive, Q_1 is biased *on*. When the control voltage goes to zero, Q_1 is *off*. At transistor saturation, the output voltage is $V_o \simeq V_s$. At cutoff, the output becomes zero. It is seen that the transistor is operating as a switch, and that the output from the gate is a series of samples of the input amplitude.

The waveforms in Figure 11-2 are drawn for a positive input signal. if the input becomes negative, as shown in Figure 11-3(a), then the transistor operates in the *inverted mode*. The emitter terminal acts as the collector, and the collector operates as the transistor emitter. This, by no means, is an efficient way to operate a transistor used for amplification. However, as a saturated switch with a large base current, the transistor performs satisfactorily in inverted mode. To ensure that the device will switch *off*, the negative swing of the control voltage must be greater than the negative

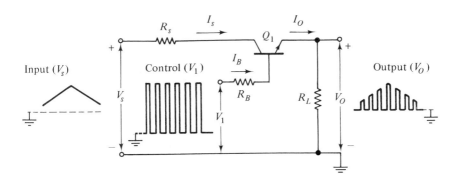

FIGURE 11-2. Bipolar transistor series sampling gate.

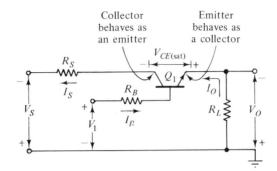

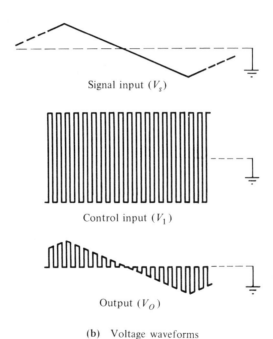

(a) Series gate with a negative signal voltage

Signal input (V_s)

Control input (V_1)

Output (V_O)

(b) Voltage waveforms

FIGURE 11-3. Series gate with negative signal voltage and transistor in *inverted* mode.

peak of the signal voltage. For a signal with positive and negative components, the circuit waveforms are as illustrated in Figure 11-3(b).

The input signal applied to a sampling gate is frequently a very low level voltage. Since the transistor saturation voltage constitutes a loss of signal amplitude [see Figure 11-4(a)], $V_{CE(\text{sat})}$ (also termed the *offset voltage*) must be maintained as small as possible. Reference to the transistor

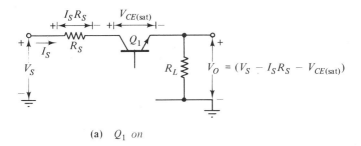

(a) Q_1 *on*

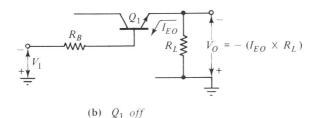

(b) Q_1 *off*

FIGURE 11-4.   Error sources in bipolar transistor series gate.

characteristics in Figure 4-2 shows that for the smallest $V_{CE(sat)}$, I_C must be kept small and I_B must be relatively large. For $I_C = 1$ mA and $I_B = 0.1$ mA, a typical $V_{CE(sat)}$ is 0.2 V. Another source of error is the emitter-base leakage current I_{EO} that flows when the device is biased *off*. I_{EO} causes an unwanted output voltage to develop across load resistance R_L [see Figure 11-4(b)]. A typical level of I_{EO} for a switching transistor is 50 nA at 25°C.

In the design of a series sampling gate, the load resistance R_L should be selected much larger than the signal source resistance R_s. This will avoid large signal currents which would cause a significant voltage drop across R_s. The signal current can be reduced to a minimum if I_B is made equal to the output current I_E. The amplitude of the control voltage should be greater than the peak signal voltage. The sampling frequency (*i.e.*, the control voltage frequency) should be several times the frequency of the signal to be sampled.

EXAMPLE 11-1

Design a transistor series gate to sample a signal with a peak amplitude of 2 V, and a source resistance of 100 Ω. Also calculate the output errors due to $V_{CE(sat)}$ and I_{EO}.

solution

$$R_L \gg R_s$$

Let

$$R_L = 100 \times R_s = 100 \times 100 \ \Omega = 10 \ k\Omega$$

When the transistor is on,

$$V_o \simeq V_s$$

$$\therefore \qquad I_o = \frac{V_s}{R_L} = \frac{2 \ V}{10 \ k\Omega} = 200 \ \mu A$$

Let

$$I_B = I_o = 200 \ \mu A$$

The control voltage $V_1 > V_s$.
Let

$$V_1 = 2 \times V_s = 2 \times 2 \ V = 4 \ V$$

$$I_B = \frac{V_1 - V_{BE} - V_o}{R_B}$$

$$200 \ \mu A = \frac{4 \ V - 0.7 \ V - 2 \ V}{R_B}$$

and

$$R_B = \frac{1.3 \ V}{200 \ \mu A} = 6.5 \ k\Omega \qquad \text{(use 6.8 } k\Omega \text{ standard value)}$$

Typically, $V_{CE(\text{sat})} = 0.2$ V, and $I_{EO} = 50$ nA.

$$\text{Error due to } V_{CE(\text{sat})} = \frac{V_{CE(\text{sat})}}{V_s} \times 100\%$$

$$= \frac{0.2 \ V}{2 \ V} \times 100\% = 10\%$$

$$\text{Error due to } I_{EO} = \frac{(I_{EO} R_L)}{V_s} \times 100\%$$

$$= \frac{50 \ nA \times 10 \ k\Omega}{2 \ V} \times 100\%$$

$$= 0.025\%$$

11-3   BIPOLAR TRANSISTOR SHUNT GATE

The series sampling gate is suitable for signals having a low source resistance. For signals with a very high source resistance, the series gate requirement that R_L be much larger than R_s is difficult to fulfill. In this case, a shunt sampling gate is most suitable.

In the shunt sampling gate (Figure 11-5), transistor Q_1 shorts the input to ground when it is switched into saturation. When Q_1 is *off*, current flows

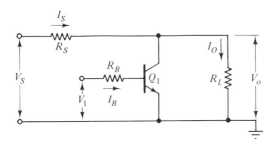

(a)   Shunt gate circuit

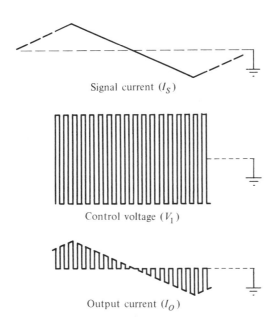

Signal current (I_S)

Control voltage (V_1)

Output current (I_O)

(b)   Current and voltage waveforms

FIGURE 11-5.   Bipolar transistor shunt gate and waveforms.

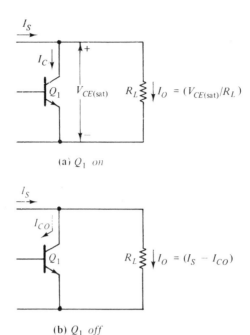

FIGURE 11-6.   Error sources in bipolar shunt gate.

from the signal source to the load resistance. Therefore, the shunt sampling gate essentially is a *current switch*, whereas the series sampling gate is a *voltage switch*. The transistor offset voltage results in a load current $V_{CE(\text{sat})}/R_L$ when the transistor is *on* [Figure 11-6(a)]. When the device is *off*, some of the signal current is lost as I_{CO} through the transistor [Figure 11-6(b)]. If the input signal becomes negative, the transistor operates in the inverted mode, as in the case of the series gate.

The load resistance for a shunt sampling gate should be selected such that $I_O R_L$ is much larger than $V_{CE(\text{sat})}$. For transistor saturation and for minimum $V_{CE(\text{sat})}$, I_B can be approximately equal to I_O. As in the case of the series gate, the sampling frequency should be at least several times the signal frequency. The transistor leakage current I_{CO} should be very much smaller than I_O.

EXAMPLE 11-2 _____

Design a transistor shunt gate to sample a signal current having a peak amplitude of 2 mA. Also, calculate the output errors due to $V_{CE(\text{sat})}$ and I_{CO}.

solution

$$I_o \simeq I_s = 2 \text{ mA}$$

Let

$$I_O R_L = 10 \times V_{CE(\text{sat})}$$
$$R_L = \frac{10 \times V_{CE(\text{sat})}}{I_O}$$
$$= \frac{10 \times 0.2 \text{ V}}{2 \text{ mA}} = 1 \text{ k}\Omega$$

Let

$$I_B \simeq I_O = 2 \text{ mA}$$

Take

$$V_1 = 4 \text{ V}$$
$$I_B = \frac{V_1 - V_{BE}}{R_B}$$
$$2 \text{ mA} = \frac{4 \text{ V} - 0.7 \text{ V}}{R_B}$$
$$R_B = \frac{3.3 \text{ V}}{2 \text{ mA}} = 1.65 \text{ k}\Omega \qquad \text{(use 1.8 k}\Omega \text{ standard value)}$$

$$\text{Error current due to } V_{CE(\text{sat})} = \frac{V_{CE(\text{sat})}}{R_L}$$

$$\text{Error due to } V_{CE(\text{sat})} = \frac{V_{CE(\text{sat})}/R_L}{I_O} \times 100\%$$

$$= \frac{0.2 \text{ V}/1 \text{ k}\Omega}{2 \text{ mA}} \times 100\% = 10\%$$

$$\text{Typical } I_{CO} = 50 \text{ nA}$$

$$\text{Error due to } I_{CO} = \frac{I_{CO}}{I_O} \times 100\%$$

$$= \frac{50 \text{ nA}}{2 \text{ mA}} \times 100\% = 0.0025\%$$

11-4   JFET SERIES GATE

A series sampling gate using an *n*-channel JFET is shown in Figure 11-7. Note that the control voltage V_1 goes from $+V_s$ to a negative level greater than the transistor pinchoff voltage V_P. When $V_1 = +V_s$, the FET is *on*. When $-V_1 > V_P$, the device is *off*. Note that because the drain terminal of the FET goes up to $+V_s$, the gate must also go up to that level for Q_1 to be correctly biased *on*. A gate resistance R_g (typically 1 MΩ) is usually included to limit any gate current that might flow. The JFET can also be operated in inverted mode, in which case the drain terminal acts as a source, and the source terminal performs the function of the drain. Inverted operation of a JFET is satisfactory only if the signal level is very small. If the signal becomes large, the (inverted) gate-channel junction

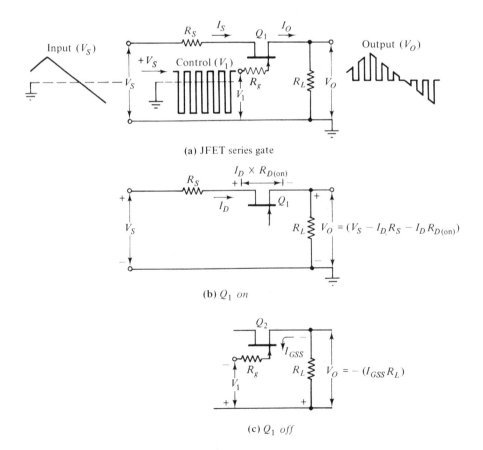

(a) JFET series gate

(b) Q_1 *on*

(c) Q_1 *off*

FIGURE 11-7.   Circuit, waveforms, and error sources for JFET series gate.

could become forward-biased, and the resultant gate current would affect the drain-source voltage.

Field effect transistors have a drain-source voltage drop of $I_D R_{D(on)}$ when biased into saturation [see Figure 11-7(b) and Sec. 4-5]. With small drain current, this drain-source voltage drop can be much smaller than the $V_{CE(sat)}$ of a bipolar transistor. A typical value of $R_{D(on)}$ for a switching FET is 30 Ω, although devices with $R_{D(on)}$ as low as 5 Ω are available. For a load current of 200 μA, as in Example 11-1, the typical FET offset voltage is (200 μA × 30 Ω) = 6 mV. This is only 0.3% of a 2 V signal, compared to the 10% loss due to the $V_{CE(sat)}$ of the bipolar transistor. When the JFET is biased *off* there is a gate-source leakage current I_{GSS}, which corresponds to I_{EO} in a bipolar transistor [Figure 11-7(c)]. Thus, I_{GSS} constitutes an unwanted load current. For a switching JFET I_{GSS} can be 0.2 nA or less, which is superior to the typical 50 nA of a bipolar device. The performance specifications for some switching JFETs are given below:

	Maximum pinchoff voltage $V_{P(max)}$	Drain-source on resistance $R_{D(on)}$	Gate-source leakage I_{GSS}	Drain-source leakage $I_{D(on)}$
2N4391	10 V	30 Ω	0.1 nA	0.1 nA
2N5433	9 V	7 Ω	0.2 nA	0.2 nA

Note that the data sheet for the 2N4391 is in Appendix 1-10.

EXAMPLE 11-3

A low-frequency signal with a peak amplitude of 1 V is applied to a voltage follower with a very low output resistance. The signal is to be

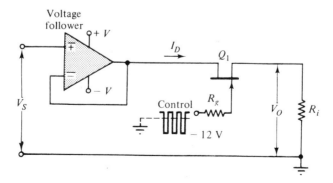

FIGURE 11-8.   FET series gate.

sampled at the output of the voltage follower and fed to a circuit with $R_i = 10$ kΩ. Design a suitable FET gate circuit and estimate the output errors.

solution

The circuit is as shown in Figure 11-8. For the 2N4391 FET, the control voltage $V_1 > (V_{P(max)} = 10$ V). Let

$$V_1 \simeq -12 \text{ V}$$

With Q_1 on,

$$I_D \simeq \frac{V_s}{R_s + R_i}$$

$$\simeq \frac{1 \text{ V}}{0 \text{ Ω} + 10 \text{ kΩ}} = 0.1 \text{ mA}$$

$$V_{DS(on)} = I_D R_{D(on)}$$

$$= 0.1 \text{ mA} \times 30 \text{ Ω} = 3 \text{ mV}$$

$$\text{Error due to } V_{DS(on)} = \frac{V_{DS(on)}}{V_s} \times 100\%$$

$$= \frac{3 \text{ mV}}{1 \text{ V}} \times 100\%$$

$$= 0.3\%$$

With Q_1 off,

$$I_o = I_{GSS} = 0.1 \text{ nA}$$

$$V_O = I_{GSS} R_i$$

$$= 0.1 \text{ nA} \times 10 \text{ kΩ} = 1 \text{ μV}$$

$$\text{Error due to } I_{GSS} = \frac{I_{GSS} R_i}{V_s} \times 100\%$$

$$= \frac{1 \text{ μV}}{1 \text{ V}} \times 100\% = 0.0001\%$$

11-5   JFET SHUNT GATE

The JFET shunt sampling gate shown in Figure 11-9(a) operates in a similar way to the bipolar shunt circuit. Like the bipolar shunt gate, the

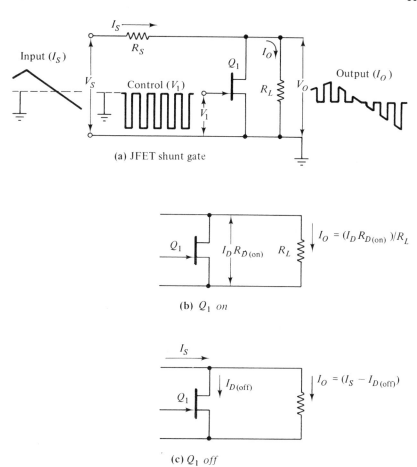

FIGURE 11-9.   Circuit, waveforms, and error sources for JFET shunt gate.

JFET shunt gate is essentially a current switch. When the FET is *on*, the output of the gate is shorted to ground. The output voltage at this time actually is $I_D R_{D(on)}$, and this produces an unwanted output current $(I_D R_{D(on)}/R_L)$ [see Figure 11-9(b)]. However, for the shunt FET gate, the unwanted output is much less than the minimum possible with a bipolar circuit. When the transistor is *off*, the drain-source leakage current $I_{D(off)}$ diverts signal current from the load [see Figure 11-9(c)]. Again, this usually is less than the corresponding bipolar leakage current.

EXAMPLE 11-4 _____

A low-frequency current with an amplitude of 0.1 mA is to be sampled and fed to the input of a circuit with $R_i = 10$ kΩ. Design a suitable FET shunt gate, and estimate the output voltage errors due to the transistor.

solution

Use a 2N4391 FET. Let the control voltage be -12 V as in Example 11-3. When Q_1 is *on*,

$$V_O = I_s R_{D(on)}$$
$$= 0.1 \text{ mA} \times 30 \text{ } \Omega = 3 \text{ mV}$$

and when Q_1 is *off*,

$$V_O = I_s R_i$$
$$= 0.1 \text{ mA} \times 10 \text{ k}\Omega = 1 \text{ V}$$

The error when Q_1 is *on* is given by

$$\frac{3 \text{ mV} \times 100}{1 \text{ V}} = 0.3\%$$

when Q_1 is *off*,

$$I_D = I_{D(off)} = 0.1 \text{ nA}$$
$$I_O = I_s - I_{D(off)}$$
$$= 0.1 \text{ mA} - 0.1 \text{ nA}$$
$$\text{Error} = \frac{0.1 \text{ nA}}{0.1 \text{ mA}} \times 100 = 0.0001\%$$

11-6   MOSFET SAMPLING GATES

MOSFETs are almost ideal devices for use as sampling gates. They have the same low $R_{D(on)}$ characteristic as JFETs, and the *enhancement mode* devices are normally *off* while the gate is at the same potential as the substrate. Figure 11-10(a) shows the circuit of a series sampling gate using an *n*-channel MOSFET. With the substrate at ground potential, the control voltage should go from 0 V to a positive voltage to switch the gate from *off* to *on*. When the input signal can be either negative or positive, the substrate should be taken to a negative bias voltage and the control voltage should start at the bias level. A MOSFET shunt sampling gate is shown in Figure 11-10(b). Here, again, the substrate terminal of the FET can be taken to a negative bias voltage, and the control voltage should start at the bias level to accommodate negative signal voltages.

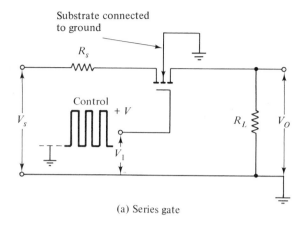

(a) Series gate

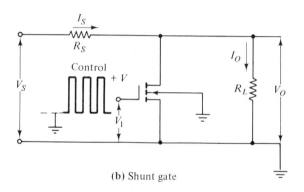

(b) Shunt gate

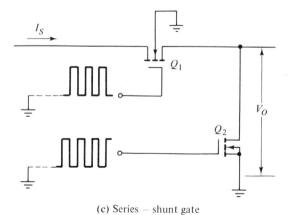

(c) Series – shunt gate

FIGURE 11-10. MOSFET sampling gate.

The circuit in Figure 11-10(c) is a series gate employing two MOSFETs. This circuit is particularly suitable where the load has a very high input resistance. When Q_1 is *on*, the signal voltage is switched to the load, and Q_2 is *off*. When Q_1 is *off*, the load voltage should be zero. With Q_2 *on* at this time, the output voltage is

$$(I_{D(off)} \text{ for } Q_1) \times (R_{D(on)} \text{ for } Q_2)$$

When typical values of 0.1 nA and 30 Ω are used, the unwanted output voltage is only 3 nV.

One problem in using MOSFETS as sampling gates is that the control voltage (applied to the gate) must always be greater than the maximum signal level. In fact, the control voltage must usually exceed the maximum signal amplitude by at least 2 V. Since there is very little voltage drop along the channel of the FET, both drain and source terminals are closely equal to the signal voltage level (as desired). Therefore, the channel of the device is always at the same potential as the signal. If the signal approaches the level of the control voltage, there may not be sufficient gate-channel bias to properly turn the device *on*. Thus, part of the signal may not be reproduced accurately at the output. The *CMOS transmission gate* overcomes this difficulty.

As shown in Figure 11-11, a CMOS transmission gate consists of two complementary MOSFETs connected in inverse parallel. The drain of Q_1 is connected to the source terminal of Q_2, and the source of Q_1 is connected to the drain of Q_2. Both substrate terminals are grounded, and each gate has its own control voltage.

Note that the control voltage waveforms in Figure 11-11 show that the *n*-channel device Q_1 has a control voltage which goes from ground to a positive voltage level. Q_2, the *p*-channel FET, has a control voltage that goes from ground to a negative level. Note also that the two control waveforms are in antiphase. When Q_1 gate is going positive, the gate of Q_2 is driven negative. This means that both devices are turned *on* and *off* simultaneously. Both present a low resistance path from input to output when *on*, and both offer a high resistance between input and output when *off*.

Now consider the signal and output waveforms illustrated in Figure 11-11. The input amplitude is shown as ± 8 V, and the control voltage on Q_1 gate is $+8$ V, while that to Q_2 is -8 V. During the time that the signal is positive, current can flow from input to output along both FET channels when the devices are biased *on*. If the positive amplitude of the signal approaches the control voltage amplitude (as illustrated), Q_1 tends to switch *off*, because there is not sufficient gate-channel voltage difference to keep it biased *on*. However, Q_2 is not affected, because its gate-channel

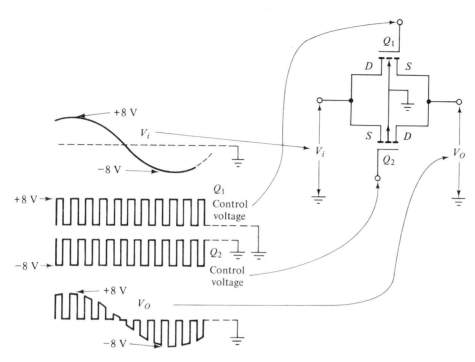

FIGURE 11-11.   CMOS transmission gate and waveforms.

voltage difference (16 V in Figure 11-11) is still easily adequate to keep it biased *on*. Thus, the signal current continues to flow via Q_2, and the output voltage remains closely equal to the input. The converse of this happens when the signal goes to a relatively large negative level. When $-V_i$ equals or exceeds the negative control voltage on the gate of Q_2 the *p*-channel device goes *off*. Now, however, the *n*-channel FET remains *on* because its gate-channel voltage is still a large positive quantity. Again, signal current finds a low resistance path (via Q_1) between input and output.

CMOS transmission gates can be purchased as integrated circuits. Usually four or more gates are contained in one IC package.

11-7 OPERATIONAL AMPLIFIER SAMPLING GATE

An operational amplifier connected as an *inverting amplifier* can be made into a sampling gate by installing a FET across its feedback resistor. The

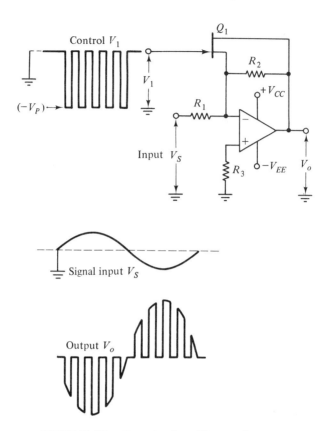

FIGURE 11-12. Operational amplifier sampling gate.

circuit is shown in Figure 11-12. When Q_1 is *off* the output voltage is

$$V_o = V_s \times \frac{R_2}{R_1}$$

When Q_1 is *on*, its channel resistance $R_{D(\text{on})}$ is in parallel with R_2, so that the output becomes

$$V_o = V_s \times \frac{R_2 \| R_{D(\text{on})}}{R_1}$$

Since $\qquad\qquad R_2 \| R_{D(\text{on})} \ll R_1$ (normally),

$$V_o \simeq 0$$

As with other sampling gates, the output waveform is a series of instantaneous samples of the input. However, in this case the output samples are an inverted version of the input. Also, the input can be amplified in the sampling process, depending upon the selection of the ratio $R_2/R1$. When Q_1 is an n-channel JFET, as illustrated, its control voltage should go down to a negative level equal to the FET maximum pinch-off voltage to ensure switch *off*, and up to ground level for switch *on*. When a MOSFET is employed, the required control voltage levels are as discussed in Section 11-6.

Because of the use of an operational amplifier, this gate has a very low output impedance. Its input impedance is equal to R_1. Design procedure for the gate simply involves designing an inverting amplifier and selection of a suitable FET.

EXAMPLE 11-5

A sampling gate using a 741 operational amplifier and a 2N4391 FET is to have a voltage gain of 10. The maximum signal voltage is $V_s = 500$ mV. Select suitable resistor values and control voltage amplitude. Also estimate the output error due to the $R_{D(\text{on})}$ of the FET.

solution

for the 741, $I_{B(\text{max})} = 500$ nA
let

$$I_1 = 100 \times I_{B(\text{max})}$$
$$= 100 \times 500 \text{ nA} = 50 \text{ }\mu\text{A}$$
$$R_1 = \frac{V_s}{I_1} = \frac{500 \text{ mV}}{50 \text{ }\mu\text{A}} = 10 \text{ k}\Omega \qquad \text{(standard value)}$$
$$R_2 = A_v \, R_1 = 10 \times 10 \text{ k}\Omega$$
$$= 100 \text{ k}\Omega \qquad \text{(standard value)}$$

Note that for a precise gain of 10, R_1 and R_2 would have to be precision resistors.

$$R_3 = R_1 \| R_2$$
$$\simeq 10 \text{ k}\Omega$$

for the 2N4391, $V_{p(\text{max})} = 10$ V (see Appendix 1-10). Therefore, the control voltage is, $V_1 \geqslant 10$ V

$$R_{D(\text{on})} = 30 \ \Omega$$

$$\textit{Zero output} = V_s \times \frac{R_{D(\text{on})} \| R_2}{R_1}$$

$$= 500 \text{ mV} \times \frac{30 \ \Omega \| 100 \text{ k}\Omega}{10 \text{ k}\Omega} = 1.5 \text{ mV}$$

$$V_o = \frac{R_2}{R_1} \times V_s$$

$$= 10 \times 500 \text{ mV}$$

$$= 5 \text{ V}$$

$$\textit{Zero error} = \frac{1.5 \text{ mV}}{5 \text{ V}} \times 100\% = 0.03\%$$

11-8   SAMPLE-AND-HOLD CIRCUIT

A *sample-and-hold circuit*, as its name implies, samples the instantaneous amplitude of a signal, and then holds the output voltage constant until the next sampling instant. The circuit, see Fig. 11-13(a), is simply a series gate with a capacitor C_1 to perform the holding function. Operational amplifiers A_1 and A_2 are connected as voltage followers (see Section 7-7) to provide high input impedance and low output impedance.

The waveforms in Figure 11-13(a) illustrate the relationship between input and output. At time t_a the instantaneous amplitude of the input is v_a. The output holds at the v_a level until time t_b, when it jumps to the input amplitude v_b. Similarly, when the input is falling, the output amplitude remains constant at v_b from t_b to t_c.

During the *sampling time* t_1 (also called the *acquisition time*). Q_1 is *on* and C_1 is charged via $R_{D(\text{on})}$, as illustrated in Figure 11-13(b). If the sampling time is

$$t_1 = 5 \text{ CR} \qquad\qquad \text{where } R \text{ is } R_{D(\text{on})},$$

the capacitor is charged to 0.993 of the input voltage. (This comes from Equation 2-2). Allowing the capacitor to charge to 0.993 of V_s results in a 0.7% error in the sampled amplitude. If $t = 7$ CR, $V_o = 0.999$ V_s, i.e., a 0.1% error.

During the holding time t_2, C_1 is partially discharged by the bias current I_{B2} flowing into A_2. The FET source-gate leakage current I_{GS} also

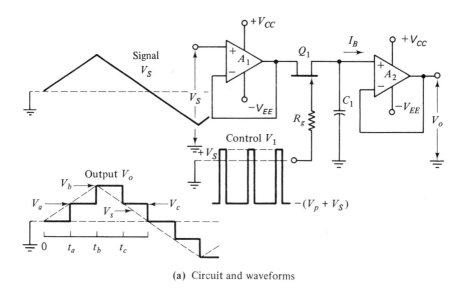

(a) Circuit and waveforms

(b) Charging circuit and capacitor voltage

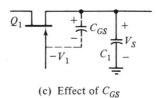

(c) Effect of C_{GS}

FIGURE 11-13. Sample-and-hold circuit.

causes some discharge of C_1. However, I_{GS} is normally very much less than I_{B2}, so it can usually be neglected. The capacitance of C_1 is calculated from the knowledge of I_{B2}, the holding time t_2, and the acceptable error due to C_1 discharge. After the value of C_1 is established, the sampling time t_1 is calculated from C_1 and the acceptable charging error.

One more source of error in the output voltage is the FET gate-source capacitance C_{GS}. When the control voltage on the gate goes to its lowest level, C_{GS} is charged to $e_c = (V_s + V_1)$. [This is illustrated in Figure 11-13(c)]. The charge on C_{GS} is removed from C_1, and thus reduces Vo. Example 11-6 demonstrated how a sample-and-hold circuit is designed, and how the various error sources affect the accuracy of the sample.

EXAMPLE 11-6 _____

A sample-and-hold circuit is to use 741 operational amplifiers and a 2N4391 FET. The signal voltage amplitude, $V_s = \pm 1$ V, is to be sampled with an accuracy of approximately 0.25%. The holding time is 500 μs. Determine the capacitor value and the minimum sampling time. Also calculate the effect of $C_{GS} = 10.5$ pF.

solution

for the 741, $I_{B(max)} = 500$ nA. Allow $I_{B(max)}$ to discharge C_1 by 0.1% during t_2.

$$\Delta V = 0.1\% \text{ of } 1 \text{ V}$$
$$= 1 \text{ mV}$$
$$C_1 = \frac{I_B \times t_2}{\Delta V}$$
$$= \frac{500 \text{ nA} \times 500 \text{ } \mu s}{1 \text{ mV}}$$
$$= 0.25 \text{ } \mu F$$

for the 2N4391, $R_{D(on)} = 30$ Ω. Allow another 0.1% error in V_o due to the sampling time t_1.

For 0.1% error, $t_1 = 7 \ C \ R_{D(on)}$
$$= 7 \times 0.25 \text{ } \mu F \times 30 \text{ } \Omega$$
$$= 52.5 \text{ } \mu s$$

For the 2N4391, $V_{p(max)} = 10$ V (see Appendix 1-10)

$$V_1 = -\left[V_{p(max)} + V_{s(peak)} \right]$$
$$= -\left[10\ V + 1\ V \right]$$
$$= -11\ V$$

Effect of $C_{GS} = 10.5$ pF:

$$V_{GS(max)} = +V_s - V_1$$
$$= 1\ V - (-11\ V)$$
$$= 12\ V$$

charge on C_{GS} is $Q = C_{GS} \times V_{GS(max)}$

$$= 10.5\ pF \times 12\ V$$
$$= 150\ pC$$

when Q is removed from C_1,

$$\Delta V_o = \frac{Q}{C_1} = \frac{150\ pC}{0.25\ \mu F}$$
$$= 600\ \mu V$$

% error due to $C_{GS} = \dfrac{\Delta V_o}{V_s} \times 100\%$

$$= \frac{600\ \mu V}{1\ V} \times 100\%$$
$$= 0.06\%$$

REVIEW QUESTIONS AND PROBLEMS

11-1 Sketch the circuit of a diode sampling gate. Show the voltage waveforms, explain the operation of the circuit, and discuss the error sources.

11-2 Repeat Problem 11-1 for a bipolar transistor series sampling gate.

11-3 Explain how a bipolar transistor series sampling gate functions when the input signal is alternately positive and negative with respect to ground.

11-4 Design a bipolar transistor series gate to sample a signal with a peak amplitude of 3 V and a source resistance of 200 Ω. Calculate the output errors due to $V_{CE(\text{sat})}$ and I_{EO}.

11-5   Repeat Problem 11-1 for a bipolar transistor shunt sampling gate.

11-6   Design a bipolar transistor shunt gate to sample a signal current with a peak amplitude of 1 mA. Calculate the output errors due to $V_{CE(\text{sat})}$ and I_{CO}.

11-7   Repeat Problem 11-1 for a JFET series sampling gate. State the precautions necessary for inverted operation of the JFET.

11-8   A signal of 1.5 V with a very low source resistance is to be sampled and passed to a circuit with $R_i = 20$ kΩ. Design a suitable JFET gate circuit, and estimate the output errors.

11-9   Repeat Problem 11-1 for a JFET shunt sampling gate.

11-10 A 200 μA signal current is to be sampled and passed to a circuit with $R_i = 15$ kΩ. Design a suitable JFET gate, and estimate the output errors.

11-11 Sketch circuits for MOSFET shunt, series and series-shunt sampling gates. Show waveforms and explain the operation of the circuits.

11-12 Compare the performances of bipolar transistors, JFETs, and MOSFETs as sampling gates.

11-13 Sketch the circuit of a CMOS transmission gate and explain its operation.

11-14 An operational amplifier and FET are to be employed as a sampling gate. Sketch the circuit and briefly explain its operation.

11-15 The circuit described for question 11-14 is to have an input voltage of $V_s = 750$ mV, and a voltage gain of 5. Design the circuit to use an operational amplifier which has $I_{B(\text{max})} = 750$ nA, and a FET with $V_{p(\text{max})} = 8$ V and $R_{D(\text{on})} = 50$ Ω. Estimate the output error due to $R_{D(\text{on})}$.

11-16 Sketch a sample-and-hold circuit using two operational amplifiers and a FET. Show all waveforms and explain how the circuit operates.

11-17 A sample-and-hold circuit is to use operational amplifiers with $I_{B(\text{max})} = 750$ nA, and a FET with $V_{p(\text{Max})} = 8$ V, $R_{D(\text{on})} = 50$ Ω, and $C_{GS} = 7$ pF. The signal voltage amplitude is $V_s = \pm 3$ V, and the sampled accuracy is to be approximately 1%. The holding time is 750 μs. Calculate the capacitor value and minimum sampling time. Also calculate the effect of C_{GS}.

CHAPTER 12

Digital
Counting

INTRODUCTION

Because the bistable multivibrator, or flip-flop, has two stable states, it can be used to count up to two. A cascade of four flip-flops can count up to sixteen. The *scale-of-16 counter* can be modified to produce a decade counter, which has an output in the form of a *binary number*. For counting in decimal form, a binary number must be converted to decimal. A further conversion stage usually is necessary to drive a numerical display. Decade counters and their numerical displays can be cascaded to construct systems for counting to hundreds, thousands, tens of thousands, etc.

12-1 FLIP-FLOPS IN CASCADE

The schematic diagram of four flip-flops (FF) connected in cascade is shown in Figure 12-1. Each flip-flop is a collector-coupled circuit, and each has symmetrical collector triggering. Negative-going input pulses are applied to FF1 via coupling capacitor C_1. Each time an input pulse is applied, FF1 will change state. The triggering circuit for FF2 is coupled via capacitor C_2 to transistor Q_2 in FF1. When Q_2 switches *off*, its collector

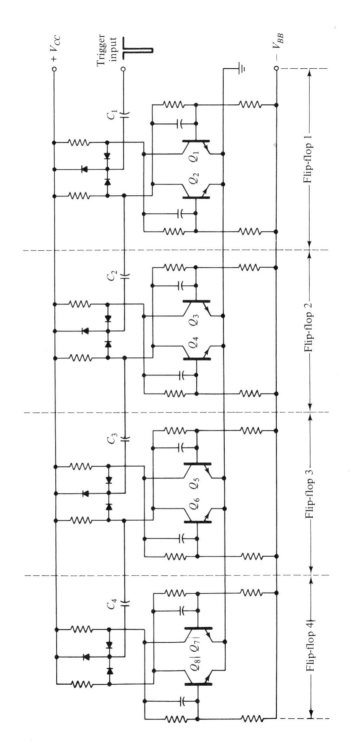

FIGURE 12-1. Cascade of four flip-flops.

voltage rises, applying a positive voltage step to C_2. Since a negative-going voltage is required to trigger these flip-flops (see Chapter 9), FF2 is not affected by the positive-going voltage. When Q_2 switches *on*, its collector voltage drops, thus applying a negative voltage step to FF2 via C_2. This negative voltage change triggers FF2. In a similar way, FF3 is triggered from FF2, and FF4 is triggered from FF3. It is seen that each flip-flop is triggered from each preceding stage.

The four-stage cascade in Figure 12-1 can have a number of combinations of flip-flop states. In Figure 12-2, the flip-flops are shown in block form with the arrowheads indicating that each is triggered from the previous stage. The state of each of the four flip-flops is best indicated by using the *binary* number system, where *0* represents a voltage at or near ground level and *1* represents a positive voltage level (see Figure 12-2). When a transistor is *on*, its collector voltage is low and is represented by *0*. An *off* transistor, on the other hand, has a high collector voltage and is designated *1*. In the decimal system, counting goes from *0* to *9*, then the next count is indicated by *0* in the first column and *1* in the next leftward column. In the binary system, the count in all columns can go only from *0* to *1*. Thus the count for *1* in both binary and decimal systems is *1*; in the

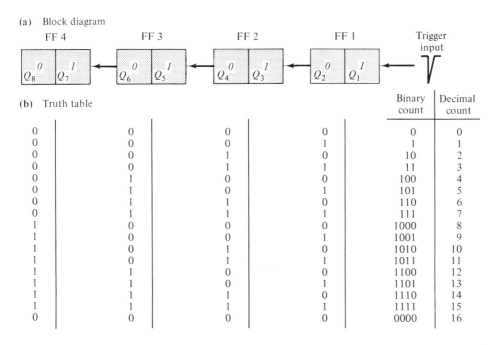

(a) Block diagram

FF 4   FF 3   FF 2   FF 1   Trigger input

| Q_8 | Q_7 | Q_6 | Q_5 | Q_4 | Q_3 | Q_2 | Q_1 | |

(b) Truth table

				Binary count	Decimal count
0	0	0	0	0	0
0	0	0	1	1	1
0	0	1	0	10	2
0	0	1	1	11	3
0	1	0	0	100	4
0	1	0	1	101	5
0	1	1	0	110	6
0	1	1	1	111	7
1	0	0	0	1000	8
1	0	0	1	1001	9
1	0	1	0	1010	10
1	0	1	1	1011	11
1	1	0	0	1100	12
1	1	0	1	1101	13
1	1	1	0	1110	14
1	1	1	1	1111	15
0	0	0	0	0000	16

FIGURE 12-2.   Block diagram and truth table for four flip-flops as a scale-of-16 counter.

binary system, the count for decimal *2* is indicated by *0* in the first column and *1* in the next leftward column. Thus, *binary 10* is equivalent to *decimal 2*. The next count in a binary system is *11* and is followed by *100*. The table of *0*'s and *1*'s showing the state of the flip-flops at each count is known as a *truth table.*

Suppose, before any pulses are applied, the state of the flip-flops is such that all even-numbered, (*i.e.*, left-hand) transistors are *on*. Reading only the even-numbered transistors (*i.e.*, in Figure 12-2) from left to right, the binary count is *0000*. At this time the decimal count is *0* and the binary count is *0*.

The first trigger pulse causes Q_1 to switch *on* and Q_2 to switch *off*. Thus, Q_2 reads as *1* (positive), and the binary count and decimal count are both *1*. The second input trigger pulse causes FF1 to change state again, so that Q_1 goes *off* and Q_2 switches *on*. When Q_2 switches *on*, a negative step is applied to FF2 triggering Q_3 *on* and Q_4 *off*. Now the binary count is *10*, and the decimal count is *2*. The third input pulse triggers Q_1 *on* and Q_2 *off* once again. This produces a positive output from FF1, which does not affect FF2. At this time, the binary count is *11*, for a decimal count of *3*. The fourth trigger pulse applied to the input, switches Q_1 *off* and Q_2 *on*. Q_2 coming *on* produces a negative step which causes Q_3 to go *off* and Q_4 to switch *on*. Q_4 switch-*on*, in turn, produces a negative voltage step which switches Q_5 *on* and Q_6 *off*. Now the binary count is read from the flip-flops as *100*, and the decimal count is *4*.

The counting process is continued with each new pulse until the maximum binary count of *1111* is reached. This occurs when 15 input pulses have been applied. The sixteenth input switches Q_2 *on* once more, producing a negative pulse which triggers Q_4 *on*. Q_4 output is a negative pulse which triggers Q_6 *on*, and Q_6 output triggers Q_8 *on*. Thus, the four flip-flops have returned to their original states, and the binary count has returned to *0000*. Including the zero condition, it is seen that the four flip-flops in cascade have 16 different states. Therefore, the circuit is termed a *scale-of*-16 *counter.*

The collector voltage levels for the scale-of-16 counter are shown as waveforms in Figure 12-3. The waveform for Q_{1C} shows that transistor Q_1 is initially *off*; its collector voltage is high and therefore is designated *1*. Q_2 is initially *on*, with its collector voltage at *0*. Each time a trigger pulse is applied, Q_1 and Q_2 change state. Q_{3C} and Q_{4C} are initially *1* and *0*, respectively, and they change state each time Q_{2C} goes from *1* to *0*, that is, when FF1 produces a negative-going output. This occurs on every second input pulse. Q_{5C} starts as *1* and Q_{6C} as *0*, and they change state only when Q_{4C} goes from *1* to *0*, which is at every fourth input pulse. Finally, the waveforms for Q_{7C} and Q_{8C} show that initially Q_{7C} is *1* and Q_{8C} is *0*, and that they change state when Q_{6C} becomes negative, that is, at every eighth

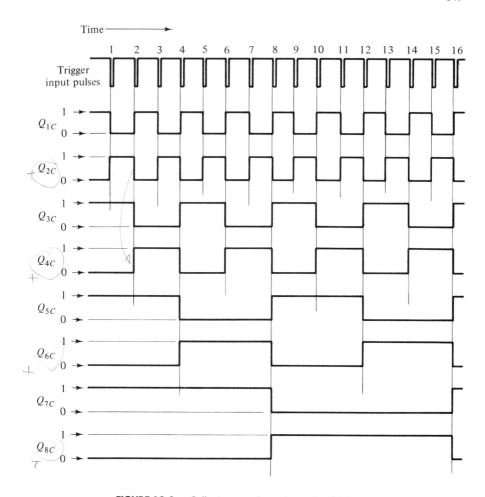

FIGURE 12-3. Collector waveforms for scale-of-16 counter.

input pulse. On the sixteenth input pulse all flip-flops change state, and the collector voltages return to their original levels.

The scale-of-16 counter actually can be used to divide the input pulse frequency by a factor of 16. Reference to the collector waveforms in Figure 12-3 shows that a negative-going voltage is produced at Q_8 collector after 16 input pulses. Another negative-going step will occur again at Q_{8C} after another 16 input pulses. Hence, the name *divide-by-16 counter* is sometimes applied to this circuit. An output taken from FF3 will produce a pulse frequency which is the input PRF divided by 8. Similarly, the output of FF2 divides the input by 4.

12-2　DECADE COUNTER

The scale-of-16 counter has many applications. However, there are also a great many instances in which a *scale-of*-10, or *decade, counter* is required. A decade counter also requires the use of a cascade of four flip-flops. Three flip-flops would count only up to seven, and then on the eighth pulse the count would revert to the *000* starting condition. This can be seen in Figure 12-2. Therefore, to produce a decade counter, a scale-of-16 must be modified to eliminate six of the sixteen states. This can be done by eliminating either the first six states or the last six states, or, perhaps, by eliminating some of the intermediate states.

When the first six states of a scale-of-16 counter are to be eliminated, the counter must always have an initial condition of *0110* (decimal 6 in Figure 12-2). To obtain this condition, transistors Q_4 and Q_6 must be in the *off* state. Q_4 and Q_6 can be reset to *off* by the asymmetrical base triggering circuit shown in Figure 12-4. (Asymmetrical base triggering is discussed in Sec. 9-5.) When Q_8 switches *on*, its collector voltage drops, providing a negative step which forward-biases D_1 and D_3, and triggers Q_6 and Q_4 *off*. Figures 12-2 and 12-3 show that Q_8 switches *on* when the sixteenth input pulse is applied. Therefore, at the end of the count of 16,

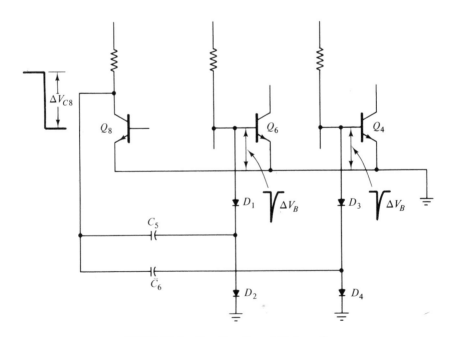

FIGURE 12-4.　Resetting Q_4 and Q_6 from Q_8.

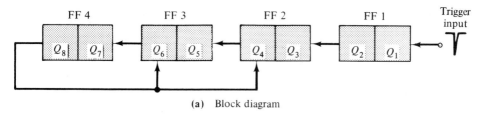

(a) Block diagram

Q_{8C}	Q_{6C}	Q_{4C}	Q_{2C}	Decimal count
0	1	1	0	0
0	1	1	1	1
1	0	0	0	2
1	0	0	1	3
1	0	1	0	4
1	0	1	1	5
1	1	0	0	6
1	1	0	1	7
1	1	1	0	8
1	1	1	1	9
Reset 0	0	0	0	10
0	1	1	0	0 =

(b) Truth table showing state of transistor collector

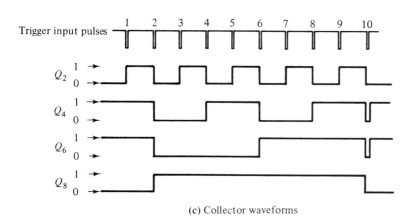

(c) Collector waveforms

FIGURE 12-5. Block diagram and truth table for four flip-flops as decade counter.

the flip-flops are set to *0110*. The block diagram, truth table, and collector waveforms, for the decade counter are shown in Figure 12-5.

In Figure 12-5(a), the line from FF4 to FF2 and FF3 indicates that these flip-flops are reset by the output from FF4. The initial state of the four flip-flips (*i.e.*, at decimal *0*) is read in Figure 12-5(b) as *0110*. The first input pulse now changes the state of FF1, causing Q_2 to switch *off*. Thus the collector of Q_2 becomes *1* (*i.e.*, high positive), and the condition of the counter is *0111*. This also is illustrated by the collector waveforms in Figure 12-5(c). The second input pulse (decimal *2*) again changes the state of FF1, this time causing Q_2 to switch *on*. The output from Q_2 is a negative step which triggers FF2, switching Q_4 *on*. This, in turn, produces a negative step which triggers FF3 from Q_6 *off* to Q_6 *on*. The output from FF3 triggers FF4. Counting continues in this way, exactly as explained for the scale-of-16 counter, until the tenth pulse. The ninth pulse sets the counter at *1111*, and the tenth pulse changes it to *0000*. However, as Q_8 switches *on*, it provides the negative output step which resets FF2 and FF3. The flip-flops have then returned to their initial conditions of *0110*, and it is seen that the circuit has only ten different states.

The waveforms in Figure 12-5(c) indicate that a negative output step is generated at Q_{8C} each time the tenth input pulse is applied. Thus, the decade counter can be employed as a *divide-by-10 counter*. Before counting begins, a decade counter (and a scale-of-16 counter) must have its flip-flops set in the correct starting condition. This can be accomplished by the manual resetting arrangement shown in Figure 12-6(a). When switch S_1 is closed, the diodes are forward-biased and the transistor bases are pulled below ground level. Thus, transistors Q_1, Q_4, Q_6, and Q_7 are switched *off*, giving the desired initial condition for the decade counter.

The flip-flops can also be reset automatically in their initial condition by the CR circuit addition in Figure 12-6(b). When the supply voltages are first switched *on*, the capacitor behaves as a short circuit. Therefore, the diode cathode voltages are at $-V$, and the transistors are biased *off*. After a brief time period, C_1 charges to $+V$ via resistor R. Now the diodes are all reverse-biased, and the reset circuit has no further effect.

Figure 12-7 shows a further modification of the circuit for resetting the flip-flops. Diode D_5 serves to isolate R and C from the rest of the reset circuit. When the cathodes of D_1 to D_4 are pulled down, D_5 is reverse-biased. D_6 and D_7, together with coupling capacitor C_2, form a triggering circuit. A negative-going voltage step applied to C_2 generates a negative pulse at the cathode of D_6. This forward-biases D_6, D_1, D_2, D_3, and D_4, causing the flip-flops to reset. Thus, as well as being reset to its starting condition when the supply is switched *on*, the counter can be reset to *zero* at any time by the application of a negative voltage step.

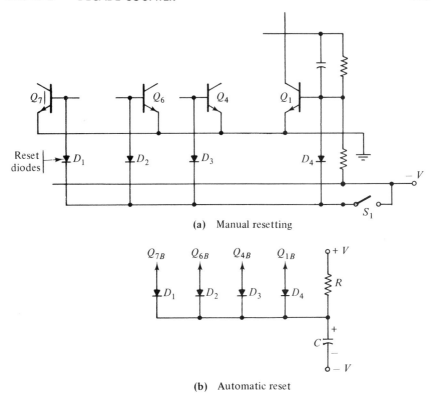

(a)   Manual resetting

(b)   Automatic reset

FIGURE 12-6.   Resetting a decade counter to zero.

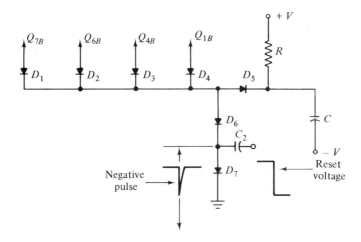

FIGURE 12-7.   Resetting to zero by pulse.

12-3 INTEGRATED CIRCUIT COUNTERS

The block diagrams and voltage waveforms for a 7493A TTL integrated circuit *binary counter* (or *scale-of*-16 *counter*) are shown in Figure 12-8. The block diagram shows that the counter consists of four *JK* flip-flops connected in cascade. The Q output of each flip-flop is connected as a trigger input to the *clock* terminal (*CK*) of the next stage. The triggering signal is applied to *input A*, which is the clock terminal of the first stage. Four outputs are available: Q_A, Q_B, Q_C, and Q_D. Separate (unidentified) *reset* terminals are provided on each flip-flop (at the bottom of each block). These are activated by *NAND* gate inputs $R_{o(1)}$ and $R_{o(2)}$. The block

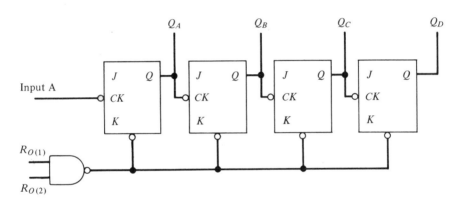

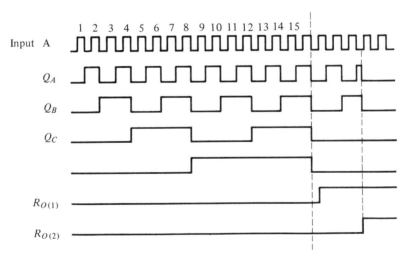

FIGURE 12-8. Block diagram and waveforms for 7493 IC binary counter.

diagram is also shown in a slightly different form on the data sheet for this counter in Appendix 1-23.

The waveforms demonstrate that Q_A changes state each time the triggering signal to input A goes negative. Similarly, Q_B changes state when Q_A goes negative, Q_C changes state when Q_B goes negative, and Q_D changes state every time Q_C goes negative. The result of this, as already explained in Sec. 12-1, is that the counter outputs have 16 different states. Note also from the waveforms that the outputs are all reset to *low* and counting ceases when *high* inputs are applied to $R_{o(1)}$ and $R_{o(2)}$.

Figure 12-9 shows the block diagram for the 7492A IC *divide-by-12 counter*. Once again, the diagram is drawn in a slightly different form from that shown on the data sheet in Appendix 1-23. The scale-of-12 differs from the scale-of-16 block diagram as follows: the Q_B output of stage 2 is connected to the J terminal of stage 3 (instead of to the clock input); the clock input of stage 3 is triggered from Q_A, which is also connected to the clock input of stage 2; the J terminal of stage 2 is connected to the $\overline{Q}$ output of stage 3.

With the arrangement described above, stage 2 output (Q_B) triggers into whatever state is applied at the J terminal. When the J input is *high*, Q_B is triggered into a *high* state when the clock terminal receives a negative-going signal. The next negative-going clock signal has no further effect on Q_B unless the J input has changed to a *low* level. (The JK flip-flops are in fact behaving as D flip-flops.) Similarly, output Q_C can be triggered into a different state only when the J input of stage 3 has changed from *high* to *low* or vice versa.

By careful consideration of the *count sequence* tables for the 92A and 93A counters in Appendix 1-23, it is found that four states of the divide-by -16 are eliminated in the divide-by-12 unit.

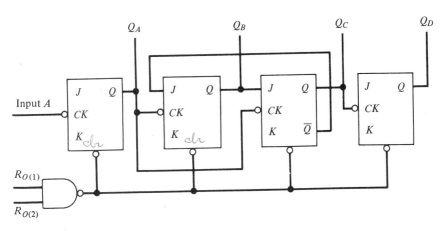

FIGURE 12-9. Block diagram for 7492A IC scale-of-12 counter.

Similarly, by the method already described in Section 12-2, and by other interconnection methods, a scale-of-16 IC counter can be employed as a decade counter.

12-4   DIGITAL DISPLAYS OR READOUTS

12-4.1   Light Emitting Diode Display

Charge carrier recombination occurs at a *pn*-junction as *electrons* cross from the *n*-side and recombine with *holes* on the *p*-side. When recombination takes place, the charge carriers give up energy in the form of heat and light. If the semiconductor material is translucent the light is emitted, and the junction is a light source, that is, a *light emitting diode* (LED).

Figure 12-10(a) shows a cross-sectional view of a typical LED. Charge carrier recombinations takes place in the *p*-type material; therefore, the *p*-region becomes the surface of the device. For maximum light emission, a metal film anode is deposited around the edge of the *p*-type material, or sometimes in a comb-shaped pattern at the center of the surface. The cathode connection for the device usually is a gold film at the bottom of the *n*-type region; this helps reflect the light to the surface. Semiconductor material used for LED manufacture is *gallium arsenide phosphide* (Ga AsP) which emits either red or yellow light, or *gallium arsenide* (Ga As) for green or red light emission.

The LED circuit symbol is shown in Figure 12-10(b). Figure 12-10(c) illustrates the arrangement of a seven-segment LED numerical display. Passing a current through the appropriate segments allows any numeral from 0 to 9 to be displayed. The actual LED device is very small, so to enlarge the lighted surface plastic *light pipes* are often employed, as illustrated in Figure 12-10(d). The typical voltage drop across a forward-biased LED is 1.2 V, and typical forward current for reasonable brightness is about 20 mA. This relatively large current requirement is a major disadvantage of LED displays. Some advantages of LEDs over other types of displays are: the ability to operate from a low voltage *dc* supply, ruggedness, rapid switching ability, and small physical size. The data sheet for a typical 7-segment LED display is shown in Appendix 1-21.

The simple transistor switch shown in Figure 12-10(b) is a suitable *on/off* control for LEDs. Q_1 is driven into saturation by input current I_B. Resistor R_C limits the current through the devices.

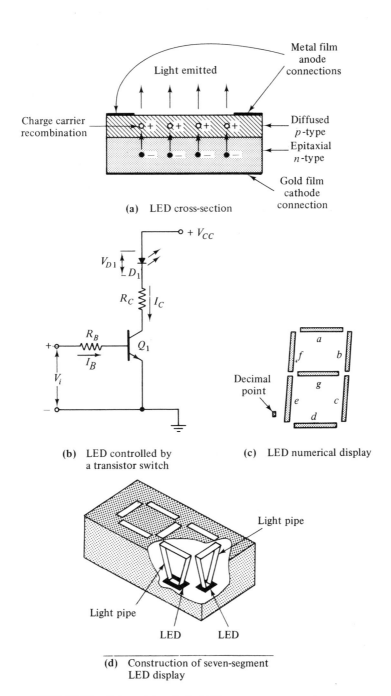

(a) LED cross-section

(b) LED controlled by a transistor switch

(c) LED numerical display

(d) Construction of seven-segment LED display

FIGURE 12-10. Light emitting diode (LED) cross section, control circuit, and seven-segment numerical display.

EXAMPLE 12-1 _____

The LED shown in Figure 12-10(b) is to have a minimum forward current of 20 mA. The diode has a forward voltage drop of 1.2 V, and transistor Q_1 has $h_{FE(min)} = 100$. Using $V_{CC} = 5$ V and $V_i = 5$ V, determine suitable values for R_C and R_B.

solution

$$V_{CC} = V_{D1} + I_C R_C + V_{CE(sat)}$$

$$R_C = \frac{V_{CC} - V_{D1} - V_{CE(sat)}}{I_C}$$

$$= \frac{5\ V - 1.2\ V - 0.2\ V}{20\ mA}$$

$$= 180\ \Omega \quad \text{(standard value)}$$

$$I_B = \frac{I_C}{h_{FE(min)}}$$

$$= \frac{20\ mA}{100} = 200\ \mu A$$

$$V_i = I_B R_B + V_{BE}$$

$$R_B = \frac{V_i - V_{BE}}{I_B}$$

$$= \frac{5\ V - 0.7\ V}{200\ \mu A} = 21.5\ k\Omega \quad \text{(use 18 k}\Omega \text{ standard value)}$$

12-4.2   Liquid Crystal Displays

Liquid crystal cell displays (LCD) usually are arranged in the same seven-segment numerical format as the LED display. There are two types of liquid crystal display, the *dynamic scattering type* and the *field effect type*. The construction of a dynamic scattering type liquid crystal cell is illustrated in Figure 12-11(a). The liquid crystal material may be one of several organic compounds which exhibit the optical properties of a crystal though they remain in liquid form. Liquid crystal is layered between glass

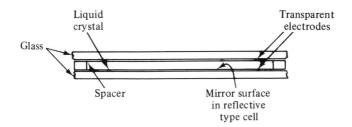

(a) Construction of liquid crystal cell

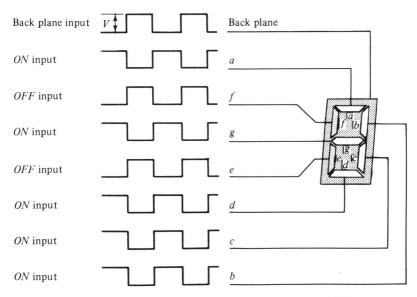

(b) Square wave drive method for liquid crystal cell 7-segment display

FIGURE 12-11.   Construction and electrical drive arrangement for liquid crystal cells.

sheets with transparent electrodes deposited on the inside faces. When a potential is applied across the cell, charge carriers flowing through the liquid disrupt the molecular alignment and produce turbulence. When not activated, the liquid crystal is transparent. When activated, the molecular turbulence causes light to be scattered in all directions, so that the cell appears quite bright. The phenomenon is termed *dynamic scattering*.

The construction of a *field effect* liquid crystal display is similar to that of the dynamic scattering type, with the exception that two thin polarizing optical filters are placed at the inside surface of each glass sheet. The liquid crystal material in the field effect cell is also a different type from that employed in the dynamic scattering cell. Known as *twisted nematic*, this liquid crystal material actually twists the light passing through the cell when the cell is not energized. This allows light to pass through the optical filters, and the cell appears bright (it can also be made to appear dark). When the cell is energized, no twisting of the light occurs and the cell remains dull.

Liquid crystal cells may be *transmittive* or *reflective*. In the *transmittive type* cell, both glass sheets are transparent, so that light from a rear source is scattered in the forward direction when the cell is activated. *The reflective type* cell has a reflecting surface on one of the glass sheets. In this case, incident light on the front surface of the cell is dynamically scattered by an activated cell. When activated, both the transmittive and reflective type cells appear quite bright even under high ambient light conditions.

Since liquid crystal cells are light reflectors or transmitters rather than light generators, they consume very small amounts of energy. The only energy required by the cell is that needed to activate the liquid crystal. The total current flow through four small 7-segment displays is typically about 25 μA for dynamic scattering cells and 300 μA for field effect cells. However, the LCD requires an ac voltage supply, either in the form of a sine wave or a square wave. This is because a direct current produces a plating of the cell electrodes, which could damage the device. A typical supply for a dynamic scattering LCD is a 30 V peak-to-peak square wave with a frequency of 60 Hz. A field effect cell typically uses 8 V peak-to-peak.

Figure 12-11(b) illustrates the square wave drive method for liquid crystal cells. The *back plane*, which is one terminal common to all cells, is supplied with a square wave. The other cell terminals each have square waves applied which are either in phase or in antiphase with the back plane square wave. Those cells with waveforms in phase with the back plane waveform [cells *e* and *f* in Figure 12-11(b)] have no voltage developed across them; therefore they are *off*. The cells with square waves in antiphase with the back plane input have an ac voltage developed across them (*e.g.*, positive square waves with 15 V peak effectively produce 30 V peak-to-peak when in antiphase). Therefore, the cells which have square wave inputs in antiphase with the back plane input are energized and appear bright.

The data sheet for the series 1603-02 liquid crystal display manufactured by Industrial Electronic Engineers, Inc., is shown in Appendix 1-22. The maximum power consumption is listed as 20 μW per segment, giving 140 μW per numeral when all seven segments are energized. Comparing this to the typical 400 mW per numeral for a LED display (see Appendix 1-21), the major advantage of liquid crystal displays is obvious. Perhaps the major disadvantage of the liquid crystal display is the decay time of

150 ms (or more). This is very slow compared to the 10 ns rise and fall times for the LED display. In fact it is so slow that the human eye can observe the fading-out of segments switching *off*. At low temperatures the response time of liquid crystal cells is considerably increased.

The series 1603-02 LCD is described in the data sheet as a *3½ decade display*. This means that the three right-hand units are complete 7-segment units while the fourth (left-hand) unit is only a single segment which indicates numeral 1 when energized. This unit is referred to as a half unit, and the entire display is then described as a $3\frac{1}{2}$ decade display. The maximum number that can be indicated by such a display is 1999.

12-4.3   Digital Indicator Tube

The basic construction of a *digital indicator tube* is shown in Figure 12-12(a), and its schematic symbol is illustrated in Figure 12-12(b). (Other names applied to this device are *cold cathode tube* and *glow tube*.) A flat metal plate with a positive voltage supply functions as an anode, and there are 10 separate wire cathodes, each in the shape of a numeral from 0 to 9.

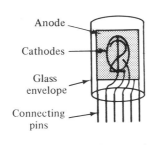

(a)   Construction

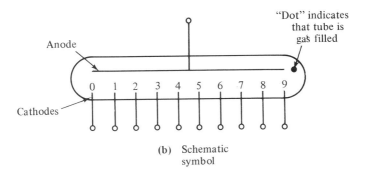

(b)   Schematic symbol

FIGURE 12-12.   Digital indicator tube, construction, and schematic symbol.

The electrodes are enclosed in a gas-filled glass envelope with connecting pins at the bottom. *Neon* gas usually is employed and it gives an orange-red glow when the tube is activated; however, other colors are available with different gases.

When a voltage is applied across the anode and one cathode, electrons are accelerated from cathode to anode. These electrons collide with gas atoms, and cause other electrons to be emitted from the gas atoms. The effect is termed *ionization by collision*. Since the ionized atoms have lost electrons, they are positively charged. Consequently, they accelerate toward the (negative) cathode, where they cause secondary electrons to be emitted when they strike. The secondary emitted electrons cause ionization and electron-atom recombination in the region close to the cathode. This results in energy being released in the form of light and produces a visible glow around the cathode. Since the cathodes are in the shape of numerals, a glowing numeral appears depending upon which cathode is energized. A transistor gate is usually employed at each cathode, so that the desired numeral can be switched *on* by a small input voltage.

The circuitry for driving digital indicator tubes is simpler than that for seven-segment devices. However, high voltages (140 V to 200 V) are required for these tubes, and in general they are much bulkier than comparable seven-segment devices.

12-4.4   Seven-segment Gas Discharge Displays

Gas discharge displays are also available in seven-segment format. Integrated circuits have been developed to drive these devices and to handle the high voltages involved. The mechanical construction of a seven-segment gas discharge display is illustrated in Figure 12-13(a). It is seen that separate cathodes are provided in seven-segment (and decimal point) form on a base. Each seven-segment group has a single anode deposited as a transparent metal film on the covering face plate. The gas is contained in the space between the anodes and cathodes, and rear connecting pins are provided for all electrodes. A *keep alive cathode* is also enclosed with each group of segments. A 50-μA current maintained through the *keep alive cathode* provides a source of ions which improves the switch-*on* speed of the display. Two circuit symbols in general use of seven-segment gas discharge displays are shown in Figures 12-13(b) and (c).

The supply voltage required to operate gas discharge displays ranges from about 140 V to 200 V, and this is the most serious disadvantage of these devices. High-voltage transistors must be employed as switches for the cathodes, and usually a separate high-voltage supply must be provided. Offsetting the disadvantage of high voltages is the fact that bright displays

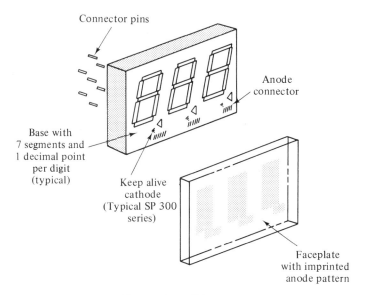

(a) Construction of 7-segment
gas discharge display
(*Courtesy of Beckman Instruments, Inc.*)

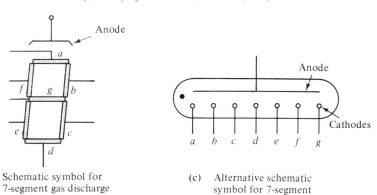

(b) Schematic symbol for
7-segment gas discharge
display

(c) Alternative schematic
symbol for 7-segment
gas discharge display

FIGURE 12-13. Mechanical construction and circuit symbols for seven-segment gas
discharge display.

can be achieved with tube currents as low as 200 μA. Thus the drain on
power supplies is minimal.

12-4.5 Fluorescent Display

The *fluorescent display* illustrated in Figure 12-14 is similar to the seven-
segment gas discharge display in that both are electron tube devices.

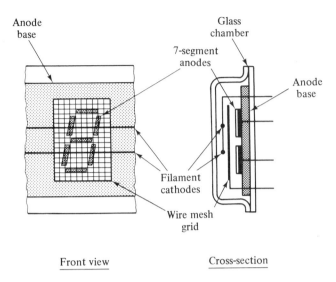

Front view   Cross-section

FIGURE 12-14.   Construction of a seven-segment-flourescent display.

However, the fluorescent display is a *vacuum* tube device with a filament type cathode, and a (wire mesh) grid to control the flow of the electrons from the cathode to the anodes (or plates). The seven electrically-separate anodes are coated with fluorescent phosphorous, so that they glow brightly (blue-green) when struck by electrons.

The major advantage of the fluorescent display is that it can operate with normal solid state supply voltage levels, ($V_{CC} = 10$ V to 40 V), whereas gas discharge devices require 140 V to 200 V supplies. The required (ac or dc) filament current is on the order of 12 mA rms for small size fluorescent displays. The grid which is typically biased to -3 V, is pulsed to approximately $+2$ V in synchronism with $+V_{CC}$ on the selected anode segments. Some grid current flows, but the total power dissipation for each seven-segment numeral is around 9 mW for small size displays.

12-5 BINARY TO DECIMAL CONVERSION

The output of a decade counter can be given in binary form if collector voltages are read as *1* when high and *0* when low. For display purposes, it is necessary to convert this binary number to decimal. Figure 12-15 shows the various binary states of a decade counter, and the circuitry required to convert each state to a decimal indication.

The *diode matrix* consists of diodes D_1 to D_{40} which have their cathodes connected to the collectors of the transistors in the decade

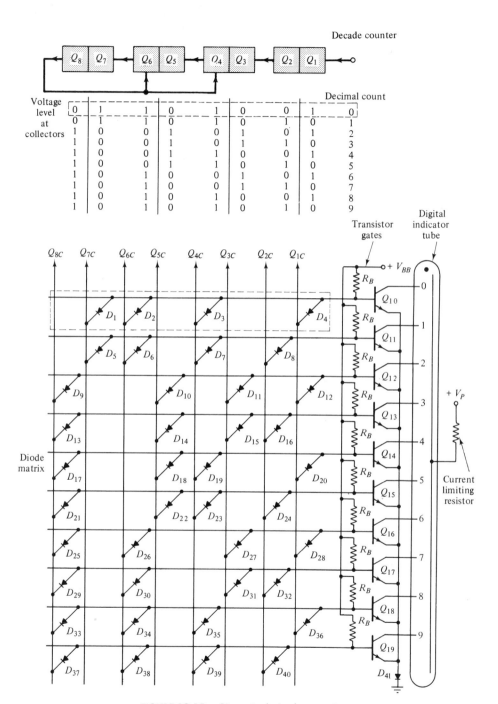

FIGURE 12-15. Binary to decimal conversion.

counter. The anodes of the diodes are connected to the bases of gate transistors Q_{10} to Q_{19}. When one transistor is switched *on*, it grounds the selected cathode in the digital indicator tube. Since the anode of the indicating tube has a positive supply, anode current flows when one of the cathodes is grounded, and the cathode glows. The transistor emitters are commoned and connected to ground via diode D_{41}. The presence of D_{41} ensures that the base voltage of each transistor has to be approximately 2 V_{BE} above ground level for it to switch *on*. Each gate transistor has four diodes connected to its base. When the cathode of one or more of these diodes is at *0* (*i.e.*, near ground level), the transistor base is held below the switching voltage. In this condition the transistor cannot switch *on*. When the cathodes of all four diodes connected to the base of any gate transistor are at *1*, the diodes are reverse-biased, and the transistor is biased *on* via base resistance R_B.

Consider the collector voltage levels and decimal count for the decade counter illustrated in Figure 12-15. For a decimal count of *0*, reading all transistor collector levels, Q_8 to Q_1 are read as *01 10 10 01*. Now, look at diodes D_1, D_2, D_3, and D_4, which are connected to the base of Q_{10}. The cathode of D_1 is connected to Q_7 collector. The collector voltage of Q_7 is at *1* (*i.e.*, positive); therefore D_1 is reverse-biased. The cathode of D_2 is connected to Q_{6C}, which is also at *1*, so D_2 is also reverse-biased. The cathode of D_3 is connected to Q_{4C}, and since Q_{4C} is at *1*, D_3 is reverse-biased. Finally, D_4 has its cathode connected to Q_{1C}, which is also at *1*. Thus all four diodes at the base of Q_{10} are reverse-biased. Base current flows from V_{BB} via R_B into the base of Q_{10}. With Q_{10} *on*, the *0* cathode in the digital tube glows, indicating that the decimal count is *0*.

For a correct *0* indication, all other gate transistors (Q_{11} to Q_{19}) must be biased *off*. To check that this is the case it is necessary to identify only one forward-biased diode at the base of each transistor. Consider diodes D_5 to D_8 at the base of Q_{11}. The cathodes of D_5, D_6, and D_7 are connected to transistor collectors which are at *1* when the decimal count is *0*. Therefore, all three are reverse-biased. D_8 cathode is connected to Q_2 collector, which is at *0*. Consequently, D_8 is forward-biased, and transistor Q_{11} is held in the *off* condition. For Q_{12}, the cathodes of D_9, D_{10}, and D_{11} are connected to transistor collectors which are at *0* while the decimal count remains *0*. Thus Q_{12} is biased *off*. Other diodes with *0* at their cathodes when the decade counter is in its decimal *0* condition are D_{13}, D_{14}, D_{15} D_{16}, D_{17}, D_{18}, D_{21}, D_{22}, D_{24}, D_{25}, D_{27}, D_{29}, D_{31}, D_{32}, D_{33}, D_{37}, D_{40}. It is seen that while the decimal count is *0*, all transistors except Q_{10} have at least one forward-biased diode at their bases. Therefore, only Q_{10} is biased *on*, and only the *0* cathode glows in the digital indicator tube.

A careful examination of the circuit conditions for any given decimal count shows that only the correct cathode is energized. All other cathodes have their transistor gates biased *off*.

EXAMPLE 12-2 _____

In Figure 12-15 identify the forward-biased and reverse-biased diodes for a decimal count of 5.

solution

For decimal 5, the transistor collectors in the decade counter read *10 01 10 10.*

At the base of transistor Q_{15}, diodes D_{21}, D_{22}, D_{23}, and D_{24} have their cathodes connected to Q_8, Q_5, Q_4, and Q_2, respectively. All these transistors have collectors at *1*; therefore, diodes D_{21} to D_{24} are reverse-biased, Q_5 is biased *on*, and cathode 5 in the digital tube glows.

Other reverse-biased diodes are D_3, D_7, D_8, D_9, D_{10}, D_{13}, D_{14}, D_{16}, D_{17}, D_{18}, D_{19}, D_{25}, D_{29}, D_{32}, D_{33}, D_{35}, D_{37}, D_{39}, D_{40}.

The forward-biased diodes and their associated transistor gates are D_1, D_2, D_4—Q_{10}; D_5, D_6—Q_{11}; D_{11}, D_{12}—Q_{12}; D_{15}—Q_{13}; D_{20}—Q_{14}; D_{26}, D_{27}, D_{28}—Q_{16}; D_{30}, D_{31}—Q_{17}; D_{34}, D_{36}—Q_{18}; D_{38}—Q_{19}.

_____

Figure 12-16 is a logic diagram for binary to decimal conversion. The flip-flop blocks have terminals as follows: *trigger input T, set S, reset R,* and outputs identified by their normal *set* conditions of *0* and *1* as shown. The triggering input pulses are applied to terminal *T* of FF1. Each succeeding flip-flop has its trigger terminal connected to the *0* output of the preceding stage. This arrangement can be compared to the cascaded flip-flops in Figure 12-1, where each stage is triggered from the second transistor in the previous stage. The *reset* terminals of FF2 and FF3 are connected to the output of FF4. This corresponds with the *reset* circuitry in Figure 12-4. In Figure 12-16 the output terminals of the decade counter are identified as Q_{1C}, Q_{2C}, etc. to show the correspondences with Figure 12-15.

Diodes D_1, D_2, D_3, and D_4 in Figure 12-15 constitute an *AND* gate. In Figure 12-16, these diodes are replaced by the *AND* gate symbol. Thus gate G_{10} represents D_1 to D_4. Also gate G_{11} represents diodes D_5 to D_8, G_{12} represents D_9 to D_{12}, etc. In Figure 12-16 the input terminals of *AND* gate G_{10} are connected to the decade counter terminals in the same configuration as D_1 to D_4 in Figure 12-15. Thus, the inputs to G_{10} are Q_{1C}, Q_{4C}, Q_{6C} and Q_{7C}. Similarly, gate G_{11} in Figure 12-16 is connected to the same decade counter terminals as D_5 to D_8 in Figure 12-15. The output of each *AND* gate is connected to the base of the appropriate transistor gate.

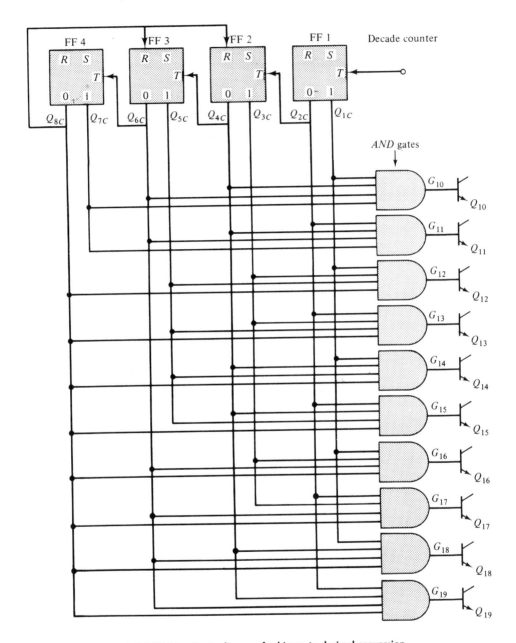

FIGURE 12-16. Logic diagram for binary to decimal conversion.

EXAMPLE 12-3

From the collector voltage levels shown in Figure 12-15 determine the input terminal connections for AND gate G_{19} in Figure 12-16.

solution

Gate G_{19} should provide an output to transistor Q_{19} only when the decimal count is 9. For G_{19} to produce an output, all its input terminals must be positive (*i.e.*, *1*). From Figure 12-15, at the decimal count of 9, transistors Q_8, Q_6, Q_4, and Q_2 all have their collectors at *1*. Therefore, the inputs of G_{19} should be connected to the *0 0 0 0* output terminals of the decade counter in Figure 12-15.

12-6   SEVEN-SEGMENT LED DISPLAY DRIVER

The binary to decimal conversion circuitry already discussed is suitable for driving a digital indicating tube. However, it is not suitable for driving a seven-segment display. Binary to decimal conversion is necessary for a seven-segment display, but in addition another diode matrix is required to convert from decimal to seven-segment format. The required diode matrix configuration is shown in Figure 12-17.

For the seven-segment display shown in Figure 12-17, the anodes of the light emitting diodes are commoned and have a positive supply voltage ($+V$). The cathodes from each segment (lettered *a* to *g*) have separate terminals. The transistor gates Q_{10} to Q_{19} of Figures 12-15 and 12-16 are shown again in Figure 12-17. When the decimal count is *0*, gate Q_{10} is *on*. When the count is *1*, gate Q_{11} is *on*, etc. For a *0* indication, LED segments *a*, *b*, *c*, *d*, *e*, and *f* should be energized. Therefore, these segments are connected via diodes D_{42} to D_{47} to the collector of transistor Q_{10}. The cathodes of the diodes are connected to the transistor collector so that they are forward-biased when Q_{10} is *on*. With this transistor *on* the seven-segment display indication is **0**.

When the decimal count is *1*, segments *b* and *c* should be energized. These segments are connected via diodes D_{48} and D_{49} to the collector of transistor Q_{11}. At the collector of transistor Q_{12}, diodes D_{50} to D_{54} connect to LED segments *a*, *b*, *d*, *e*, and *g*. When transistor Q_{12} is *on*, these segments are energized and display the numeral 2, as shown in the figure. At this time, only diodes D_{50} to D_{54} are conducting. No other diodes are conducting because only transistor Q_{12} is *on*. Figure 12-17 shows the LED segments that are energized for each decimal count.

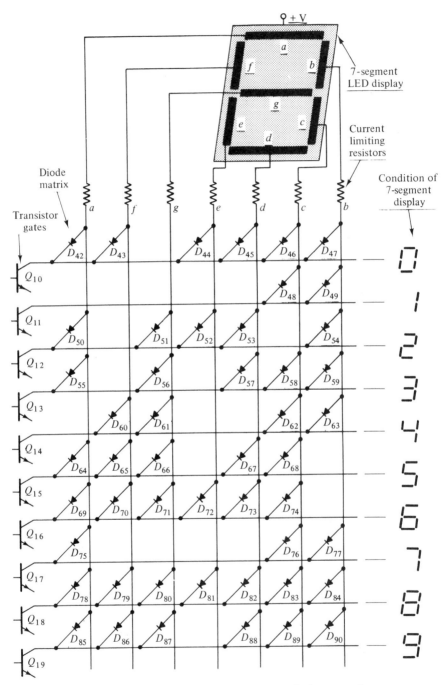

FIGURE 12-17. Decimal to seven-segment display conversion.

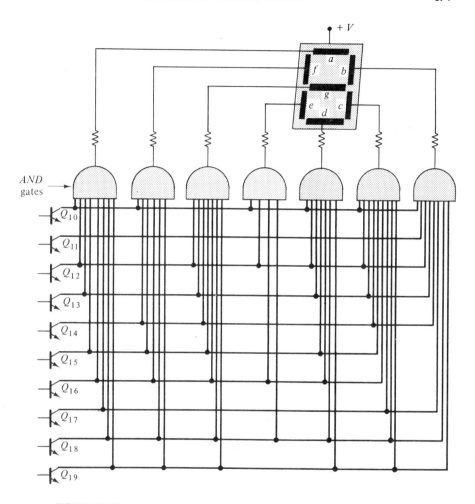

FIGURE 12-18.   Logic diagram for decimal to seven-segment display conversion.

Consideration of the diode matrix in Figure 12-17 reveals that the diodes connected to each LED segment constitute an *AND* gate. For segment *e*, for example, diodes D_{44}, D_{52}, D_{72}, and D_{81} provide a *low* level at the segment cathode when any one of the diodes has a low input level. The segment cathode is *high* only when the inputs to all four diodes are *high*. The diode matrix can be replaced by a group of *AND* gates, as shown in the logic diagram of Figure 12-18. The *AND* gate inputs for each segment are the same as the cathode connections for the diodes associated with the segments in Figure 12-17.

EXAMPLE 12-4 _____

Determine the input terminal connections for the *AND* gate connected to segment *b* in Figure 12-18.

solution

Consideration of the 0 to 9 display arrangements shows that segment *b* must be energized for display of numerals 0, 1, 2, 3, 4, 7, 8, 9. Therefore, the *AND* gate input for segment *b* must be connected to transistors Q_{10}, Q_{11}, Q_{12}, Q_{13}, Q_{14}, Q_{17}, Q_{18}, and Q_{19}.

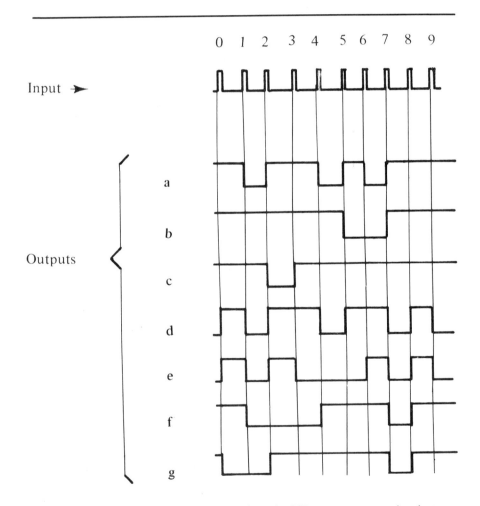

FIGURE 12-19. Input and output waveforms for BCD-to-seven-segment decoder/
driver.

Integrated circuits which convert directly from binary to seven-segment displays are, of course, available. Such IC's are usually listed as *BCD-to-seven-segment decoder/drivers* by their manufacturers. BCD are the initials for *binary-coded-decimal*, which is the form of output obtained from four cascaded flip-flops.

The input/output waveforms for a BCD-to-seven-segment decoder/driver are shown in Figure 12-19. The output waveforms identified as: *a, b, c, d, e, f, g*, refer to each of the segments of the display device, (see the seven-segment device in Figure 12-17). For the waveforms shown, the segments are energized when the outputs are *high*. Thus, at the count of zero; outputs *a, b, c, d, e* and *f* are *high*, and output *g* is *low*. This would display the numeral *0*. Similarly, at the count of *3*; only outputs *a, b, c, d* and *g* are *high*. Examination of the output waveforms at each stage of the input count, shows that the displayed numerals change consecutively from *0* to *9*.

12-7 SCALE-OF-10,000 COUNTER

One decade counter together with a seven-segment display and the necessary binary to seven-segment conversion circuitry can be employed to count from 0 to 9. Each time the tenth input pulse is applied, the display goes from 9 to 0 again. When this occurs, the output of the final transistor in the decade counter goes from *1* to *0* (see Figure 12-5). This is the only time that the final transistor produces a negative-going output, and this output can be used to trigger another decade counter.

Consider the block diagram of the scale-of-10,000 counter shown in Figure 12-20. The system consists of four complete decade counters and displays. Starting from *0*, all four counters can be set at their normal starting conditions. This gives an indication of *0000*. The first 9 input pulses register only on the first (right-hand side) display. On the tenth input pulse, the first display goes to *0*, and a negative-going pulse output from the first decade counter triggers the second decade counter. The display of the second counter now registers *1*, so that the complete display reads *0010*. The counter has counted to *10*, and has also registered *10* on the display system.

The next nine input pulses cause the first counter to go from *0* to *9*, so that the display reads *0019* on the nineteenth pulse. The twentieth pulse causes the first display to go to *0* again. At this time, the final transistor in the first decade counter puts out another negative pulse, which again triggers the second decade counter. The total display now reads *0020*, which indicates the fact that 20 pulses have been applied to the input of the first decade counter. It is seen that the second decade counter and display is counting *tens* of input pulses.

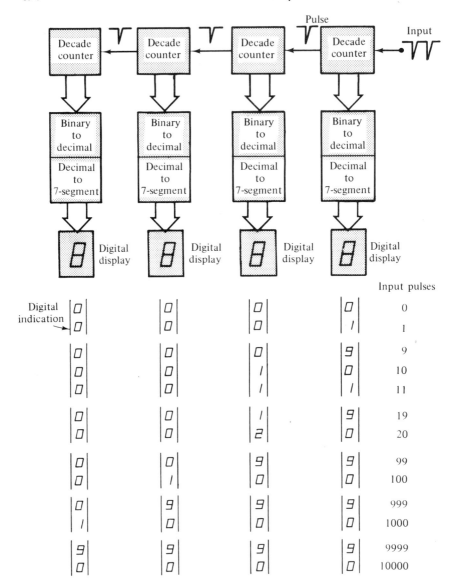

FIGURE 12-20.   Scale-of-10,000 counter.

Counting continues as described until the display indicates *0099* after the ninety-ninth input pulse. The one-hundredth input pulse causes the first two displays (from the right) to go to *0*. The second decade counter emits a negative pulse at this time, which triggers the third decade counter. Therefore, the count reads *100*. It is seen that the system shown in Figure 12-20 can count to a maximum of *9999*. One more pulse causes the display to return to its initial *0000* condition. To increase the maximum count to *99999*, it is necessary to add one more decade counter, together with a display, and binary to seven-segment conversion circuitry.

12-8   COUNTER CONTROLS

A simple system for switching the counter input pulses *on* and *off* is shown in Figure 12-21. The pulses to be counted are applied to one input of an *AND* gate. The voltage level at the other input of the gate is controlled by the output of a flip-flop. The input triggering pulses pass through the gate to the counter only when the flip-flop output is at its *1* level, that is, when it is high. The flip-flop can be reset to *1* output by switching a negative input voltage to the *reset R* terminal. The *manual start* control shown in the figure is provided for this purpose. A connection to the *reset* input of each decade counter (see Figure 12-7) ensures that the counting circuits return to *0* condition before counting begins. The *manual stop* control provides a negative voltage which returns the flip-flop to its original *set* condition. This applies a *0* to the *AND* gate input, and thus stops the pulses to be counted from passing through.

The flip-flop can also be triggered, and consequently counting can be started by means of a negative *start counting* pulse applied to the *reset*

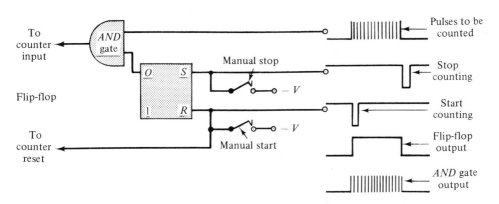

FIGURE 12-21.   Starting and stopping a counter.

terminal as shown in Figure 12-21. Similarly, a negative pulse applied to the *set* terminal of the flip-flop interrupts the passage of pulses to the counter. The waveforms in the figure show that counting pulses pass through the *AND* gate only during the time interval between application of *start-counting* and *stop-counting* pulses.

REVIEW QUESTIONS AND PROBLEMS

12-1   Sketch the complete circuitry for four cascaded flip-flops. Briefly explain the triggering process.

12-2   Sketch a block diagram for a scale-of-16 counter. Reading only the left-hand transistors in each flip-flop, prepare a truth table showing the state of the counter after each input pulse from 0 to 16. Explain the procedure.

12-3   Sketch the waveforms of collector voltages for every transistor in a scale-of-16 counter for input pulses from 0 to 16. Briefly explain each waveform change.

12-4   Briefly explain how a scale-of-16 counter can be converted to a decade counter. Identify the flip-flops which must be reset in the process, and show which of 16 states of the counter are eliminated.

12-5   Sketch circuitry employed for resetting flip-flops when a scale-of-16 is converted to a decade counter. Briefly explain.

12-6   Sketch the block diagram of a decade counter. Prepare a table that shows the state of the counter after each input pulse from 0 to 10. Also show the collector waveforms for the even-numbered transistors. Explain the waveforms and the states of the counter.

12-7   Identify the transistors in a decade counter that should be reset to *0* before counting commences. Sketch suitable circuitry for (a) manual resetting, (b) automatic resetting, (c) resetting by pulse input. Briefly explain each circuit.

12-8   Sketch the logic block diagram for a 7493 integrated circuit divide-by-16 counter. Sketch waveforms showing the counter output voltage levels after each input pulse. Explain the operation of the counter.

12-9   Refer to the block diagram for the 7492A IC decade counter, and to the count sequence table for this IC in Appendix 1-23. Sketch waveforms showing the relationship between triggering and output signals for the counter.

12-10 Explain the operation of a light-emitting diode. Sketch the cross section of an LED, and identify and explain all component parts of the device.

12-11 Discuss the current and voltage requirements of an LED. Sketch the circuit symbol for the device, and show how it is employed in a seven-segment numerical display.

12-12 Three series-connected LEDs are to have 15 mA passed through them from a -9 V supply when energized. A *pnp* transistor with $h_{FE(min)} = 75$ is to be used as a switch. The input voltage to the transistor is -9 V. Sketch an appropriate circuit and determine the resistor values required.

12-13 Explain the operation of a liquid crystal cell. Sketch the cross section of a liquid crystal cell and identify and explain each component part.

12-14 Explain *transmittive type* and *reflective type* liquid crystal cells. Discuss the voltage and current requirements for liquid crystal displays, and show how square waves are employed to drive a seven-segment liquid crystal display.

12-15 Explain the construction of 7-segment fluorescent displays, and discuss its characteristics.

12-16 Using sketches, explain the construction of a digital indicator tube. Also sketch the schematic symbol for the device and explain its operation.

12-17 Using sketches, explain the operation of a seven-segment gas discharge display. Discuss the current and voltage requirements for the display, and explain the function of the *keep-alive cathode*. Sketch two schematic symbols in general use for seven-segment gas discharge displays.

12-18 Show how a diode matrix may be employed for binary to decimal conversion. Sketch the complete circuit of the diode matrix, transistor gates, and digital indicator tube.

12-19 For the diode matrix in the binary to decimal conversion circuitry sketched for Problem 12-18, identify the diodes that are forward-biased and those that are reverse-biased at a decimal count of 6 and at a decimal count of 3.

12-20 Sketch a complete logic diagram for binary to decimal conversion. Briefly explain how the system functions.

12-21 Sketch a complete diode matrix for driving a seven-segment display from a digital input. Explain the operation of the circuitry, and identify the segments of the display that are energized for each decimal input.

12-22 Sketch a complete logic diagram for decimal to seven-segment display conversion. Briefly explain how the system functions.

12-23 Draw the block diagram of a scale-of-1,000 counter. Explain the operation of the system.

12-24 Draw the block diagram of a control system that will start and stop a counter manually and by means of input pulses. Show the waveforms of input, and control voltages. Explain the operation of the system.

CHAPTER 13

Digital Frequency Meters and Digital Voltmeters

INTRODUCTION

If a pulse waveform is fed to the input of a digital counter for a time period of exactly one second, the counter indicates the frequency of the waveform. Suppose the counter registers 1000 at the end of a second; then the frequency of the input is 1000 *pulses per second*. Essentially, a digital frequency meter is a digital counter combined with an accurate timing system. The timing system usually is such that the input frequency is sampled repeatedly. This necessitates the use of a *latch*, which keeps the display constant while the frequency remains unchanged.

A dc voltage can be converted to a frequency which is directly proportional to the voltage. Then this frequency can be measured by a digital frequency meter and the output is read as a voltage. Several methods of converting from voltage to frequency are available.

13-1 TIMING SYSTEM

A block diagram and voltage waveform of a typical timing circuit for a digital frequency meter are shown in Figure 13-1. The source of the time interval over which input pulses are counted is a very accurate crystal-controlled oscillator usually referred to as a *clock source*. The crystal is often enclosed in a constant temperature *oven* to maintain a stable oscillation frequency.

The output frequency from the final flip-flop of a decade counter is exactly one-tenth of the input triggering frequency (see Chapter 12). This means that the time period of the output waveform is exactly ten times the time period of the input waveform. The 1 MHz output from the crystal oscillator in Figure 13-1 has a time period of 1 μs, and the output waveform from the first decade counter has a time period of 10μs. The time period of the output from the second decade counter is 100 μs, and that from the third is 1 ms, etc. With all six decade counters, the available time periods are 1 μs, 10 μs, 100 μs, 1 ms, 10 ms, 100 ms, and 1 second.

When the counting circuits in a digital frequency meter are triggered for a period of 1 second, the display registers the input frequency directly. A count of 1000 cycles over the 1 second period represents a frequency of 1000 Hz or 1 kHz, a 5000 display indicates 5000 Hz, etc. These figures are more easily read when a decimal point is placed after the first numeral and the output is identified in kilohertz. Thus a display of 1.000 is 1 kHz, 5.000 is 5 kHz, and 5.473 is 5.473 kHz. In an LED display, the decimal point is created by use of a single suitably placed light-emitting diode. Also, a *kHz* indication usually is displayed.

Now consider the effect of using the 100-ms time period to control the counting circuits. A display of 1000 now means 1000 cycles per 100 ms. This is 10,000 cycles per second or *10 kHz*. When the time period is switched from 1 second to 100 ms, the decimal point also is switched from the first numeral to a position after the second numeral. The 10.00 display is now read as *10.00 kHz;* 50.00 is read as *50.00 kHz*, etc.

When the time period is switched to 10 ms, the decimal point is moved to a new position after the third numeral on the display. A 100.0 display now becomes *100.0 kHz*. Since this is the result of 1000 cycles of input counted over a period of 10 ms, the actual input frequency is (1000/ 10 ms), that is, 100 kHz. With 1 ms time period, a display of 1000 indicates 1000 cycles during 1 ms, or 1 MHz. The decimal point is now moved back to its original position after the first numeral, and a *MHz* indication is displayed. Therefore, the display of 1.000 with a 1 ms time base is read as 1.000 MHz. If the 100 μs and 10 μs time periods are used, the decimal point is again moved so that the 1000 indication becomes 10.00 MHz and 100.0 MHz, respectively.

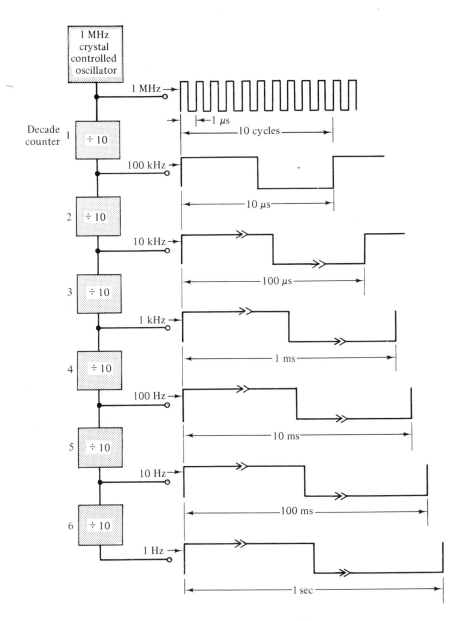

FIGURE 13-1. Time base generation for digital frequency meter.

EXAMPLE 13-1 _____

A 3.5 kHz sine wave is applied to a digital frequency meter. The time base is derived from a 1 MHz clock generator frequency divided by decade counters. Determine the meter indication when the time base uses (a) six decade counters and (b) four decade counters.

solution

(a) When six decade counters are used:

$$\text{Time base frequency} = f_1 = \frac{1 \text{ MHz}}{10^6} = 1 \text{ Hz}$$

$$\text{Time base} = t_1 = \frac{1}{f_1} = \frac{1}{1 \text{ Hz}} = 1 \text{ sec}$$

Cycles of input

counted during t_1 = Input frequency $\times t_1$

$$= 3.5 \text{ kHz} \times 1 \text{ sec}$$

$$= 3,500$$

Thus, the display indication is 3500.

(b) When four decade counters are used:

$$\text{Time base frequency} = f_2 = \frac{1 \text{ MHz}}{10^4} = 100 \text{ Hz}$$

$$\text{Time base} = t_2 = \frac{1}{f_2} = \frac{1}{100 \text{ Hz}} = 10 \text{ ms}$$

Cycles of input

counted during t_2 = Input frequency $\times t_2$

$$= 3.5 \text{ kHz} \times 10 \text{ ms}$$

$$= 35$$

Thus, the display indication is 0035.

13-2 LATCH CIRCUITS

If the display devices in a digital frequency meter are controlled directly from the counting circuits, the display changes rapidly as the count progresses from zero. Suppose the input pulses are counted over a period of 1 second, and then the count is held constant for 1 second. The display alternates between being constant for one second and continuously changing for the next second. Therefore, the display is quite difficult to read, and the difficulty is increased when shorter time periods are employed for counting. To overcome this problem, *latch* circuits are employed.

A *latch* isolates the display devices from the counting circuits while counting is in progress. At the end of the counting time, a signal to the latch causes the display to change to the decimal equivalent of the final condition of the counting circuits. If the input frequency is constant, as is usually the case, the displayed count remains constant. The counting circuits continue to sample the input frequency, and the *latches* are repeatedly triggered to check the displays against each final state of the counting circuits.

Figure 13-2 illustrates the use of *JK* flip-flops as latch circuits. The input (*J* and *K*) terminals of the flip-flops are connected to the collectors

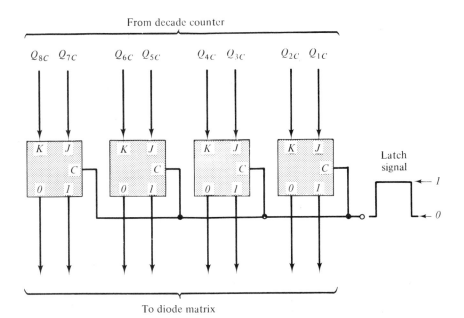

FIGURE 13-2. JK flip-flops operated as latch circuits.

of the transistors in the decade counter. The flip-flop outputs are fed to the binary-to-decimal diode matrix (Figure 12-15). The latching signal is applied to the *clock* input terminal c of each JK flip-flop.

The theory of the JK flip-flop is explained in Sec. 9-10 and a timing diagram is shown in Figure 9-16. When triggered, the flip-flop outputs assume the level of the J and K input terminals. The output then remains constant until another trigger (*i.e*, latch) input is applied. If the input levels have changed, the output will change. If the input levels are the same as before, the flip-flop outputs remain unaltered. Thus, the latch circuits sample the output levels of the decade counter, and pass these levels to the diode matrix. Since sampling occurs only when a latch signal is applied, the inputs to the diode matrix remain constant during the time interval between latch signals. Therefore, the numerical display devices also remain in a constant state during this time, and the counting circuits can go from zero to maximum count without affecting the display.

A *display enable* control which open-circuits the supply voltage to the display devices can sometimes be used instead of a latch. The display is simply switched *off* during the counting time and *on* during the noncounting time. The (normally constant) displayed numerals are thus switched *on* and *off* continuously, with no display occurring during the counting time. When the display time and counting time are brief enough, the *on/off* frequency of the display is so high that the human eye sees only a constant display.

13-3 DIGITAL FREQUENCY METER

The block diagram of a digital frequency meter is shown in Figure 13-3, and the voltage waveforms for the system are illustrated in Figure 13-4. The input signal which is to have its frequency measured is first amplified and then fed to a Schmitt trigger circuit. Amplification ensures that the signal amplitude is large enough to trigger the Schmitt circuit, and the Schmitt circuit produces a square wave output of the same frequency as the input. A square wave is required for triggering the counting circuits. Before it gets to the counting circuits, however, the square wave must pass through an *AND* gate.

The square wave passes to the counting circuits only when output 1 from the flip-flop is at logic *1* (*i.e.*, positive). The flip-flop changes state each time a negative-going output is received from the *timer*. Therefore, when $T = 1$ sec (see Figure 13-4) the flip-flop output is alternately at level *1* for one second and at level *0* for one second. Consequently the *AND* gate is alternately *on* for one second and *off* for one second. During the time that the *AND* gate is *on*, the Schmitt output triggers the counting circuits.

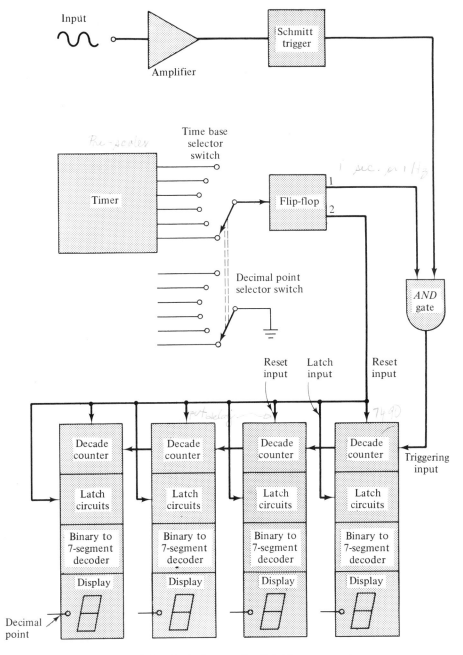

Input

Amplifier

Schmitt
trigger

Time base
selector
switch

Pre-scaler

Timer

Flip-flop

1

2

1 sec. or 1 Hz

Decimal point
selector switch

AND
gate

Reset
input

Latch
input

Reset
input

put delay

74 90

Decade
counter

Decade
counter

Decade
counter

Decade
counter

Triggering
input

Latch
circuits

Latch
circuits

Latch
circuits

Latch
circuits

Binary to
7-segment
decoder

Binary to
7-segment
decoder

Binary to
7-segment
decoder

Binary to
7-segment
decoder

Display

Display

Display

Display

Decimal
point

FIGURE 13-3. Block diagram of a digital frequency meter.

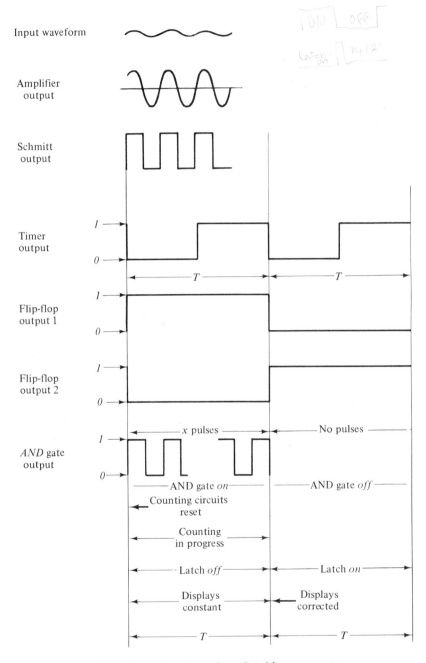

Input waveform

Amplifier
output

Schmitt
output

Timer
output

Flip-flop
output 1

Flip-flop
output 2

AND gate
output

x pulses — No pulses

AND gate *on* — AND gate *off*

Counting circuits
reset

Counting
in progress

Latch *off* — Latch *on*

Displays
constant

Displays
corrected

FIGURE 13-4. Waveforms for a digital frequency meter.

The exact number of input pulses are counted during that time and, as already discussed, when $T = 1$ second the count is a measure of the input frequency. The timer has six (or more) available output time periods over which counting can take place. The desired time period is selected by means of a switch, as shown in Figure 13-3. A separate decimal point selector switch moves with the time base selector.

Output 2 from the flip-flop is an antiphase to output 1 (see Figure 13-4). This waveform is employed for resetting the counting circuits, and for opening and closing the latches. At the beginning of the counting time, output 2 from the flip-flop is a negative-going voltage. This triggers the reset circuitry in each decade counter. Since flip-flop output 2 is at logic *0* during the counting time, its application to the latching circuits ensures that each latch is *off*. That is, during the counting time, nothing passes through the latching circuits. At the end of the counting time, the wave-form fed to the latch inputs goes to logic *1*. This triggers each latch *on* so that the conditions of the displays are corrected, if necessary, to reflect the states of the counting circuits. During the latch *on* time, the *AND* gate is *off* and no counting occurs. Therefore, once corrected, the displays remain constant. The displays remain constant also during the counting time, since the latch circuits are *off*.

The digital frequency meter can be used to measure the frequency of a signal with almost any waveform. Also, it can be employed for accurate measurement of time periods, and for determining the ratio of two frequencies.

13-4 DUAL SLOPE INTEGRATOR FOR DIGITAL VOLTMETER

In the dual slope integration type of digital voltmeter, an integrating circuit is used to generate a ramp over a timed interval. The slope of the ramp is proportional to the voltage to be measured. Consequently, the ramp amplitude is also proportional to the input voltage. The integrator then is reversed and discharged at a constant rate proportional to an accurate reference voltage. When the time for the ramp to go to zero is measured digitally, the numerical display can be read as the input voltage.

A dual slope integrating circuit is shown in Figure 13-5. The voltage V_i to be measured is applied to the input terminal of a voltage follower in order to present a high input impedance. The voltage follower output is switched via a FET(Q_1) to the input of a Miller integrator. (The voltage follower and Miller integrator are introduced in Chapter 7.) An accurate current source is also connected to the input of the Miller integrator, and the integrator output level is monitored by a *zero crossing detector*. The

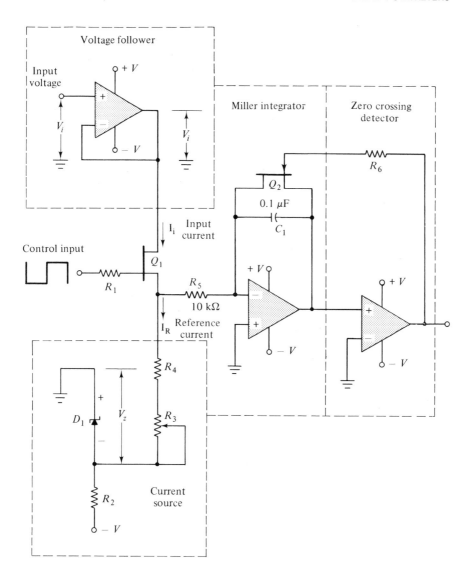

FIGURE 13-5. Dual slope integrator.

zero crossing detector is merely a high gain operational amplifier, which gives a large positive output when the input voltage is slightly above ground, and a large negative output when the input is slightly below ground.

The square wave (control) input to Q_1 provides the time interval over which integration occurs. This square wave is generated by using decade counters to divide the output frequency of a *clock source*. When the input to Q_1 is negative, the FET is biased *off* and V_i is isolated from the integrator. During this time, the reference current is fed into the Miller integrator circuit. The level of this reference current is

$$I_R = \frac{-V_{z1}}{R_3 + R_4 + R_5}$$

The direction of the current is such that C_1 tends to charge positively on the right-hand side, so that the output of the Miller integrator tends to be positive. However, when the output of the Miller circuit becomes slightly positive, the zero crossing detector generates a large positive output. This biases FET Q_2 *on*, and Q_2 short-circuits C_1. Therefore, at the

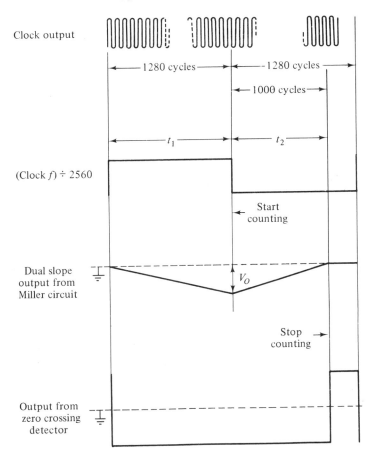

FIGURE 13-6.   Waveforms for a dual slope integrator.

end of the negative half of the square wave input to Q_1, capacitor C_1 is short-circuited and the Miller circuit output is held close to ground level.

Transistor Q_1 switches *on* when the square wave input becomes positive. This action connects voltage V_i to resistance R_5, and provides an input current, $I_i = V_i/R_5$, to the Miller circuit. Capacitor C_1 now charges with negative polarity on the right-hand side and this produces a negative-going output from the Miller circuit (Figure 13-6). Consequently, the zero crossing detector has a large negative output, which biases transistor Q_2 *off*; thus permitting C_1 to charge. The output from the Miller circuit is a linear negative ramp voltage, which continues during the positive portion of the square wave input to Q_1. Since I_i is directly proportional to V_i, the slope of the ramp is also proportional to V_i. Also, the time duration t_1 of the positive input voltage is a constant. This means that the ramp amplitude V_o is directly proportional to V_i.

When the square wave input again becomes negative, Q_1 switches *off* and the reference current I_R commences to flow once more. I_R discharges C_1 so that the Miller circuit output now becomes a positive ramp (Figure 13-6). The positive ramp continues until it arrives at ground level. Then the zero crossing detector provides an output which switches Q_2, *on*, discharges C_1, and holds it in short circuit once again.

The time t_2 for the ramp to discharge to zero now is directly proportional to the input voltage. Time t_2 is measured by starting the counting circuits at the negative-going edge of the square wave input to Q_1, and stopping them at the positive-going edge of the output from the zero crossing detector.

EXAMPLE 13-2

The dual slope integrator in Figure 13-5 has a square wave input with each half-cycle equivalent to 1280 clock pulses (see Figure 13-6). The output frequency from the clock is 200 kHz. If 1000 pulses during time t_2 are to represent an input of $V_i = 1$ V, determine the required level of reference current.

solution

$$I_i = \frac{V_i}{R_5}$$

For $V_i = 1$ V,

$$I_i = \frac{1 \text{ V}}{10 \text{ k}\Omega}$$

$$= 100 \ \mu\text{A}$$

Clock frequency $= 200$ kHz

$$T = \frac{1}{f} = \frac{1}{200 \text{ kHz}}$$

$$= 5\mu\text{s}$$

If t_1 is the time duration of 1280 clock pulses:

$$t_1 = 5 \ \mu\text{s} \times 1280$$

$$= 6.4 \text{ ms}$$

I_i is applied to the integrator input for a time period t_1. Since

$$C = \frac{It}{V}$$

$$\text{Ramp voltage } V_o = \frac{I_i t_1}{C_1} = \frac{100 \ \mu\text{A} \times 6.4 \text{ ms}}{0.1 \ \mu\text{F}} = 6.4 \text{ V}$$

If t_2 is the time duration of 1000 clock pulses,

$$t_2 = 5 \ \mu\text{s} \times 1000$$

$$= 5 \text{ ms}$$

and I_R must discharge C_1 in time period t_2.

$$\therefore \quad I_R = \frac{C_1 V_o}{t_2}$$

$$= \frac{0.1 \ \mu\text{F} \times 6.4 \text{ V}}{5 \text{ ms}}$$

$$= 128 \ \mu\text{A}$$

One of the most important advantages of the dual slope integration method is that small drifts in the clock frequency have little or no effect on the accuracy of measurements. Consider the following example: Let clock frequency $= f$; then time period of one cycle of clock frequency $= T = 1/f$.

The time duration of 1280 clock pulses, t_1, is $1280 \times T$.

$$V_o = \frac{I_i t_1}{C_1}$$

$$= \frac{100\ \mu\mathrm{A} \times 1280T}{C_1}$$

$$t_2 = \frac{C_1 V_o}{I_R} = \frac{C_1}{128\ \mu\mathrm{A}} \times \frac{100\ \mu\mathrm{A} \times 1280T}{C_1}$$

$$= 1000T$$

The number of clock pulses during t_2 is given by

$$\frac{t_2}{\text{Time period of clock pulses}} = \frac{1000T}{T} = 1000$$

It is seen that when the clock frequency drifts, the digital measurement of voltage is unaffected.

13-5 DIGITAL VOLTMETER (DVM)

Figure 13-7 shows a block diagram of a DVM system employing dual slope integration. In this particular system, the clock generator has a frequency of 200 kHz. The 200 kHz is divided by a decade counter and two divide-by-16 counters as shown, giving a frequency of approximately 78 Hz. This (78 Hz) is the square wave which controls the integrator, as explained in Sec. 13-4. The 200 kHz clock signal, the 78 Hz square wave, and the integrator output are all fed to input terminals of an *AND* gate.

The 200 kHz clock output acts as a triggering signal to the counting circuitry when the other two inputs to the *AND* gate are high. This occurs during time t_2, as illustrated in Figure 13-6. Note that the output of the zero crossing detector and the 78 Hz square wave must be inverted before being applied to the *AND* gate. The integrator output (i.e., zero crossing detector output) is also used to reset the counting circuits and to control the latch. The counting circuits are reset at the beginning of time period t_1. Counting commences at the start of t_2. The latch is switched *on* at the end of t_2 in order to set the displays according to the counting circuits.

The range selector is adjusted to suit the input voltage. An input of less than 1 V is applied directly to the integrator, and a decimal point is selected so that the display can indicate a maximum of 0.9999 V. An input voltage between 1 V and 10 V is first potentially divided by 10 and applied to the integrator again as a voltage less than 1 V. In this case the decimal

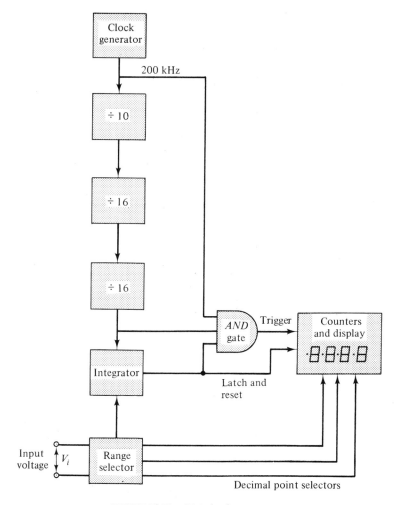

FIGURE 13-7. Digital voltmeter system.

point is selected so that the display can indicate a maximum of 9.999 V. An input voltage between 10 V and 100 V is reduced by a factor of 100 before passing to the integrator. Decimal point selection now allows the meter to indicate a maximum of 99.99 V.

REVIEW QUESTIONS AND PROBLEMS

13-1 Sketch the block diagram of a timing system for a digital frequency meter. Also sketch the output waveforms and carefully explain the operation of the system.

13-2   A crystal-controlled oscillator with an output frequency of 100 MHz is available for use in the timing circuit of a digital voltmeter. Draw a block diagram to show how time intervals of 100 μs, 1 ms, 10 ms, and 100 ms can be obtained.

13-3   Discuss the numerical display obtained with a digital frequency meter when the time base is 1 second and the input frequency is 3 kHz. Explain how the time base and the display must be altered when the frequency goes to 30 kHz, 300 kHz, and 3 MHz.

13-4   A digital frequency meter uses a time base derived by decade counters from a 10 MHz source. Determine the display indication produced by a 1.5 kHz input when the time base uses five decade counters. Also determine the number of decade counters required for the display to read 1500.

13-5   Sketch a block diagram showing JK flip-flops employed as latching circuits. Carefully explain the operation of the latch, and the effect that it has on the numerical display.

13-6   Sketch the complete block diagram of a digital frequency meter. Also sketch the voltage waveforms that occur throughout the system. Explain the operation of the system.

13-7   Sketch the circuit of a *dual slope integrator*. Also sketch the circuit waveforms, and carefully explain the operation of the integrator.

13-8   Show that the accuracy of the dual slope integration method is not affected by small drifts in clock frequency.

13-9   The Miller circuit in a dual slope integrator (as in Figure 13-5) has an input resistance $R_5 = 15$ kΩ and a capacitor $C_1 = 0.1$ μF. The input voltage is 1 V, and each half-cycle of the square wave input is equivalent to 1500 cycles of the clock frequency. The clock frequency is 400 kHz. Determine the required reference current if the 1 V input is to be represented by a count of 1000.

13-10   Sketch the block diagram of a digital voltmeter using dual slope integration. Carefully explain the operation of the system.

CHAPTER 14

Pulse Modulation and Multiplexing

INTRODUCTION

Information can be transmitted, recorded, or otherwise processed in the form of pulses. The technique used to do this may be by *pulse amplitude modulation, pulse duration modulation, pulse position modulation, or pulse code modulation.* Although it is the most complicated of all four methods, pulse code modulation can be the most accurate and most efficient technique. Several separate pulse-modulated signals can be transmitted or recorded on one channel by *time division multiplexing.* In this process, pulse signals are inserted in the spaces between other pulse signals. The circuits involved in pulse modulation and demodulation, and in coding and decoding time-multiplexed information are, in general, those that have been studied in previous chapters.

14-1   TYPES OF PULSE MODULATION

The instantaneous amplitude of a signal may be measured (or sampled) at regular intervals, and the measured amplitudes converted to pulses. The pulses may then be transmitted, recorded, or otherwise processed. A low-frequency alternating signal and four types of pulse modulation by which the signal may be represented are illustrated in Figure 14-1.

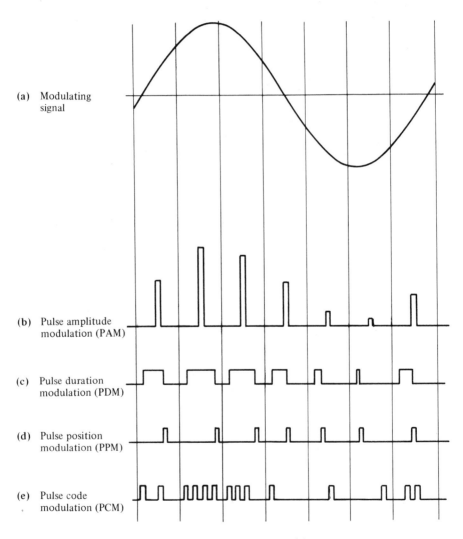

FIGURE 14-1.   Types of pulse modulation.

Pulse amplitude modulation (PAM) is the simplest type of pulse modulation. As the name implies, the amplitude of each pulse is made proportional to the instantaneous amplitude of the modulating signal [Figure 14-1(b)]. The largest pulse represents the greatest positive signal amplitude sampled, while the smallest pulse represents the largest negative sample. The time duration of each pulse may be quite short, and the time interval between pulses may be relatively long. If a radio frequency is pulse-amplitude-modulated instead of simply being amplitude-modulated, much less power is required for the transmission of information because the transmitter actually is switched *off* between pulses. This is one advantage of pulse modulation.

In *pulse duration modulation* (PDM), also termed *pulse width modulation*, the pulses have a constant amplitude and a variable time duration. The time duration (or width) of each pulse is proportional to the instantaneous amplitude of the modulating signal [Figure 14-1(c)]. In this case, the narrowest pulse represents the most negative sample of the original signal, and the widest pulse represents the largest positive sample. When PDM is applied to radio transmission, the carrier frequency has a constant amplitude, and the transmitter *on*-time is carefully controlled. In some circumstances PDM can be more accurate than PAM. One example of this is in magnetic tape recording, where pulse widths can be recorded and reproduced with less error than pulse amplitudes.

Pulse position modulation (PPM) [Figure 14-1(d)] is more efficient than PAM or PDM for radio transmission. In PPM, all pulses have the same constant amplitude and narrow pulse width. The position in time of the pulses is made to vary in proportion to the amplitude of the modulating signal. Note that in Figure 14-1(d) each PPM pulse occurs just at the end of a PDM pulse in Figure 14-1(c). Thus, the pulses near the right-hand side of the sampling time period represent the largest positive signal sample, and those toward the left-hand side correspond to the most negative samples of the original signal. PPM uses less power than PDM and, essentially, has all the advantages of PDM. One disadvantage of PPM is that the demodulation process to recover the original signal is more difficult than with PDM.

Pulse code modulation (PCM), illustrated in Figure 14-1(e), is the most complicated type of pulse modulation. However, PCM can be the most accurate and the most efficient of the four methods. In certain circumstances, it may be the only type of pulse modulation that can be employed. In PCM, each amplitude sample of the original modulating signal is converted to a *binary number* (see Chapter 12). The binary number is then represented by a group of pulses, the presence of a pulse indicating *1* and the absence of a pulse indicating *0*. The *four-bit* code illustrated in Figure 14-1(e) can represent only sixteen discrete levels of signal amplitude.

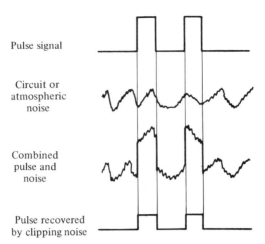

Pulse signal

Circuit or
atmospheric
noise

Combined
pulse and
noise

Pulse recovered
by clipping noise

FIGURE 14-2.   Effect of noise on pulse signals.

Thus it is far from accurate. Accuracy can be improved by increasing the number of *bits* (*i.e.*, pulses) employed. A seven-bit code, for example, can represent 128 discrete levels of signal amplitude, or to an accuracy of better than 1%. The process of converting the signal to standard amplitudes which are to be represented by the binary code is termed *quantizing*, and the error introduced by this process is referred to as the *quantizing error*.

For all four pulse modulation methods the sampling frequency is determined by the highest signal frequency that must be processed. It can be shown that if samples are taken at a rate greater than twice the signal frequency, then the original signal can be recovered. However, in practice it is normal to sample at a minimum rate of about ten times the highest signal frequency. For audio, voice transmission, for example, with a "high" frequency of 3 kH, the sampling frequency might be 30 kHz.

Another major advantage of pulse modulation is illustrated in Figure 14-2. When radio signals are very weak, they may be almost completely lost in circuit or atmospheric noise. If the modulation method is PDM, PPM, or PCM, the signals can be recovered simply by clipping off the noise. For this, PCM gives the best results, since it is necessary only to determine whether each pulse is present or absent.

In the following sections, modulation and demodulation methods are explained for PAM, PDM, and PPM. Before PCM techniques can be understood, *time division multiplexing* methods must be studied.

14-2   PAM MODULATION AND DEMODULATION

A process for producing a pulse amplitude modulated waveform is illustrated in Figure 14-3. The block diagram of Figure 14-3(a) shows a

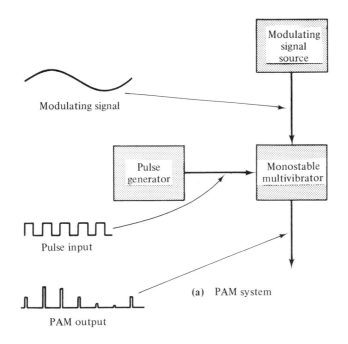

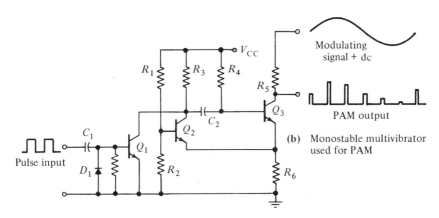

FIGURE 14-3.   PAM modulating system and use of a monostable multivibrator for PAM.

pulse generator triggering a monostable multivibrator at a sampling frequency. The output pulses from the multivibrator are made to increase and decrease in amplitude by the modulating signal. Figure 14-3(b) shows the circuit of a monostable multivibrator (Chapter 8) with its load resistance R_5 supplied from the modulating signal source. When Q_3 is *on*, the output voltage is the saturation level of Q_3 collector. When Q_3 is switched

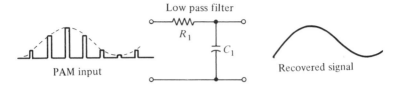

FIGURE 14-4.   PAM demodulation.

off for the pulse time, the output voltage (*i.e.*, the pulse amplitude) is equal to the modulating signal level. The actual voltage applied as a supply to R_5 must have a dc component as well as the ac modulating signal. This is necessary to ensure correct operation of Q_3.

Demodulation of PAM is accomplished simply by passing the amplitude-modulated pulses through a low-pass filter. This process is illustrated in Figure 14-4. The PAM waveform consists of the fundamental modulating frequency, and a number of high-frequency components which give the pulses their shape. The resistance R_1 and capacitance C_1 shown in Figure 14-4 form a potential divider. At low frequencies, the impedance of C_1 is very much larger than R_1. Consequently, low-frequency signals (*i.e.*, the fundamental) are passed with very little attenuation. At high frequencies, the impedance of C_1 becomes quite small, and the signals experience severe attenuation. Thus, the filter output is the signal frequency, with perhaps a very small pulse frequency component. If necessary, more than one stage of filtering can be employed to remove the pulse frequency completely.

14-3 PDM MODULATION AND DEMODULATION

In pulse duration modulation, the signal samples must be converted to pulses which have a time duration directly proportional to the amplitude of the samples. One method of producing PDM, shown in Figure 14-5, uses a free-running ramp generator and a *voltage comparator*. The operation of the voltage comparator is explained in Sec. 6-8.

Consider the circuit and waveforms illustrated in Figure 14-5. The modulating signal is capacitor-coupled via C_1 to the noninverting input terminal of the comparator. Therefore, the voltage at that terminal is a dc level (provided by R_1 and R_2) with the ac signal superimposed. The free-running ramp generator produces a ramp waveform output at the desired sampling frequency. This is directly coupled to the inverting input terminal of the comparator. When the ramp is at its zero level, the comparator inverting input terminal voltage is below the level at the noninverting input terminal. In this condition the comparator output is at

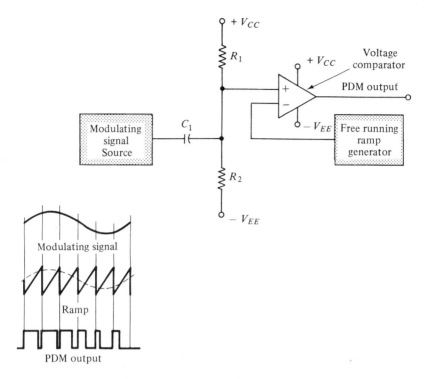

FIGURE 14-5.   PDM modulating system.

its extreme positive voltage level. The ramp voltage grows linearly and eventually becomes equal to the voltage at the noninverting terminal of the comparator. When this occurs, the comparator output switches rapidly from its extreme positive level to its extreme negative voltage level. When the ramp voltage returns to zero, the inverting input voltage is once again below the level of the noninverting input, and the comparator output returns to its extreme positive voltage level.

It is seen that the output from the comparator is a series of positive pulses. Each pulse commences at the instant the ramp waveform returns to zero volts, and ends when the ramp level coincides with the signal voltage. When the signal is at its highest level, the ramp takes its longest time to reach equality with the noninverting input voltage. Therefore, the output pulses at this time are of the longest duration. At the instant that the signal is at its lowest level, the ramp takes the shortest time to arrive at the same voltage as that at the noninverting terminal. Consequently, the width of the output pulses is a minimum at this instant.

Demodulation of PDM waves can be accomplished by first converting each duration-modulated pulse into an amplitude-modulated pulse. Then, filtering can be employed to recover the original modulating signal. The

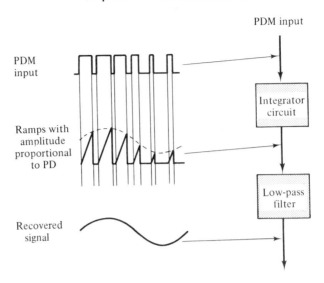

FIGURE 14-6.   **PDM demodulating system.**

block diagram and waveforms for such a PDM demodulation system are shown in Figure 14-6. The PDM wave is applied to an integrator which generates a ramp-type output. The integrator output ramp always increases linearly at the same rate (*i.e.*, for constant amplitude pulses). The ramp commences at the start of each input pulse and finishes at the end of the pulse. Consequently, the ramp peak value is proportional to the pulse width. Since the pulse widths are made proportional to the instantaneous samples of the original signal voltage, the ramp peaks represent amplitude samples of the original signal. The integrator output waveform now is fed to the low-pass filter, which removes the high-frequency components and passes the low-frequency signal waveform.

A Miller integrator circuit (Figure 7-14) is suitable for use with the PDM demodulation system described above. The circuit, as shown, requires negative input pulses. Thus the positive PDM waveform in Figure 14-6 would first have to be passed through an inverter circuit [*e.g.*, Figures 5-10(b) or (c)] before being applied to the integrator.

14-4 PPM MODULATION AND DEMODULATION

The simplest form of modulation process for pulse position modulation is a PDM system with the addition of a monostable multivibrator (see Figure 14-7). The monostable is arranged so that it is triggered by the trailing edges of the PDM pulses. Thus, the monostable output is a series of

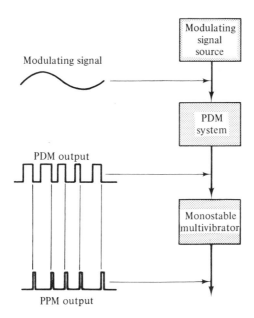

Modulating signal

PDM output

PPM output

FIGURE 14-7.   PPM modulating system.

constant-width constant-amplitude pulses which vary in position according to the original signal amplitude.

For demodulation of PPM, a PDM waveform is first constructed by triggering an *RS flip-flop*, as shown in Figure 14-8. The flip-flop is triggered into its *set* condition by the leading edges of a square wave which must be synchronized with the original signal source. Synchronization is necessary so that the leading edges of the square wave coincide with the leading edges of the PDM wave that was employed to generate the PPM pulses (Figure 14-7). The flip-flop in Figure 14-8 is *reset* by the leading edge of the PPM pulses. The output of the flip-flop is now a PDM wave which may be demodulated by the process illustrated in Figure 14-6.

One of the most difficult requirements for PPM demodulation is synchronization of the square wave for triggering the flip-flop. Several alternatives are available. PPM pulses may be generated at both the leading and lagging edges of the PDM waveform in Figure 14-7. For demodulation, the leading edge pulses are applied to the *set* terminal of the RS flip-flop. In this case, a square wave generator is not required. However, some method of identifying the leading edge pulses must be employed. A more efficient system is to include periodic synchronizing pulses. For example, every fiftieth pulse might be a synchronizing pulse which corrects any frequency drift in the square wave generator. Again, some means of identifying the synchronizing pulses is essential. This could be done by

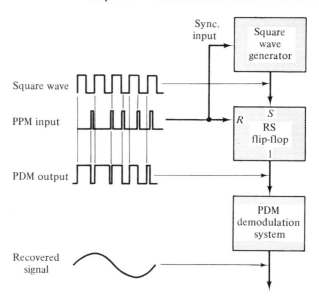

FIGURE 14-8. PPM demodulation.

making all synchronizing pulses negative while the other pulses are positive. Alternatively, the synchronizing pulses could be made larger or wider than the other PPM pulses. Perhaps the best method of identifying a sychronizing pulse is to precede it with a longer-than-normal space. Methods for generating such a space during the modulation process and for identifying it during demodulation are similar to those employed in *time division multiplexing*, the subject of the next section.

14-5   TIME DIVISION MULTIPLEXING

14-5.1   TDM Waveforms

Suppose a 2.5 kHz signal is sampled ten times in every cycle. The samples occur at time intervals of one-tenth of the time period of the waveform, that is, at 40 μs intervals. If PDM is employed, and the maximum pulse width is made less than 10 μs, a 30 μs time interval (or space) is left between pulses. If other signals are sampled at the same rate, the additional pulse samples might be included in the 30 μs space. This process, known as *time division multiplexing* (TDM), is illustrated in Figure 14-9.

    Channel 1 in Figure 14-9 is a series of PDM samples, with the first pulse shown commencing at time t_1. Channel 2 is another series of PDM

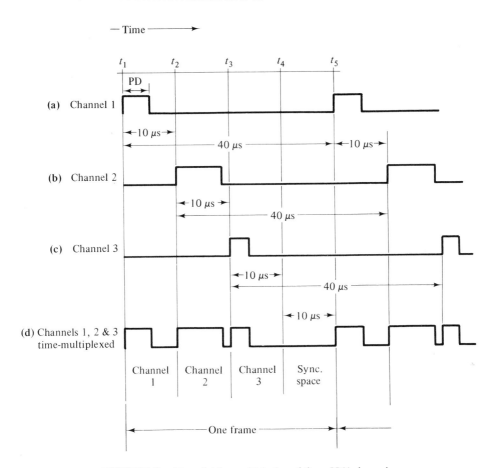

FIGURE 14-9.   Time division multiplexing of three PDM channels.

pulses with the first pulse shown commencing at t_2, 10 μs after t_1. The second pulse in channel 2 occurs 40 μs after t_2 (10 μs after t_5). For channel 3, the first pulse shown starts at t_3, which is 20 μs after t_1 and 10 μs after t_2. As in the case of the other channels, there is a 40 μs time interval between commencement of each pulse in channel 3. When the pulses are time-multiplexed, as shown in Figure 14-9(d), three channels of information are contained in the waveform. This waveform now may be recorded on a single magnetic track, transmitted on a single radio frequency, or otherwise processed.

Each channel in the time-multiplexed waveform [Figure 14-9(d)] is allocated a 10 μs time period. Therefore, the maximum pulse width must be less than 10 μs, so that the end of one pulse can be distinguished from the beginning of the next. At each sampling, the three channels occupy a total time period of 30 μs. This leaves a 10 μs time interval, or one unoccupied channel space, between the end of the three pulses and commencement of the next three. This time period is deliberately left clear so that it may be used for synchronization in the demodulation process. Channel 1 can easily be identified (during demodulation) as the first information pulse after the *sync. space*. Channels 2 and 3 can then be identified in sequence after channel 1. The series of three (or more) information pulses together with the synchronizing space usually is termed *one frame* of the TDM waveform.

PDM is not the only pulse modulation process that lends itself to time division multiplexing. Figure 14-10 shows typical waveforms of PAM, PPM, and PCM signals in time-multiplexed form. Note that the PPM demodulation synchronization problem discussed in Sec. 14-4 is solved by the TDM synchronization arrangements. Instead of using a space between pulses, synchronization may be accomplished by means of a negative pulse, or, perhaps, a wider or taller pulse than normal. The number of channels that may be time-multiplexed is by no means limited to three.

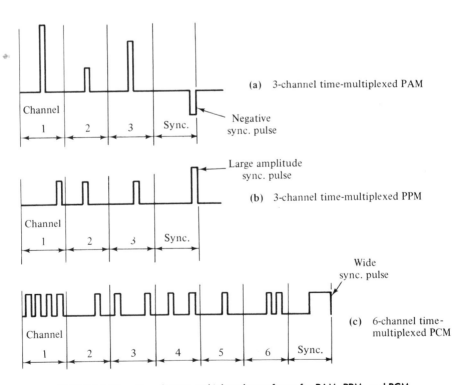

FIGURE 14-10.   Time division multiplexed waveforms for PAM, PPM, and PCM.

The maximum number of channels is determined by the highest frequency to be sampled and the time period that must be allocated to each channel.

14-5.2   Ring Counter

A *ring counter* is the circuit employed in a TDM coding system to select the signals to be sampled in the correct repetitive sequence. Triggered by input pulses, the circuit switches through a number of states equal to one more than the number of TDM channels. For a three-channel system, the ring counter must have four states, that is, three channels plus the synchronizing space. Ring counters usually are constructed of flip-flops and diode gates. The binary-to-decimal conversion system shown in Figure 12-15 could be employed as a ten-state ring counter. In this case, the outputs (to the transistor gates) would control nine signal channels and a synchronizing space.

Figure 14-11 shows a four-state ring counter made up of two flip-flops and a diode matrix. The operation of the circuit is similar to the binary-to-decimal conversion system explained in Chapter 12. In the initial condition, Q_1 and Q_3 are *off*, and the output voltage at their collectors is *high* and is represented as logic *1*. Q_2 and Q_4 are *on*, so their *low* collector voltage level is represented as logic *0*. This means that the voltage levels at the cathodes of D_1 and D_2 are *high*, and that the output of terminal 1 is *high*. All other output terminals have at least one diode biased *on*, holding the output voltage *low*. When the first toggle pulse is applied, the states of the flip-flops change to Q_4 on Q_3 *off*, Q_2 *off*, Q_1 *on*. The cathode of D_2 has a low voltage applied so that the output of terminal 1 is *low*. Also D_3 and D_4 have *high* voltages at their cathodes, which produces a *high* output at terminal 2. While the input count remains at *1*, diodes D_5, D_6, and D_7 have *low* cathode voltages and thus terminals 3 and 4 are *low*. Continuing through input pulse counts 2 and 3, it is seen that output terminals 3 and 4 switch *high* in turn, while all other outputs become *low*. After the count of four, the next input pulse switches the circuit back to *high* at output terminal 1. Then the sequence is repeated. The *ring counter* derives its name from the fact that it toggles in a repetitive or *ring* sequence from *state 1 to state 4* and back through *state 1* again for the four-stage ring counter. A synchronizing input terminal also can be provided as was explained in Chapter 12, so that a *sync. pulse* can set the circuit in its initial state.

The voltage waveforms in Figure 14-11 help to illustrate the operation of the ring counter. It is seen that in the initial state, only output terminal 1 is *high*. When the first toggle pulse is applied, output terminal 2 becomes *high*, and 1 becomes *low*. The third and fourth toggle pulses, respectively, switch outputs 3 and 4 to *high*, while all others go to *low*. On receipt of the

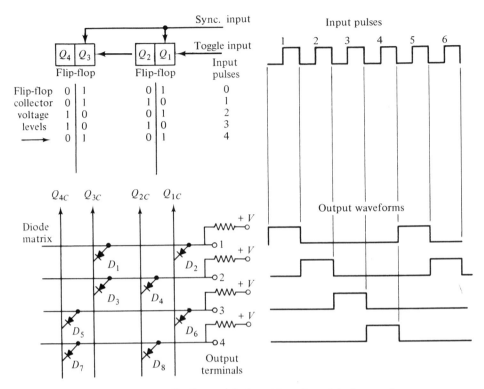

FIGURE 14-11. Flip-flop and diode matrix as four-state ring counter.

fifth pulse, output 1 becomes *high* again, and the sequence is repeated as the input pulses continue.

14-5.3 TDM Coding System

The block diagram and waveforms for the time-multiplexing of three signals are shown in Figure 14-12. A time division multiplexing system usually is referred to as a TDM *coding system*. The system for separating the waveform into individual channels is termed a TDM *decoding system*. If pulse modulation is performed concurrently with time-multiplexing, a single modulation system can be used for all channels.

In Figure 14-12, a clock generator produces a square waveform to toggle the ring counter at the desired frequency. The ring counter outputs switch the sampling gates *on* and *off* in the correct sequence. (Sampling gates are discussed in Chapter 11.) When one output from the ring counter is *high*, all others are *low*. Thus, one sampling gate is *on*, and all others are *off*. The time periods during which a sampling gate is in the *on* and *off*

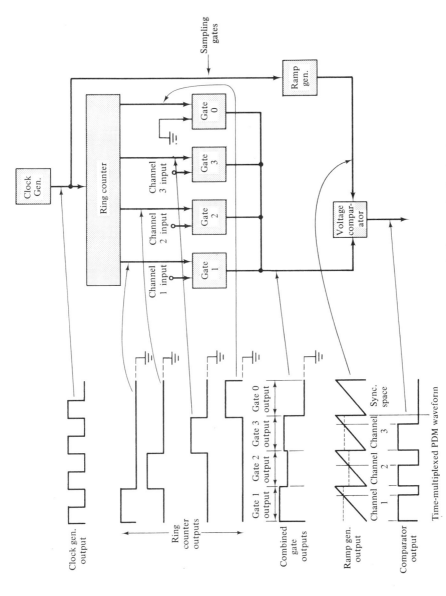

FIGURE 14-12. Three-channel time division multiplexing and pulse duration modulation system.

states is determined by the clock generator frequency and the number of channels. For example, if the clock generator output is a 1 kHz square wave, each gate is *on* for 1 ms. For the three-channel system shown, each gate is *off* for three time periods of the clock generator (*i.e.*, for 3 ms).

Each signal to be sampled is applied to the input terminal of a sampling gate. The input of gate 0, the synchronizing gate, is grounded, so that during the sync. time its output is zero volts. The output of all four gates is "commoned." Thus, as each gate switches *on* in turn, amplitude samples of the signal waveforms are time-multiplexed into a single waveform. The combined output of the gates (see Figure 14-12) is in the form of a pulse-amplitude-modulated waveform with no spaces between pulses, and with a sync. space at the end of each set of samples.

The waveform from the gates is converted to PDM by applying it to one input of a voltage comparator. The other input terminal of the comparator has a ramp generator output applied to it, as shown in Figure 14-12. The ramp generator has a linear output and is triggered by the negative-going edges of the clock generator output, so that the ramp commences at the beginning of the output from each sampling gate. The voltage comparator output becomes high when the ramp output becomes zero. When the ramp amplitude is equal to the amplitude of the signal being sampled, the comparator output goes to zero. (The process of converting amplitude samples to PDM was explained in Sec. 14-3.) During the sync. space, the comparator input (from the gates) remains at zero; thus the output level from the comparator does not switch positively when the ramp goes to zero. The output waveform from the system is seen to be time-multiplexed duration-modulated pulses with intervening sync. spaces.

14-5.4 TDM Decoding System

Before demodulation, time-multiplexed signals must be *decoded*, or separated into individual channels. Figure 14-13 shows the block diagram and waveforms of a system for decoding three-channel time-multiplexed PDM signals.

The waveform to be decoded is applied simultaneously to an integrator circuit, the toggle input of a ring counter, and to one input on each of three *AND* gates. The function of the integrator, and the Schmitt trigger circuit which follows it, is to detect the sync. space in the TDM waveform. A suitable integrator circuit for this situation is the ramp generator shown in Figure 7-3(b). During the time that the input pulses are applied, the ramp generator output remains at zero. When the pulses are absent the ramp output grows linearly, then rapidly returns to zero again at commencement of the next input pulse. Thus, the integrator output is a series of small ramps generated during the space time. During the synchronizing

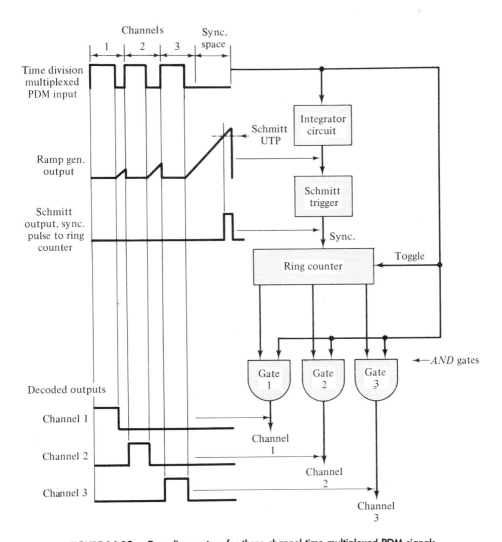

FIGURE 14-13.   Decoding system for three-channel time-multiplexed PDM signals.

space, the integrator generates a larger-than-usual output. When this output arrives at the upper trigger point of the Schmitt circuit, the Schmitt produces an output pulse which sets the ring counter to its synchronized state. The next input pulse (*i.e.*, from channel 1) toggles the ring counter to its channel 1 state.

In the channel 1 state, the ring counter provides a positive output to *AND* gate 1, and a zero output to gates 2 and 3. At this time, channel 1 pulse is present at one input of all three *AND* gates. Only gate 1 has a positive voltage at both input terminals. Therefore, only gate 1 produces an output during the application of the pulse from channel 1. The output pulse from gate 1 commences at the beginning of channel 1, and ends at the end of the channel 1 PDM pulse. At the commencement of channel 2 (*i.e.*, on receipt of the next positive-going pulse edge) the ring counter toggles to its channel 2 state. Therefore, *AND* gate 2 now has positive voltage levels at both input terminals, while *AND* gates 1 and 3 have zero levels applied from the ring counter. Thus, gate 2 produces the channel 2 PDM pulse at its output terminals. Similarly, at commencement of channel 3, the ring counter toggles to its channel 3 state. *AND* gate 3 output then becomes positive for the duration of the PDM pulse in the channel 3 portion of the input waveform. It is seen that the decoding system separates the TDM wave into the individual channels. Now, each channel can be demodulated to recover the original modulating signals.

14-6 PCM MODULATION AND DEMODULATION

14-6.1 PCM Modulation System

PCM signals are essentially time division multiplexed pulses; therefore, modulation and demodulation techniques for PCM are similar to TDM methods. The block diagram of a PCM modulation system is shown in Figure 14-14, and the waveforms for the system are illustrated in Figure 14-15.

The PCM system in Figure 14-14 employs a *shift register*. This is simply a cascade of flip-flops similar to the arrangement illustrated in Figures 12-1 and 12-2. The shift register is toggled by input pulses, and counts the pulses in binary form. An output is taken from one collector of each flip-flop. The five-stage shift register can count to binary *11111*, which is equivalent to decimal *31*.

The input signal voltage which is to be converted to PCM (V_s in Fig. 14-14) is applied to one input of a voltage comparator, and a ramp generator output is applied to the other input of the comparator. This is similar to the PDM method, in which the signal amplitude sample is converted to a time period. In Figure 14-15, the comparator output is shown as having a time duration t. To convert this time period to a five-bit binary number, the shift register is toggled by the clock generator during the time period. Once the shift register has settled at the binary equivalent of the signal sample, the ring counter is triggered to *read* the state of the shift register and produce a PCM output.

Suppose the clock generator frequency f_1 is 32 kHz. This frequency is divided by a factor of 32 to produce $f_2 = (f_1/32) = 1$ kHz (see Figure

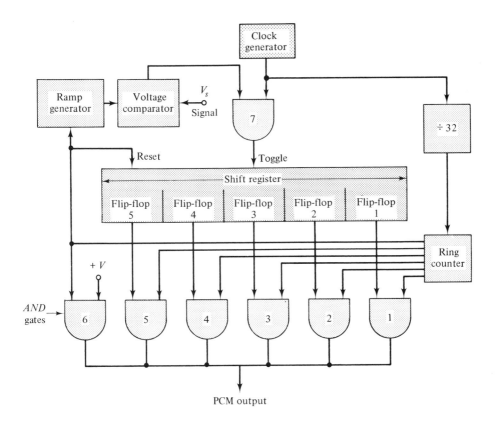

FIGURE 14-14.   PCM modulating system.

14-15). (Frequency dividing techniques are discussed in Chapter 13.) The time period of f_2 is 1 ms, and this is used to trigger the ring counter. As shown by the waveforms in Figure 14-15, the ring counter provides a 1 ms pulse to *AND* gate 6. This pulse also goes to the ramp generator and to the *reset* input terminal of the shift register. Therefore, at commencement of time period t_1 (Figure 14-15) the shift register is set to its zero condition, and the ramp generator output is triggered from its maximum output voltage level to zero. When the ramp generator output becomes zero, it causes the output of the voltage comparator to switch from zero to its maximum voltage level. Thus, the input (from the comparator) to *AND* gate 7 is positive, and the clock generator pulses (*i.e.*, *not* divided by 32) pass through to toggle the shift register.

At the commencement of time period t_1, the ramp voltage also starts to grow linearly from the zero level. When the ramp amplitude becomes equal to the instantaneous amplitude of the signal voltage V_s, the comparator output switches to zero once again. This means that no more

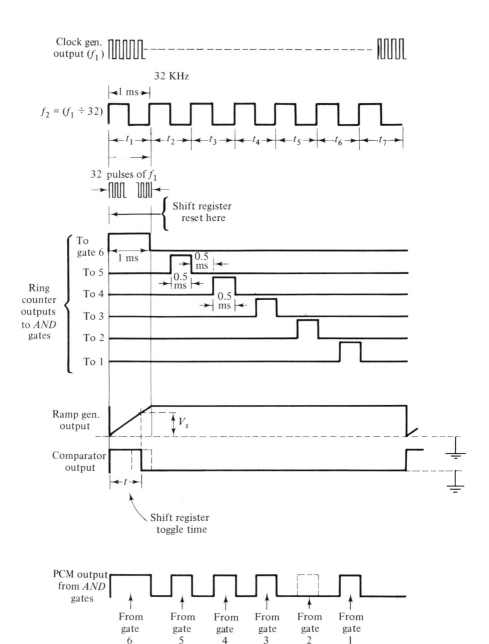

FIGURE 14-15. Waveforms for PCM modulating system.

clock generator pulses can pass through AND gate 7. At this time, the state of the shift register represents the binary equivalent of the signal voltage amplitude. Since no more toggle pulses arrive, the shift register state remains constant.

The 1 ms ring counter output pulse that is applied to the ramp generator during time t_1 is also applied to AND gate 6. Since the other input to the gate is at $+V$, the output of gate 6 is a positive 1 ms pulse. During time period t_2, the ring counter provides a 0.5 ms pulse to AND gate 5. The other input to AND gate 5 is derived from FF5 in the shift register. If the output of FF5 is *high*, gate 5 produces a 0.5 ms output pulse, as shown. If the flip-flop output is *low*, there is no pulse present at this point in the PCM output waveform. Through time periods t_3, t_4, t_5, and t_6, the ring counter switches *on* gates 4, 3, 2, and 1, respectively, to sample the shift register voltage levels at each flip-flop. Thus the PCM output waveform is produced with pulses that represent the binary equivalent of the signal sample as measured during t_1. At the end of t_6, a gap (t_7) is generated by the ring counter holding gates 1 to 6 *off*, then the long synchronizing pulse t_1 commences again, and the sampling and coding process is repeated.

14-6.2 PCM Decoding and Demodulation System

For decoding and demodulation, PCM signals are first processed through a TDM decoding system to separate the bits into different channels. The process, illustrated in Figure 14-16(a), shows a TDM decoding system with five output channels for a five-bit PCM system. Several techniques are available for converting the decoded PCM back to analog voltage samples. The method illustrated is termed *digital-to-analog conversion by weighted resistors*. The operational amplifier, connected as shown, is termed a *summing amplifier*.

In Figure 14-16(b) it may be seen that each pulse in a five-bit PCM code has a binary and a decimal equivalent. The decimal equivalent of bit 5 is 16 and that of bit 4 is 8, etc. The resistors R_1 to R_5 are *weighted* (or selected) to have values inversely proportional to the decimal equivalent of the bit number applied to them. Thus, if $R_1 = 16$ kΩ for bit 1, which has a decimal equivalent of 1, then $R_5 = 1$ kΩ for bit 5 with a decimal equivalent of 16. Similarly, $R_2 = 8$ kΩ for bit 2 (the decimal equivalent is 2), and $R_4 = 2$ kΩ for bit 4 (the decimal equivalent is 8). Note also that $R_6 = 1.6$ kΩ.

Consider the effect on the summing amplifier when bit 5 is present, and all other bits are absent. Let the amplitude of all bits be $V = 1$ V. Recall, from the previous study of operational amplifiers in Chapters 5, 6, and 7, that with the noninverting terminal grounded the inverting terminal

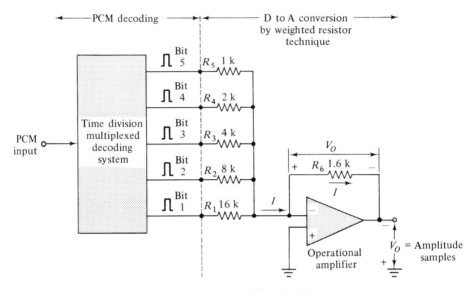

(a)   PCM decoding and demodulation

	Binary number represented by bits	Decimal equivalent
Bit 5	1 0 0 0 0	16
Bit 4	1 0 0 0	8
Bit 3	1 0 0	4
Bit 2	1 0	2
Bit 1	1	1

(b)   Binary and decimal equivalents of bits in 5-bit PCM

FIGURE 14-16.   PCM decoding and demodulation system.

is always very close to ground potential. Therefore, with the input voltage applied at one end of R_5 and a *virtual ground* at the other end, the amplifier input current is

$$I = \frac{V_i}{R_5}$$

Also in previous studies it was shown that the actual current flow into the amplifier terminal is near zero. Consequently, all of the input current effectively flows through R_6. Since one end of R_6 is at the virtual ground input terminal and the other end is at the output terminal, the output voltage is the voltage drop across R_6. Thus, the output voltage is

$$V_o = IR_6 = V_i \times (R_6/R_5)$$
$$= 1\ \text{V} \times (1.6\ \text{k}\Omega/1\ \text{k}\Omega)$$
$$= 1.6\ \text{V}$$

Thus, bit 5 is converted to 1.6 V and since the decimal equivalent of bit 5 is 16, this seems to make sense.

When only bit 1 is present, the output becomes

$$V_o = V_i \times (R_6/R_1)$$
$$= 1\ \text{V} \times (1.6\ \text{k}\Omega/16\ \text{k}\Omega)$$
$$= 0.1\ \text{V}$$

This is one-sixteenth of the output produced from bit 5. Thus the two outputs are in correct proportion, since bit 5 has a decimal equivalent of 16, and bit 1 has a decimal equivalent of 1.

The combined output produced when bits 5 and 1 are present is

$$V_o = V_i(R_6/R_5) + V_i(R_6/R_1)$$
$$= V_i R_6 \times (1/R_5 + 1/R_1)$$
$$= 1\ \text{V} \times 1.6\ \text{k}\Omega \times (1/1\ \text{k}\Omega + 1/16\ \text{k}\Omega)$$
$$= 1.7\ \text{V}$$

The decimal equivalent of 10001 (i.e., bit 5 and bit 1) is 17; therefore, the *summing amplifier* has summed the bits and converted the input voltages from digital form to analog form. Any combination of input bits may be considered, and the summing amplifier output calculated. It is always found that the amplifier output is directly proportional to the decimal equivalent of the binary number represented by the bits. After conversion to amplitude samples, low-pass filtering can be employed to recover the original signal. Note that the output voltage samples from the summing amplifier are negative. These can be converted to positive voltage levels by the use of another inverting amplifier.

REVIEW QUESTIONS AND PROBLEMS

14-1 Sketch waveforms to illustrate the various types of pulse modulation. Identify and briefly explain each type of modulation.

14-2   Discuss the performance of the various types of pulse modulation with respect to noise.

14-3   Show how a monostable multivibrator may be employed for pulse amplitude modulation. Briefly explain.

14-4   Sketch a block diagram for a PAM modulating system. Show the waveforms, and explain the system.

14-5   Sketch and explain the process for demodulating PAM signals.

14-6   Draw a block diagram for a PDM modulating system. Show the circuit of the voltage comparator, and the system waveforms. Explain the modulation process.

14-7   Sketch a system block diagram and waveform for demodulation of PDM signals. Explain the process.

14-8   Using illustrations, explain the process of producing a PPM waveform.

14-9   Explain the process of demodulating a PPM waveform. Sketch the appropriate block diagram and voltage waveforms. Also, discuss the *synchronizing* problem which occurs with PPM and state the possible solutions.

14-10   Explain *time division multiplexing* and discuss its advantages. Sketch waveforms for time-multiplexed signals using PAM, PDM, PPM, and PCM.

14-11   Draw a block diagram and diode circuit to show how three flip-flops and a diode matrix can be employed as an eight-state ring counter. Also sketch the input and output waveforms, and explain how the ring counter functions.

14-12   Draw a block diagram for a four-channel TDM system which includes pulse duration modulation. Show the waveforms throughout, and explain the operation of the system.

14-13   Draw a block diagram for a decoding system for four-channel time-multiplexed PDM signals. Show the system waveforms and explain the operation of the system.

14-14   A signal having a maximum frequency of 5 kHz is to be pulse-code -modulated with not more than 2% quantizing error. Determine the number of pulses required to represent each sample, and the number of samples required per second. Explain.

14-15   Draw a block diagram of a PCM modulating system that would be suitable for the waveform described in problem 14-14. Sketch appropriate waveforms and explain the system process.

14-16   Explain the process of decoding and demodulating a 5-bit PCM signal. Sketch the appropriate circuitry and identify the level of output for each input pulse.

Data Sheets

TYPES 1N914, 1N914A, 1N914B, 1N915, 1N916, 1N916A, 1N916B and 1N917
DIFFUSED SILICON SWITCHING DIODES

- **Extremely Stable and Reliable High-Speed Diodes**

mechanical data

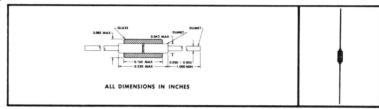

ALL DIMENSIONS IN INCHES

absolute maximum ratings at 25°C ambient temperature (unless otherwise noted)

		1N914	1N914A	1N914B	1N915	1N916	1N916A	1N916B	1N917	Unit
V_R	Reverse Voltage at − 65 to + 150°C	75	75	75	50	75	75	75	30	v
I_o	Average Rectified Fwd. Current	75	75	75	75	75	75	75	50	ma
I_o	Average Rectified Fwd. Current at + 150°C	10	10	10	10	10	10	10	10	ma
i_f	Recurrent Peak Fwd. Current	225	225	225	225	225	225	225	150	ma
$i_{f(surge)}$	Surge Current, 1 sec	500	500	500	500	500	500	500	300	ma
P	Power Dissipation	250	250	250	250	250	250	250	250	mw
T_A	Operating Temperature Range	− 65 to + 175								°C
T_{stg}	Storage Temperature Range	200								°C

maximum electrical characteristics at 25°C ambient temperature (unless otherwise noted)

		1N914	1N914A	1N914B	1N915	1N916	1N916A	1N916B	1N917	Unit
BV_R	Min Breakdown Voltage at 100 μa	100	100	100	65	100	100	100	40	v
I_R	Reverse Current at V_R	5	5	5	5	5	5	5		μa
I_R	Reverse Current at − 20 v	0.025	0.025	0.025		0.025	0.025	0.025		μa
I_R	Reverse Current at − 20 v at 100°C	3	3	3	5	3	3	3	25	μa
I_R	Reverse Current at − 20 v at + 150°C	50	50	50		50	50	50		μa
I_R	Reverse Current at − 10 v				0.025				0.05	μa
I_R	Reverse Current at − 10 v at 125°C									μa
I_F	Min Fwd Current at $V_F = 1$ v	10	20	100	50	10	20	30	10	ma
V_F	at 250 μa								0.64	v
V_F	at 1.5 ma								0.74	v
V_F	at 3.5 ma								0.83	v
V_F	at 5 ma		0.72	0.73			0.73			v
V_F	Min at 5 ma				0.60					v
C	Capacitance at $V_R = 0$	4	4	4	4	2	2	2	2.5	pf

operating characteristics at 25°C ambient temperature (unless otherwise noted)

		1N914	1N914A	1N914B	1N915	1N916	1N916A	1N916B	1N917	Unit
t_{rr}	Max Reverse Recovery Time	**4	**4	**4	°10	**4	**4	**4	°3	nsec
		°8	°8	°8		°8	°8	°8		nsec
V_f	Fwd Recovery Voltage (50 ma Peak Sq. wave, 0.1 μsec pulse width, 10 nsec rise time, 5 kc to 100 kc rep. rate)	2.5	2.5	2.5	2.5	2.5	2.5	2.5	2.5	v

* Trademark of Texas Instruments

° Lumatron (10 ma I_F, 10 ma I_R, recover to 1 ma)

** EG&G (10 ma I_F, 6v V_R, recover to 1 ma)

*Courtesy of Texas Instruments, Incorporated

APPENDIX 1-2*

1N4001thru 1N4007

CATHODE

CASE 59

$$I_O = 1 \text{ A}$$
$$V_R - \text{to } 1000 \text{ V}$$

Low-current, passivated silicon rectifiers in subminiature void-free, flame-proof silicone polymer case. Designed to operate under military environmental conditions.

MAXIMUM RATINGS (At 60 cps Sinusoidal, Input, Resistive or Inductive Load)

Rating	Symbol	1N4001	1N4002	1N4003	1N4004	1N4005	1N4006	1N4007	Unit
Peak Repetitive Reverse Voltage DC Blocking Voltage	$V_{RM(rep)}$ V_R	50	100	200	400	600	800	1000	Volts
RMS Reverse Voltage	V_r	35	70	140	280	420	560	700	Volts
Average Half-Wave Rectified Forward Current (75°C Ambient) (100°C Ambient)	I_O	1000 750	1000 750	1000 750	1000 750	1000 750	1000 750	1000 750	mA mA
Peak Surge Current 25°C (1/2 Cycle Surge, 60 cps) Peak Repetitive Forward Current	$I_{FM(surge)}$ $I_{FM(rep)}$	30 10	30 10	30 10	30 10	30 10	30 10	30 10	Amps Amps
Operating and Storage Temperature Range	T_J, T_{stg}	-65 to + 175							°C

ELECTRICAL CHARACTERISTICS

Characteristic	Symbol	Rating	Unit
Maximum Forward Voltage Drop (1 Amp Continuous DC, 25°C)	V_F	1.1	Volts
Maximum Full-Cycle Average Forward Voltage Drop (Rated Current @ 25°C)	$V_{F(AV)}$	0.8	Volts
Maximum Reverse Current @ Rated DC Voltage (25°C) (100°C)	I_R	0.01 0.05	mA
Maximum Full-Cycle Average Reverse Current (Max Rated PIV and Current, as Half-Wave Rectifier, Resistive Load, 100°C)	$I_{R(AV)}$	0.03	mA

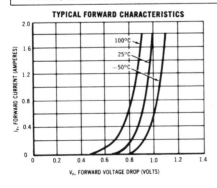

TYPICAL FORWARD CHARACTERISTICS

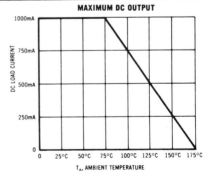

MAXIMUM DC OUTPUT

*Courtesy of Motorola, Inc.

TYPES 1N746 THRU 1N759, 1N746A THRU 1N759A
SILICON VOLTAGE REGULATOR DIODES

3.3 TO 12 VOLTS • 400 mw

GUARANTEED DYNAMIC ZENER IMPEDANCE

Available in 5% and 10% tolerances
-65 to 175°C operation & storage

*electrical characteristics at 25°C free-air temperature (unless otherwise noted)

PARAMETER		V_Z Zener Breakdown Voltage				α_Z Temperature Coefficient of Breakdown Voltage	Z_Z Small-Signal Breakdown Impedance	I_R Static Reverse Current	
TEST CONDITIONS		$I_{ZT} = 20$ ma				$I_{ZT} = 20$ ma	$I_{ZT} = 20$ ma, $I_{zt} = 1$ ma	$V_R = 1$ v	$V_R = 1$ v, $T_A = 150$°C
LIMIT →	NOM	1N746–1N759		1N746A–1N759A		TYP	MAX	MAX	MAX
		MIN	MAX	MIN	MAX				
UNIT →	v	v	v	v	v	% / °C	Ω	μa	μa
1N746	3 3	2 97	3 63	3 135	3 465	−0 062	28	10	30
1N747	3 6	3 24	3 96	3 420	3 780	−0 055	24	10	30
1N748	3 9	3 51	4 29	3 705	4 095	−0 049	23	10	30
1N749	4 3	3 87	4 73	4 085	4 515	−0 036	22	2	30
1N750	4 7	4 23	5 17	4 465	4 935	−0 018	19	2	30
1N751	5 1	4 59	5 61	4 845	5 355	−0 008	17	1	20
1N752	5 6	5 04	6 16	5 320	5 880	+0 006	11	1	20
1N753	6 2	5 58	6 82	5 890	6 510	+0 022	7	0 1	20
1N754	6 8	6 12	7 48	6 460	7 140	+0 035	5	0 1	20
1N755	7 5	6 75	8 25	7 125	7 875	+0 045	6	0 1	20
1N756	8 2	7 38	9 02	7 790	8 610	+0 052	8	0 1	20
1N757	9 1	8 19	10 01	8 645	9 555	+0 056	10	0 1	20
1N758	10 0	9 00	11 00	9 500	10 500	+0 060	17	0 1	20
1N759	12 0	10 80	13 20	11 400	12 000	+0 060	30	0 1	20

*absolute maximum ratings

Average Rectified Forward Current at (or below) 25°C Free-Air Temperature . . . . 230 ma

Average Rectified Forward Current at 150°C Free-Air Temperature . . . . 85 ma

Continuous Power Dissipation at (or below) 50°C Free-Air Temperature . . . . 400 mw

Continuous Power Dissipation at 150°C Free-Air Temperature . . . . 100 mw

Operating Free-Air Temperature Range . . . . −65°C to 175°C

Storage Temperature Range . . . . −65°C to 175°C

*Indicates JEDEC registered data

*Courtesy of Texas Instruments, Incorporated

2N3903 (SILICON)
2N3904

$V_{CB} = 60$ V
$I_C = 200$ mA
$C_{ob} = 4.0$ pf (max)

CASE 29
(TO-92)

NPN silicon annular transistors, designed for general purpose switching and amplifier applications, features one-piece, injection-molded plastic package for high reliability. The 2N3903 and 2N3904 are complementary with types 2N3905 and 2N3906, respectively.

MAXIMUM RATINGS (T_A = 25°C unless otherwise noted)

Characteristic	Symbol	Rating	Unit
Collector-Base Voltage	V_{CB}	60	Vdc
Collector-Emitter Voltage	V_{CEO}	40	Vdc
Emitter-Base Voltage	V_{EB}	6	Vdc
Collector Current	I_C	200	mAdc
Total Device Dissipation @ T_A = 60°C	P_D	210	mW
Total Device Dissipation @ T_A = 25°C Derate above 25°C	P_D	310 2.81	mW mW/°C
Thermal Resistance, Junction to Ambient	θ_{JA}	0.357	°C/mW
Junction Operating Temperature	T_J	135	°C
Storage Temperature Range	T_{stg}	-55 to +135	°C

ELECTRICAL CHARACTERISTICS (T_A = 25°C unless otherwise noted)

Characteristic	Symbol	Min	Max	Unit
OFF CHARACTERISTICS				
Collector-Base Breakdown Voltage (I_C = 10 μAdc, I_E = 0)	BV_{CBO}	60	—	Vdc
Collector-Emitter Breakdown Voltage* (I_C = 1 mAdc)	BV_{CEO}^*	40	—	Vdc
Emitter-Base Breakdown Voltage (I_E = 10 μAdc, I_C = 0)	BV_{EBO}	6	—	Vdc
Collector Cutoff Current (V_{CE} = 40 Vdc, V_{OB} = 3 Vdc)	I_{CEX}	—	50	nAdc
Base Cutoff Current (V_{CE} = 40 Vdc, V_{OB} = 3 Vdc)	I_{BL}	—	50	nAdc

*Pulse Test: Pulse Width = 300 μsec, Duty Cycle = 2%. V_{OB} = Base Emitter Reverse Bias

*Courtesy of Motorola, Inc.

ELECTRICAL CHARACTERISTICS (continued)

Characteristic		Symbol	Min	Max	Unit
ON CHARACTERISTICS					
DC Current Gain * (I_C = 0.1 mAdc, V_{CE} = 1 Vdc) 2N3903 2N3904		h_{FE}*	20 40	— —	—
(I_C = 1.0 mAdc, V_{CE} = 1 Vdc) 2N3903 2N3904			35 70	— —	
(I_C = 10 mAdc, V_{CE} = 1 Vdc) 2N3903 2N3904			50 100	150 300	
(I_C = 50 mAdc, V_{CE} = 1 Vdc) 2N3903 2N3904			30 60	— —	
(I_C = 100 mAdc, V_{CE} = 1 Vdc) 2N3903 2N3904			15 30	— —	
Collector-Emitter Saturation Voltage* (I_C = 10 mAdc, I_B = 1 mAdc)		$V_{CE(sat)}$*	—	0.2	Vdc
(I_C = 50 mAdc, I_B = 5 mAdc)			—	0.3	
Base-Emitter Saturation Voltage* (I_C = 10 mAdc, I_B = 1 mAdc)		$V_{BE(sat)}$*	0.65	0.85	Vdc
(I_C = 50 mAdc, I_B = 5 mAdc)			—	0.95	
SMALL SIGNAL CHARACTERISTICS					
High Frequency Current Gain 2N3903 (I_C = 10 mA, V_{CE} = 20 V, f = 100 mc) 2N3904		$|h_{fe}|$	2.5 3.0	— —	—
Current-Gain—Bandwidth Product 2N3903 (I_C = 10 mA, V_{CE} = 20 V, f = 100 mc) 2N3904		f_T	250 300	— —	mc
Output Capacitance (V_{CB} = 5 Vdc, I_E = 0, f = 100 kc)		C_{ob}	—	4	pf
Input Capacitance (V_{OB} = 0.5 Vdc, I_C = 0, f = 100 kc)		C_{ib}	—	8	pf
Small Signal Current Gain 2N3903 (I_C = 1.0 mA, V_{CE} = 10 V, f = 1 kc) 2N3904		h_{fe}	50 100	200 400	—
Voltage Feedback Ratio 2N3903 (I_C = 1.0 mA, V_{CE} = 10 V, f = 1 kc) 2N3904		h_{re}	0.1 0.5	5.0 8.0	$\times 10^{-4}$
Input Impedance 2N3903 (I_C = 1.0 mA, V_{CE} = 10 V, f = 1 kc) 2N3904		h_{ie}	0.5 1.0	8 10	Kohms
Output Admittance (I_C = 1.0 mA, V_{CE} = 10 V, f = 1 kc) Both Types		h_{oe}	1.0	40	μmhos
Noise Figure (I_C = 100 μA, V_{CE} = 5 V, R_g = 1 Kohms, 2N3903 Noise Bandwidth = 10 cps to 15.7 kc) 2N3904		NF	— —	6 5	db
SWITCHING CHARACTERISTICS					
Delay Time	V_{CC} = 3 Vdc, V_{OB} = 0.5 Vdc, I_C = 10 mAdc, I_{B1} = 1 mA	t_d	—	35	nsec
Rise Time		t_r	—	35	nsec
Storage Time	V_{CC} = 3 Vdc, I_C = 10 mAdc, 2N3903 I_{B1} = I_{B2} = 1 mAdc 2N3904	t_s	— —	175 200	nsec
Fall Time		t_f	—	50	nsec

*Pulse Test: Pulse Width = 300 μsec, Duty Cycle = 2% V_{OB} = Base Emitter Reverse Bias

2N3905 (SILICON)
2N3906

$V_{CB} = 40$ V
$I_C = 200$ mA
$C_{ob} = 4.5$ pf (max)

CASE 29
(TO-92)

PNP silicon annular transistor, designed for general purpose switching and amplifier applications, features one-piece, injection-molded plastic package for high reliability. The 2N3905 and 2N3906 are complementary with types 2N3903 and 2N3904, respectively.

MAXIMUM RATINGS (T$_A$ = 25°C unless otherwise noted)

Characteristic	Symbol	Rating	Unit
Collector-Base Voltage	V_{CB}	40	Vdc
Collector-Emitter Voltage	V_{CEO}	40	Vdc
Emitter-Base Voltage	V_{EB}	5	Vdc
Collector Current	I_C	200	mAdc
Total Device Dissipation @ T$_A$ = 60°C	P_D	210	mW
Total Device Dissipation @ T$_A$ = 25°C Derate above 25°C	P_D	310 2.81	mW mW/°C
Thermal Resistance, Junction to Ambient	θ_{JA}	0.357	°C/mW
Junction Operating Temperature	T_J	135	°C
Storage Temperature Range	T_{stg}	-55 to +135	°C

ELECTRICAL CHARACTERISTICS (T$_A$ = 25°C unless otherwise noted)

Characteristic	Symbol	Min	Max	Unit
OFF CHARACTERISTICS				
Collector-Base Breakdown Voltage (I_C = 10 μAdc, I_E = 0)	BV_{CBO}	40	—	Vdc
Collector-Emitter Breakdown Voltage* (I_C = 1 mAdc)	BV_{CEO}*	40	—	Vdc
Emitter-Base Breakdown Voltage (I_E = 10 μAdc, I_C = 0)	BV_{EBO}	5	—	Vdc
Collector Cutoff Current (V_{CE} = 40 Vdc, V_{OB} = 3 Vdc)	I_{CEX}	—	50	nAdc
Base Cutoff Current (V_{CE} = 40 Vdc, V_{OB} = 3 Vdc)	I_{BL}	—	50	nAdc

*Pulse Test: Pulse Width = 300 μsec, Duty Cycle = 2% V_{OB} = Base Emitter Reverse Bias

*Courtesy of Motorola, Inc.

ELECTRICAL CHARACTERISTICS (continued)

Characteristic		Symbol	Min	Max	Unit
ON CHARACTERISTICS					
DC Current Gain*		h_{FE}*		—	—
(I_C = 0.1 mAdc, V_{CE} = 1 Vdc)	2N3905		30	—	
	2N3906		60	—	
(I_C = 1.0 mAdc, V_{CE} = 1 Vdc)	2N3905		40	—	
	2N3906		80	—	
(I_C = 10 mAdc, V_{CE} = 1 Vdc)	2N3905		50	150	
	2N3906		100	300	
(I_C = 50 mAdc, V_{CE} = 1 Vdc)	2N3905		30	—	
	2N3906		60	—	
(I_C = 100 mAdc, V_{CE} = 1 Vdc)	2N3905		15	—	
	2N3906		30	—	
Collector-Emitter Saturation Voltage*		$V_{CE(sat)}$*			Vdc
(I_C = 10 mAdc, I_B = 1 mAdc)			—	0.25	
(I_C = 50 mAdc, I_B = 5 mAdc)			—	0.4	
Base-Emitter Saturation Voltage*		$V_{BE(sat)}$*			Vdc
(I_C = 10 mAdc, I_B = 1 mAdc)			0.65	0.85	
(I_C = 50 mAdc, I_B = 5 mAdc)			—	0.95	
SMALL SIGNAL CHARACTERISTICS					
High-Frequency Current Gain	2N3905	$|h_{fe}|$	2.0	—	—
(I_C = 10 mAdc, V_{CE} = 20 Vdc, f = 100 mc)	2N3906		2.5	—	
Current-Gain—Bandwidth Product	2N3905	f_T	200	—	mc
(I_C = 10 mAdc, V_{CE} = 20 Vdc, f = 100 mc)	2N3906		250	—	
Output Capacitance		C_{ob}		4.5	pf
(V_{CB} = 5 Vdc, I_E = 0, f = 100 kc)			—		
Input Capacitance		C_{ib}		10	pf
(V_{OB} = 0.5 Vdc, I_C = 0, f = 100 kc)			—		
Small Signal Current Gain	2N3905	h_{fe}	50	200	—
(I_C = 1.0 mA, V_{CE} = 10 V, f = 1 kc)	2N3906		100	400	
Voltage Feedback Ratio	2N3905	h_{re}	0.1	5	$\times 10^{-4}$
(I_C = 1.0 mA, V_{CE} = 10 V, f = 1 kc)	2N3906		1.0	10	
Input Impedance	2N3905	h_{ie}	0.5	8	Kohms
(I_C = 1.0 mA, V_{CE} = 10 V, f = 1 kc)	2N3906		2.0	12	
Output Admittance	2N3905	h_{oe}	1.0	40	μmhos
(I_C = 1.0 mA, V_{CE} = 10 V, f = 1 kc)	2N3906		3.0	60	
Noise Figure		NF			db
(I_C = 100 μA, V_{CE} = 5 V, R_g = 1 Kohms,	2N3905		—	5.0	
Noise Bandwidth = 10 cps to 15.7 kc)	2N3906		—	4.0	
SWITCHING CHARACTERISTICS					
Delay Time	V_{CC} = 3 Vdc, V_{OB} = 0.5 Vdc,	t_d	—	35	nsec
Rise Time	I_C = 10 mAdc, I_{B1} = 1 mA	t_r	—	35	nsec
Storage Time	V_{CC} = 3 Vdc, I_C = 10 mAdc,	t_s 2N3905	—	200	nsec
		2N3906	—	225	
Fall Time	I_{B1} = I_{B2} = 1 mAdc	t_f 2N3905	—	60	nsec
		2N3906	—	75	

*Pulse Test: PW = 300 μsec, Duty Cycle = 2% V_{OB} = Base-Emitter Reverse Bias

APPENDIX 1-6*

FOR EXTREMELY LOW-LEVEL, LOW-NOISE, HIGH-GAIN, SMALL-SIGNAL AMPLIFIER APPLICATIONS

- Guaranteed h_{FE} at 10 µa, $T_A = -55°C$ and $25°C$
- Guaranteed Low-Noise Characteristics at 10 µa
- Usable at Collector Currents as Low as 1 µa
- Very High Reliability
- 2N929 and 2N930 Also Are Available to MIL-S-19500/253 (Sig C)

*mechanical data

THE COLLECTOR IS IN ELECTRICAL CONTACT WITH THE CASE

ALL JEDEC TO-18 DIMENSIONS AND NOTES ARE APPLICABLE

ALL DIMENSIONS ARE IN INCHES UNLESS OTHERWISE SPECIFIED

*absolute maximum ratings at 25°C free-air temperature (unless otherwise noted)

Collector-Base Voltage	45 v
Collector-Emitter Voltage (See Note 1)	45 v
Emitter-Base Voltage	5 v
Collector Current	30 ma
Total Device Dissipation at (or below) 25°C Free-Air Temperature (See Note 2)	300 mw
Total Device Dissipation at (or below) 25°C Case Temperature (See Note 3)	600 mw
Operating Collector Junction Temperature	175°C
Storage Temperature Range	−65°C to + 200°C

NOTES: 1. This value applies when the base-emitter diode is open-circuited.
    2. Derate linearly to 175°C free-air temperature at the rate of 2.0 mw/C°.
    3. Derate linearly to 175°C case temperature at the rate of 4.0 mw/C°.

*Indicates JEDEC registered data

*Courtesy of Texas Instruments, Incorporated

TYPES 2N929, 2N930
N-P-N PLANAR SILICON TRANSISTOR

*electrical characteristics at 25°C free-air temperature (unless otherwise noted)

PARAMETER		TEST CONDITIONS	2N929		2N930		UNIT
			MIN	MAX	MIN	MAX	
BV$_{CEO}$	Collector-Emitter Breakdown Voltage	I$_C$ = 10 ma, I$_B$ = 0, (See Note 4)	45		45		v
BV$_{EBO}$	Emitter-Base Breakdown Voltage	I$_E$ = 10 na I$_C$ = 0	5		5		v
I$_{CBO}$	Collector Cutoff Current	V$_{CB}$ = 45 v, I$_E$ = 0		10		10	na
I$_{CES}$	Collector Cutoff Current (See Note 5)	V$_{CE}$ = 45 v, V$_{BE}$ = 0		10		10	na
		V$_{CE}$ = 45 v, V$_{BE}$ = 0, T$_A$ = 170°C		10		10	μa
I$_{CEO}$	Collector Cutoff Current	V$_{CE}$ = 5 v, I$_B$ = 0		2		2	na
I$_{EBO}$	Emitter Cutoff Current	V$_{EB}$ = 5 v, I$_C$ = 0		10		10	na
h$_{FE}$	Static Forward Current Transfer Ratio	V$_{CE}$ = 5 v, I$_C$ = 10 μa	40	120	100	300	
		V$_{CE}$ = 5 v, I$_C$ = 10 μa, T$_A$ = −55°C	10		20		
		V$_{CE}$ = 5 v, I$_C$ = 500 μa	60		150		
		V$_{CE}$ = 5 v, I$_C$ = 10 ma, (See Note 4)		350		600	
V$_{BE}$	Base-Emitter Voltage	I$_B$ = 0.5 ma, I$_C$ = 10 ma, (See Note 4)	0.6	1.0	0.6	1.0	v
V$_{CE(sat)}$	Collector-Emitter Saturation Voltage	I$_B$ = 0.5 ma, I$_C$ = 10 ma, (See Note 4)		1.0		1.0	v
h$_{ib}$	Small-Signal Common-Base Input Impedance	V$_{CB}$ = 5 v, I$_E$ = −1 ma, f = 1 kc	25	32	25	32	ohm
h$_{rb}$	Small-Signal Common-Base Reverse Voltage Transfer Ratio	V$_{CB}$ = 5 v, I$_E$ = −1 ma, f = 1 kc	0	6.0 x 10^{-4}	0	6.0 x 10^{-4}	
h$_{ob}$	Small-Signal Common-Base Output Admittance	V$_{CB}$ = 5 v, I$_E$ = −1 ma, f = 1 kc	0	1.0	0	1.0	μmho
h$_{fe}$	Small-Signal Common-Emitter Forward Current Transfer Ratio	V$_{CE}$ = 5 v, I$_C$ = 1 ma, f = 1 kc	60	350	150	600	
|h$_{fe}$|	Small-Signal Common-Emitter Forward Current Transfer Ratio	V$_{CE}$ = 5 v, I$_C$ = 500 μa, f = 30 mc	1.0		1.0		
C$_{ob}$	Common-Base Open-Circuit Output Capacitance	V$_{CB}$ = 5 v, I$_E$ = 0, f = 1 mc		8		8	pf

*operating characteristics at 25°C free-air temperature

PARAMETER		TEST CONDITIONS	2N929	2N930	UNIT
			MAX	MAX	
NF	Average Noise Figure	V$_{CE}$ = 5 v, I$_C$ = 10 μa, R$_G$ = 10 kΩ Noise Bandwidth 10 cps to 15.7 kc	4	3	db

NOTES: 4. These parameters must be measured using pulse techniques. PW = 300 μsec, Duty Cycle ≤ 2%.
  5. I$_{CES}$ may be used in place of I$_{CBO}$ for circuit stability calculations.
* Indicates JEDEC registered data

TYPES 2N4418 AND 2N4419
N-P-N EPITAXIAL PLANAR SILICON TRANSISTORS

SILECT† TRANSISTORS FOR HIGH-SPEED SWITCHING APPLICATIONS
- 2N4418 Electrically Similar to the 2N2369
- Rugged, One-Piece Construction with Standard TO-18 100-mil Pin Circle

*switching characteristics at 25°C free-air temperature

PARAMETER		TEST CONDITIONS†		2N4418 MAX	2N4419 MAX	UNIT
t_d	Delay Time	$I_C = 10$ mA, $I_{B(1)} = 1$ mA, $V_{BE(off)} = 0$, $R_L = 280$ Ω, See Figure 1		10	10	ns
t_r	Rise Time			12	14	ns
t_{on}	Turn-on Time			20	22	ns
t_s	Storage Time	$I_C = 10$ mA, $I_{B(1)} = 1$ mA, $I_{B(2)} = -1$ mA, $R_L = 280$ Ω, See Figure 2		12	14	ns
t_f	Fall Time			14	16	ns
t_{off}	Turn-off Time			22	28	ns
t_s	Storage Time	$I_C = 10$ mA, $I_{B(1)} = 10$ mA, $I_{B(2)} = -10$ mA, See Figure 3		18	20	ns

†Voltage and current values shown are nominal; exact values vary slightly with transistor parameters.

*absolute maximum ratings at 25°C free-air temperature (unless otherwise noted)

	2N4418	2N4419
Collector-Base Voltage	40 V	30 V
Collector-Emitter Voltage (See Note 1)	40 V	30 V
Collector-Emitter Voltage (See Note 2)	15 V	12 V
Emitter-Base Voltage	4.5 V	4.5 V
Continuous Collector Current	← 200 mA →	
Peak Collector Current (See Note 3)	← 500 mA →	
Continuous Device Dissipation at (or below) 25°C Free-Air Temperature (See Note 4)	← 360 mW→	
Continuous Device Dissipation at (or below) 25°C Lead Temperature (See Note 5)	←500 mW→	
Storage Temperature Range	−65°C to 150°C	
Lead Temperature 1/16 Inch from Case for 10 Seconds	← 260°C →	

*electrical characteristics at 25°C free-air temperature (unless otherwise noted)

PARAMETER		TEST CONDITIONS	2N4418 MIN	2N4418 MAX	2N4419 MIN	2N4419 MAX	UNIT
$V_{(BR)CBO}$	Collector-Base Breakdown Voltage	$I_C = 10$ μA, $I_E = 0$	40		30		V
$V_{(BR)CEO}$	Collector-Emitter Breakdown Voltage	$I_C = 10$ mA, $I_B = 0$, See Note 6	15		12		V
$V_{(BR)CES}$	Collector-Emitter Breakdown Voltage	$I_C = 10$ μA, $V_{BE} = 0$	40		30		V
$V_{(BR)EBO}$	Emitter-Base Breakdown Voltage	$I_E = 10$ μA, $I_C = 0$	4.5		4.5		V
I_{CBO}	Collector Cutoff Current	$V_{CB} = 20$ V, $I_E = 0$		0.4		0.4	μA
		$V_{CB} = 20$ V, $I_E = 0$, $T_A = 70°C$		3		3	μA
I_{EBO}	Emitter Cutoff Current	$V_{EB} = 3$ V, $I_C = 0$		20		25	nA
h_{FE}	Static Forward Current Transfer Ratio	$V_{CE} = 1$ V, $I_C = 10$ mA, See Note 6	40	120	30		
		$V_{CE} = 2$ V, $I_C = 100$ mA, See Note 6	20				
V_{BE}	Base-Emitter Voltage	$I_B = 1$ mA, $I_C = 10$ mA	0.72	0.87	0.72	0.87	V
$V_{CE(sat)}$	Collector-Emitter Saturation Voltage	$I_B = 1$ mA, $I_C = 10$ mA		0.25		0.25	V
$|h_{fe}|$	Small-Signal Common-Emitter Forward Current Transfer Ratio	$V_{CE} = 10$ V, $I_C = 10$ mA, f = 100 MHz	5		4		
C_{cb}	Collector-Base Capacitance	$V_{CB} = 5$ V, $I_E = 0$, f = 1 MHz, See Note 7		4		4	pF

NOTES: 1. This value applies when the base-emitter diode is short-circuited.
2. These values apply between 0 and 200 mA collector current when the base-emitter diode is open-circuited. Maximum rated voltage and 200 mA collector current may be simultaneously applied provided the time of application is 10 μs or less and the duty cycle is 2% or less.
3. This value applies for $t_p ≤ 10$ μs and duty cycle ≤ 2%.
4. Derate linearly to 150°C free-air temperature at the rate of 2.88 mW/deg.

5. Derate linearly to 150°C lead temperature at the rate of 4 mW/deg. Lead temperature is measured on the collector lead 1/16 inch from the case.
6. These parameters must be measured using pulse techniques. $t_p = 300$ μs, duty cycle ≤ 2%.
7. C_{cb} is measured using three-terminal measurement techniques with the emitter guarded.

*Indicates JEDEC registered data
†Trademark of Texas Instruments
‡Patent Pending

*Courtesy of Texas Instruments, Incorporated

TYPES 2N4856, 2N4857, 2N4858, 2N4859, 2N4860, 2N4861
N-CHANNEL EPITAXIAL PLANAR SILICON FIELD-EFFECT TRANSISTORS

SYMMETRICAL N-CHANNEL FIELD-EFFECT TRANSISTORS
FOR HIGH-SPEED COMMUTATOR AND CHOPPER APPLICATIONS
2N4859 Formerly TIXS41

- **Low $r_{ds(on)}$: 25 Ω Max (2N4856, 2N4859)**
- **Low $I_{D(off)}$: 0.25 nA Max**

*mechanical data

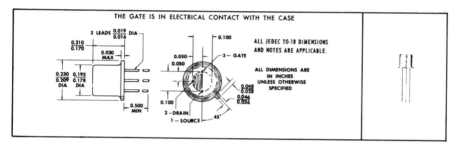

THE GATE IS IN ELECTRICAL CONTACT WITH THE CASE

ALL JEDEC TO-18 DIMENSIONS AND NOTES ARE APPLICABLE.

ALL DIMENSIONS ARE IN INCHES UNLESS OTHERWISE SPECIFIED

3 — GATE
2 — DRAIN
1 — SOURCE

*absolute maximum ratings at 25°C free-air temperature (unless otherwise noted)

	2N4856 2N4857 2N4858	2N4859 2N4860 2N4861
Drain-Gate Voltage	40 V	30 V
Drain-Source Voltage	40 V	30 V
Reverse Gate-Source Voltage	–40 V	–30 V
Forward Gate Current	← 50 mA →	
Continuous Device Dissipation at (or below) 25°C Free-Air Temperature (See Note 1)	← 360 mW →	
Storage Temperature Range	–65°C to 200°C	
Lead Temperature ⅟₁₆ Inch from Case for 10 Seconds	← 300°C →	

NOTE 1: Derate linearly to 175°C free-air temperature at the rate of 2.4 mW/deg.
*Indicates JEDEC registered data

*Courtesy of Texas Instruments, Incorporated

*electrical characteristics at 25°C free-air temperature (unless otherwise noted)

PARAMETER		TEST CONDITIONS	2N4856 MIN	2N4856 MAX	2N4857 MIN	2N4857 MAX	2N4858 MIN	2N4858 MAX	2N4859 MIN	2N4859 MAX	2N4860 MIN	2N4860 MAX	2N4861 MIN	2N4861 MAX	UNIT
$V_{(BR)GSS}$	Gate-Source Breakdown Voltage	$I_G = -1\ \mu A$, $V_{DS} = 0$	−40		−40		−40		−30		−30		−30		V
I_{GSS} Gate Reverse Current		$V_{GS} = -20$ V, $V_{DS} = 0$		−0.25		−0.25		−0.25							nA
		$V_{GS} = -20$ V, $V_{DS} = 0$, $T_A = 150°C$		−0.5		−0.5		−0.5							μA
		$V_{GS} = -15$ V, $V_{DS} = 0$								−0.25		−0.25		−0.25	nA
		$V_{GS} = -15$ V, $V_{DS} = 0$, $T_A = 150°C$								−0.5		−0.5		−0.5	μA
$I_{D(off)}$ Drain Cutoff Current		$V_{DS} = 15$ V, $V_{GS} = -10$ V		0.25		0.25		0.25		0.25		0.25		0.25	nA
		$V_{DS} = 15$ V, $V_{GS} = -10$ V, $T_A = 150°C$		0.5		0.5		0.5		0.5		0.5		0.5	μA
$V_{GS(off)}$	Gate-Source Cutoff Voltage	$V_{DS} = 15$ V, $I_D = 0.5$ nA	−4	−10	−2	−6	−0.8	−4	−4	−10	−2	−6	−0.8	−4	V
I_{DSS}	Zero-Gate-Voltage Drain Current	$V_{DS} = 15$ V, $V_{GS} = 0$, See Note 2	50		20	100	8	80	50		20	100	8	80	mA
$V_{DS(on)}$ Drain-Source On-State Voltage		$I_D = 20$ mA, $V_{GS} = 0$		0.75						0.75					V
		$I_D = 10$ mA, $V_{GS} = 0$				0.50						0.50			V
		$I_D = 5$ mA, $V_{GS} = 0$						0.50						0.50	V
$r_{ds(on)}$	Small-Signal Drain-Source On-State Resistance	$V_{GS} = 0$, $I_D = 0$, $f = 1$ kHz		25		40		60		25		40		60	Ω
C_{iss}	Common-Source Short-Circuit Input Capacitance	$V_{GS} = -10$ V, $V_{DS} = 0$, $f = 1$ MHz		18		18		18		18		18		18	pF
C_{rss}	Common-Source Short-Circuit Reverse Transfer Capacitance	$V_{GS} = -10$ V, $V_{DS} = 0$, $f = 1$ MHz		8		8		8		8		8		8	pF

*switching characteristics at 25°C free-air temperature

PARAMETER		TEST CONDITIONS		2N4856 2N4859 MAX	2N4857 2N4860 MAX	2N4858 2N4861 MAX	UNIT
$t_{d(on)}$	Turn-On Delay Time	$V_{DD} = 10$ V,	$I_{D(on)}† = $ { 20 mA (2N4856, 2N4859) 10 mA (2N4857, 2N4860) 5 mA (2N4858, 2N4861)	6	6	10	ns
t_r	Rise Time	$V_{GS(on)} = 0$,		3	4	10	ns
t_{off}	Turn-Off Time	See Figure 1	$V_{GS(off)} = $ { −10 V (2N4856, 2N4859) −6 V (2N4857, 2N4860) −4 V (2N4858, 2N4861)	25	50	100	ns

NOTE 2: This parameter must be measured using pulse techniques. $t_p \approx 100$ ms, duty cycle $\leq 10\%$.

*Indicates JEDEC registered data

†These are nominal values; exact values vary slightly with transistor parameters.

APPENDIX 1-9*

2N**5457** (SILICON)
2N**5458**
2N**5459**

Silicon N-channel junction field-effect transistors depletion mode (Type A) designed for general-purpose audio and switching applications.

CASE 29 (5)
(TO-92)

Drain and source may be
interchanged.

MAXIMUM RATINGS

Rating	Symbol	Value	Unit
Drain-Source Voltage	V_{DS}	25	Vdc
Drain-Gate Voltage	V_{DG}	25	Vdc
Reverse Gate-Source Voltage	$V_{GS(r)}$	25	Vdc
Gate Current	I_G	10	mAdc
Total Device Dissipation @ T_A= 25°C	P_D [2]	310	mW
Derate above 25°C		2.82	mW/°C
Operating Junction Temperature	T_J [2]	135	°C
Storage Temperature Range	T_{stg} [2]	–65 to +150	°C

ELECTRICAL CHARACTERISTICS (T_A = 25°C unless otherwise noted)

Characteristic		Symbol	Min	Typ	Max	Unit
OFF CHARACTERISTICS						
Gate-Source Breakdown Voltage (I_G = -10μAdc, V_{DS} = 0)		BV_{GSS}	25	—	—	Vdc
Gate Reverse Current (V_{GS} = -15 Vdc, V_{DS} = 0)		I_{GSS}	—	—	1.0	nAdc
(V_{GS} = -15 Vdc, V_{DS} = 0, T_A = 100°C)			—	—	200	
Gate-Source Cutoff Voltage (V_{DS} = 15 Vdc, I_D = 10 nAdc)	2N5457	$V_{GS(off)}$	0.5	—	6.0	Vdc
	2N5458		1.0	—	7.0	
	2N5459		2.0	—	8.0	
Gate-Source Voltage (V_{DS} = 15 Vdc, I_D = 100 μAdc)	2N5457	V_{GS}	—	2.5	—	Vdc
(V_{DS} = 15 Vdc, I_D = 200 μAdc)	2N5458		—	3.5	—	
(V_{DS} = 15 Vdc, I_D = 400 μAdc)	2N5459		—	4.5	—	
ON CHARACTERISTICS						
Zero-Gate-Voltage Drain Current [1] (V_{DS} = 15 Vdc, V_{GS} = 0)	2N5457	I_{DSS}	1.0	3.0	5.0	mAdc
	2N5458		2.0	6.0	9.0	
	2N5459		4.0	9.0	16	
DYNAMIC CHARACTERISTICS						
Forward Transfer Admittance [1] (V_{DS} = 15 Vdc, V_{GS} = 0, f = 1 kHz)	2N5457	$\|y_{fs}\|$	1000	3000	5000	μmhos
	2N5458		1500	4000	5500	
	2N5459		2000	4500	6000	
Output Admittance [1] (V_{DS} = 15 Vdc, V_{GS} = 0, f = 1 kHz)		$\|y_{os}\|$	—	10	50	μmhos
Input Capacitance (V_{DS} = 15 Vdc, V_{GS} = 0, f = 1 MHz)		C_{iss}	—	4.5	7.0	pF
Reverse Transfer Capacitance (V_{DS} = 15 Vdc, V_{GS} = 0, f = 1 MHz)		C_{rss}	—	1.5	3.0	pF

[1] Pulse Test: Pulse Width ≤ 630 ms; Duty Cycle ≤ 10%

[2] Continuous package improvements have enhanced these guaranteed Maximum Ratings as follows: P_D = 1.0 W @ T_C = 25°C, Derate above 25°C – 8.0 mW/°C, T_J = –65 to +150°C, θ_{JC} = 125° C/W.

*Courtesy of Motorola, Inc.

2N**4391** (SILICON)
2N**4392**
2N**4393**

SILICON N-CHANNEL
JUNCTION FIELD-EFFECT TRANSISTORS

Depletion Mode (Type A) Junction Field-Effect Transistors designed for chopper and high-speed switching applications.

- Low Drain-Source "On" Resistance —
  $r_{ds(on)}$ = 30 Ohms (Max) @ f = 1.0 kHz (2N4391)

- Low Gate Reverse Current —
  I_{GSS} = 0.1 nAdc (Max) @ V_{GS} = 20 Vdc

- Guaranteed Switching Characteristics

N-CHANNEL
JUNCTION FIELD-EFFECT
TRANSISTORS
(Type A)

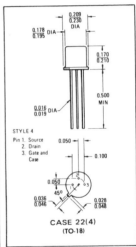

CASE 22(4)
(TO-18)

MAXIMUM RATINGS

Rating	Symbol	Value	Unit
Drain-Source Voltage	V_{DS}	40	Vdc
Drain-Gate Voltage	V_{DG}	40	Vdc
Gate-Source Voltage	V_{GS}	40	Vdc
Forward Gate Current	$I_{G(f)}$	50	mAdc
Total Device Dissipation @ T_C = 25°C Derate above 25°C	P_D	1.8 10	Watts mW/°C
Operating Junction Temperature Range	T_J	-65 to +175	°C
Storage Temperature Range	T_{stg}	-65 to +200	°C

FIGURE 1   SWITCHING TIMES TEST CIRCUIT

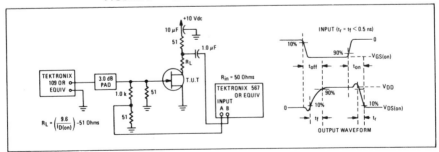

*Courtesy of Motorola, Inc.

2N4391, 2N4392, 2N4393 (continued)

Characteristic		Symbol	Min	Max	Unit
OFF CHARACTERISTICS					
Gate-Source Breakdown Voltage ($I_G = 1.0 \mu Adc$, $V_{DS} = 0$)		$V_{(BR)GSS}$	40	-	Vdc
Gate-Source Forward Voltage ($I_G = 1.0$ mAdc, $V_{DS} = 0$)		$V_{GS(f)}$	-	1.0	Vdc
Gate-Source Voltage ($V_{DS} = 20$ Vdc, $I_D = 1.0$ nAdc)	2N4391 2N4392 2N4393	V_{GS}	4.0 2.0 0.5	10 5.0 3.0	Vdc
Gate Reverse Current ($V_{GS} = 20$ Vdc, $V_{DS} = 0$) ($V_{GS} = 20$ Vdc, $V_{DS} = 0$, $T_A = 150°C$)		I_{GSS}	- -	0.1 0.2	nAdc μAdc
Drain-Cutoff Current ($V_{DS} = 20$ Vdc, $V_{GS} = 12$ Vdc) ($V_{DS} = 20$ Vdc, $V_{GS} = 7.0$ Vdc) ($V_{DS} = 20$ Vdc, $V_{GS} = 5.0$ Vdc) ($V_{DS} = 20$ Vdc, $V_{GS} = 12$ Vdc, $T_A = 150°C$) ($V_{DS} = 20$ Vdc, $V_{GS} = 7.0$ Vdc, $T_A = 150°C$) ($V_{DS} = 20$ Vdc, $V_{GS} = 5.0$ Vdc, $T_A = 150°C$)	2N4391 2N4392 2N4393 2N4391 2N4392 2N4393	$I_{D(off)}$	- - - - - -	0.1 0.1 0.1 0.2 0.2 0.2	nAdc μAdc
ON CHARACTERISTICS					
Zero-Gate Voltage Drain Current (1) ($V_{DS} = 20$ Vdc, $V_{GS} = 0$)	2N4391 2N4392 2N4393	I_{DSS}	50 25 5.0	150 75 30	mAdc
Drain-Source "ON" Voltage ($I_D = 12$ mAdc, $V_{GS} = 0$) ($I_D = 6.0$ mAdc, $V_{GS} = 0$) ($I_D = 3.0$ mAdc, $V_{GS} = 0$)	2N4391 2N4392 2N4393	$V_{DS(on)}$	- - -	0.4 0.4 0.4	Vdc
Static Drain-Source "ON" Resistance ($I_D = 1.0$ mAdc, $V_{GS} = 0$)	2N4391 2N4392 2N4393	$r_{DS(on)}$	- - -	30 60 100	Ohms
SMALL-SIGNAL CHARACTERISTICS					
Drain-Source "ON" Resistance ($V_{GS} = 0$, $I_D = 0$, $f = 1.0$ kHz)	2N4391 2N4392 2N4393	$r_{ds(on)}$	- - -	30 60 100	Ohms
Input Capacitance ($V_{DS} = 20$ Vdc, $V_{GS} = 0$, $f = 1.0$ MHz)		C_{iss}	-	14	pF
Reverse Transfer Capacitance ($V_{DS} = 0$, $V_{GS} = 12$ Vdc, $f = 1.0$ MHz) ($V_{DS} = 0$, $V_{GS} = 7.0$ Vdc, $f = 1.0$ MHz) ($V_{DS} = 0$, $V_{GS} = 5.0$ Vdc, $f = 1.0$ MHz)	2N4391 2N4392 2N4393	C_{rss}	- - -	3.5 3.5 3.5	pF
SWITCHING CHARACTERISTICS					
Turn-On Time (See Figure 1) ($I_{D(on)} = 12$ mAdc) ($I_{D(on)} = 6.0$ mAdc) ($I_{D(on)} = 3.0$ mAdc)	2N4391 2N4392 2N4393	t_{on}	- - -	15 15 15	ns
Rise Time (See Figure 1) ($I_{D(on)} = 12$ mAdc) ($I_{D(on)} = 6.0$ mAdc) ($I_{D(on)} = 3.0$ mAdc)	2N4391 2N4392 2N4393	t_r	- - -	5.0 5.0 5.0	ns
Turn-Off Time (See Figure 1) ($V_{GS(off)} = 12$ Vdc) ($V_{GS(off)} = 7.0$ Vdc) ($V_{GS(off)} = 5.0$ Vdc)	2N4391 2N4392 2N4393	t_{off}	- - -	20 35 50	ns
Fall Time (See Figure 1) ($V_{GS(off)} = 12$ Vdc) ($V_{GS(off)} = 7.0$ Vdc) ($V_{GS(off)} = 5.0$ Vdc)	2N4391 2N4392 2N4393	t_f	- - -	15 20 30	ns

(1) Pulse Test: Pulse Width $\leq 100 \mu$s, Duty Cycle ≤ 1.0%.

APPENDIX 1-11*

µA741
FREQUENCY-COMPENSATED OPERATIONAL AMPLIFIER
FAIRCHILD LINEAR INTEGRATED CIRCUITS

GENERAL DESCRIPTION — The µA741 is a high performance monolithic Operational Amplifier constructed using the Fairchild Planar* epitaxial process. It is intended for a wide range of analog applications. High common mode voltage range and absence of "latch-up" tendencies make the µA741 ideal for use as a voltage follower. The high gain and wide range of operating voltage provides superior performance in integrator, summing amplifier, and general feedback applications.

- NO FREQUENCY COMPENSATION REQUIRED
- SHORT CIRCUIT PROTECTION
- OFFSET VOLTAGE NULL CAPABILITY
- LARGE COMMON-MODE AND DIFFERENTIAL VOLTAGE RANGES
- LOW POWER CONSUMPTION
- NO LATCH UP

ABSOLUTE MAXIMUM RATINGS

Supply Voltage	
Military (741)	±22 V
Commercial (741C)	±18 V
Internal Power Dissipation (Note 1)	
Metal Can	500 mW
DIP	670 mW
Mini DIP	310 mW
Flatpak	570 mW
Differential Input Voltage	±30 V
Input Voltage (Note 2)	±15 V
Storage Temperature Range	
Metal Can, DIP, and Flatpak	−65°C to +150°C
Mini DIP	−55°C to +125°C
Operating Temperature Range	
Military (741)	−55°C to +125°C
Commercial (741C)	0°C to +70°C
Lead Temperature (Soldering)	
Metal Can, DIP, and Flatpak (60 seconds)	300°C
Mini DIP (10 seconds)	260°C
Output Short Circuit Duration (Note 3)	Indefinite

EQUIVALENT CIRCUIT

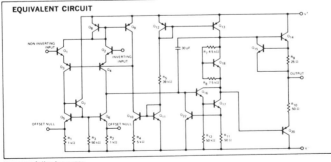

Notes on following pages.

*Courtesy of Fairchild Semiconductors

CONNECTION DIAGRAMS

8-LEAD METAL CAN
(TOP VIEW)
PACKAGE OUTLINE 5B

Note: Pin 4 connected to case

ORDER INFORMATION
TYPE	PART NO.
741	741HM
741C	741HC

14-LEAD DIP
(TOP VIEW)
PACKAGE OUTLINE 6A

ORDER INFORMATION
TYPE	PART NO.
741	741DM
741C	741DC

10-LEAD FLATPAK
(TOP VIEW)
PACKAGE OUTLINE 3F

ORDER INFORMATION
TYPE	PART NO.
741	741FM

8-LEAD MINIDIP
(TOP VIEW)
PACKAGE OUTLINE 9T

ORDER INFORMATION
TYPE	PART NO.
741C	741TC

*Planar is a patented Fairchild process.

ELECTRICAL CHARACTERISTICS (V_S = ±15 V, T_A = 25°C unless otherwise specified)

PARAMETERS (see definitions)		CONDITIONS	MIN.	TYP.	MAX.	UNITS
Input Offset Voltage		$R_S \leqslant 10$ kΩ		1.0	5.0	mV
Input Offset Current				20	200	nA
Input Bias Current				80	500	nA
Input Resistance			0.3	2.0		MΩ
Input Capacitance				1.4		pF
Offset Voltage Adjustment Range				±15		mV
Large Signal Voltage Gain		$R_L \geqslant 2$ kΩ, V_{OUT} = ±10 V	50,000	200,000		
Output Resistance				75		Ω
Output Short Circuit Current				25		mA
Supply Current				1.7	2.8	mA
Power Consumption				50	85	mW
Transient Response (Unity Gain)	Risetime	V_{IN} = 20 mV, R_L = 2 kΩ, $C_L \leqslant 100$ pF		0.3		μs
	Overshoot			5.0		%
Slew Rate		$R_L \geqslant 2$ kΩ		0.5		V/μs

The following specifications apply for −55°C ≤ T_A ≤ +125°C:

Input Offset Voltage		$R_S \leqslant 10$ kΩ		1.0	6.0	mV
Input Offset Current	T_A = +125°C			7.0	200	nA
	T_A = −55°C			85	500	nA
Input Bias Current	T_A = +125°C			0.03	0.5	μA
	T_A = −55°C			0.3	1.5	μA
Input Voltage Range			±12	±13		V
Common Mode Rejection Ratio		$R_S \leqslant 10$ kΩ	70	90		dB
Supply Voltage Rejection Ratio		$R_S \leqslant 10$ kΩ		30	150	μV/V
Large Signal Voltage Gain		$R_L \geqslant 2$ kΩ, V_{OUT} = ±10 V	25,000			
Output Voltage Swing	$R_L \geqslant 10$ kΩ		±12	±14		V
	$R_L \geqslant 2$ kΩ		±10	±13		V
Supply Current	T_A = +125°C			1.5	2.5	mA
	T_A = −55°C			2.0	3.3	mA
Power Consumption	T_A = +125°C			45	75	mW
	T_A = −55°C			60	100	mW

TYPICAL PERFORMANCE CURVES FOR 741

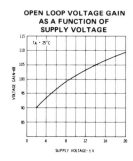

OPEN LOOP VOLTAGE GAIN
AS A FUNCTION OF
SUPPLY VOLTAGE

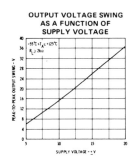

OUTPUT VOLTAGE SWING
AS A FUNCTION OF
SUPPLY VOLTAGE

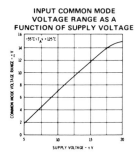

INPUT COMMON MODE
VOLTAGE RANGE AS A
FUNCTION OF SUPPLY VOLTAGE

TYPES 2N3980, 2N4947 THRU 2N4949
P-N PLANAR UNIJUNCTION SILICON TRANSISTORS

PLANAR UNIJUNCTION TRANSISTORS SPECIFICALLY CHARACTERIZED
FOR A WIDE RANGE OF MILITARY, SPACE, AND INDUSTRIAL APPLICATIONS:

2N3980 for General-Purpose UJT Applications
2N4947 for High-Frequency Relaxation-Oscillator Circuits
2N4948 for Thyristor (SCR) Trigger Circuits
2N4949 for Long-Time-Delay Circuits

● **Planar Process Ensures Extremely Low Leakage, High Performance for Low Driving Currents, and Greatly Improved Reliability**

*mechanical data

Package outline is same as JEDEC TO-18 except for lead position. All TO-18 registration notes also apply to this outline.

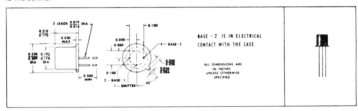

*absolute maximum ratings at 25°C free-air temperature (unless otherwise noted)

Emitter—Base-Two Reverse Voltage . . . . −30 V
Interbase Voltage . . . . See Note 1
Continuous Emitter Current . . . . 50 mA
Peak Emitter Current (See Note 2) . . . . 1 A
Continuous Device Dissipation at (or below) 25°C Free-Air Temperature (See Note 3) . . . . . 360 mW
Storage Temperature Range . . . . −65°C to 200°C
Lead Temperature 1/16 Inch from Case for 10 Seconds . . . . 260°C

*electrical characteristics at 25°C free-air temperature (unless otherwise noted)

PARAMETER		TEST CONDITIONS	2N3980		2N4947		2N4948		2N4949		UNIT
			MIN	MAX	MIN	MAX	MIN	MAX	MIN	MAX	
r_{BB}	Static Interbase Resistance	$V_{B2-B1} = 3\,V$, $\quad I_E = 0$	4	8	4	9.1	4	12	4	12	kΩ
$\alpha_{r/BB}$	Interbase Resistance Temperature Coefficient	$V_{B2-B1} = 3\,V$, $\quad I_E = 0$, $T_A = -65°C$ to $100°C$, $\quad$ See Note 4	0.4	0.9	0.1	0.9	0.1	0.9	0.1	0.9	%/deg
η	Intrinsic Standoff Ratio	$V_{B2-B1} = 10\,V$, $\quad$ See Figure 1	0.68	0.82	0.51	0.69	0.55	0.82	0.74	0.86	
$I_{B2(mod)}$	Modulated Interbase Current	$V_{B2-B1} = 10\,V$, $\quad I_E = 50\,mA$, See Note 5	12		12		12		12		mA
I_{EB2O}	Emitter Reverse Current	$V_{EB2} = -30\,V$, $\quad I_{B1} = 0$		−10		−10		−10		−10	nA
		$V_{EB2} = -30\,V$, $\quad I_{B1} = 0$, $\quad T_A = 125°C$		−1		−1		−1		−1	μA
I_P	Peak-Point Emitter Current	$V_{B2-B1} = 25\,V$		2		2		2		1	μA
$V_{EB1(sat)}$	Emitter — Base-One Saturation Voltage	$V_{B2-B1} = 10\,V$, $\quad I_E = 50\,mA$, See Note 5		.3		3		3		3	V
I_V	Valley-Point Emitter Current	$V_{B2-B1} = 20\,V$	1	10	4		2		2		mA
V_{OB1}	Base-One Peak Pulse Voltage	See Figure 2	6		3		6		3		V

*Courtesy of Texas Instruments, Incorporated

438

APPENDIX 1-13*

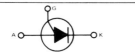

SILICON PROGRAMMABLE UNIJUNCTION TRANSISTORS

. . . designed to enable the engineer to "program" unijunction characteristics such as R_{BB}, η, I_V, and I_P by merely selecting two resistor values. Application includes thyristor-trigger, oscillator, pulse and timing circuits. These devices may also be used in special thyristor applications due to the availability of an anode gate. Supplied in an inexpensive TO-92 plastic package for high-volume requirements, this package is readily adaptable for use in automatic insertion equipment.

- Programmable — R_{BB}, η, I_V and I_P.
- Low On-State Voltage — 1.5 Volts Maximum @ I_F = 50 mA
- Low Gate to Anode Leakage Current — 10 nA Maximum
- High Peak Output Voltage — 11 Volts Typical
- Low Offset Voltage — 0.35 Volt Typical (R_G = 10 k ohms)

SILICON PROGRAMMABLE UNIJUNCTION TRANSISTORS

40 VOLTS
375 mW

ELECTRICAL CHARACTERISTICS (T_A = 25°C unless otherwise noted)

Characteristic	Figure	Symbol	Min	Typ	Max	Unit
*Peak Current	2,9,11	I_P				µA
(V_S = 10 Vdc, R_G = 1.0 MΩ) 2N6027			–	1.25	2.0	
2N6028			–	0.08	0.15	
(V_S = 10 Vdc, R_G = 10 k ohms) 2N6027			–	4.0	5.0	
2N6028			–	0.70	1.0	
*Offset Voltage	1	V_T				Volts
(V_S = 10 Vdc, R_G = 1.0 MΩ) 2N6027			0.2	0.70	1.6	
2N6028			0.2	0.50	0.6	
(V_S = 10 Vdc, R_G = 10 k ohms) (Both Types)			0.2	0.35	0.6	
*Valley Current	1,4,5,	I_V				µA
(V_S = 10 Vdc, R_G = 1.0 MΩ) 2N6027			–	18	50	
2N6028			–	18	25	
(V_S = 10 Vdc, R_G = 10 k ohms) 2N6027			70	270	–	
2N6028			25	270	–	
(V_S = 10 Vdc, R_G = 200 Ohms) 2N6027			1.5	–	–	mA
2N6028			1.0	–	–	
Gate to Anode Leakage Current	–	I_{GAO}				nAdc
(V_S = 40 Vdc, T_A = 25°C, Cathode Open)			–	1.0	10	
(V_S = 40 Vdc, T_A = 75°C, Cathode Open)			–	3.0	–	
Gate to Cathode Leakage Current	–	I_{GKS}	–	5.0	50	nAdc
(V_S = 40 Vdc, Anode to Cathode Shorted)						
*Forward Voltage (I_F = 50 mA Peak)	1,6	V_F	–	0.8	1.5	Volts
*Peak Output Voltage	3,7	V_O	6.0	11	–	Volts
(V_B = 20 Vdc, C_C = 0.2 µF)						
Pulse Voltage Rise Time	3	t_r	–	40	80	ns
(V_B = 20 Vdc, C_C = 0.2 µF)						

*Indicates JEDEC Registered Data

*Courtesy of Motorola, Inc.

SCHMITT-TRIGGER POSITIVE-NAND GATES AND INVERTERS WITH TOTEM-POLE OUTPUTS

recommended operating conditions

		54 FAMILY			SERIES 54		
		74 FAMILY			SERIES 74		
			'13			'14	
		MIN	NOM	MAX	MIN	NOM	MAX
Supply voltage, V_{CC}	54 Family	4.5	5	5.5	4.5	5	5.5
	74 Family	4.75	5	5.25	4.75	5	5.25
High-level output current, I_{OH}				−800			−800
Low-level output current, I_{OL}				16			16
Operating free-air temperature, T_A	54 Family	−55		125	−55		125
	74 Family	0		70	0		70

electrical characteristics over recommended operating free-air temperature range (unless otherwise noted)

PARAMETER		TEST FIGURE	TEST CONDITIONS[†]		SERIES 54 SERIES 74					
					'13			'14		
					MIN	TYP[‡]	MAX	MIN	TYP[‡]	MAX
V_{T+}	Positive-going threshold voltage	8	V_{CC} = 5 V		1.5	1.7	2	1.5	1.7	2
V_{T-}	Negative-going threshold voltage	9	V_{CC} = 5 V		0.6	0.9	1.1	0.6	0.9	1.1
	Hysteresis ($V_{T+}-V_{T-}$)	8, 9	V_{CC} = 5 V		0.4	0.8		0.4	0.8	
V_I	Input clamp voltage	3	V_{CC} = MIN, I_I = §				−1.5			−1.5
V_{OH}	High-level output voltage	9	V_{CC} = MIN, V_I = V_T_min, I_{OH} = MAX	54 Family	2.4	3.4		2.4	3.4	
				74 Family	2.4	3.4		2.4	3.4	
V_{OL}	Low-level output voltage	8	V_{CC} = MIN, V_I = $V_{T+}max$, I_{OL} = MAX			0.2	0.4		0.2	0.4
I_{T+}	Input current at positive-going threshold	8	V_{CC} = 5 V, V_I = V_{T+}			−0.65			−0.43	
I_{T-}	Input current at negative-going threshold	9	V_{CC} = 5 V, V_I = V_{T-}			−0.85			−0.56	
I_I	Input current at maximum input voltage	4	V_{CC} = MAX, V_I = 5.5 V				1			1
I_{IH}	High-level input current	4	V_{CC} = MAX, V_I = 2.4 V				40			40
				V_I = 2.7 V						
I_{IL}	Low-level input current	5	V_{CC} = MAX, V_{IL} = 0.4 V		−1	−1.6		−0.8	−1.2	
				V_{IL} = 0.5 V						
I_{OS}	Short-circuit output current♦	6	V_{CC} = MAX		−18		−55	−18		−55
I_{CC}	Supply current	Total, output high	7	V_{CC} = MAX		14	23		22	36
		Total, output low				20	32		39	60
		Average per gate		V_{CC} = 5 V, 50% duty cycle		8.5			5.1	

[†] For conditions shown as MIN or MAX, use the appropriate value specified under recommended operating conditions.
[‡] All typical values are at V_{CC} = 5 V, T_A = 25°C.
§ I_I = −12 mA for SN54'/SN74' and −18 mA for 'S132.
♦ Not more than one output should be shorted at a time, and for 'S132, duration of output short-circuit should not exceed one second.

switching characteristics, V_{CC} = 5 V, T_A = 25°C

TYPE	TEST CONDITIONS	t_{PLH} (ns) Propagation delay time, low-to-high-level output		t_{PHL} (ns) Propagation delay time, high-to-low-level output	
		TYP	MAX	TYP	MAX
'13	C_L = 15 pF, R_L = 400 Ω	18	27	15	22
'14, '132		15	22	15	22
'S132	C_L = 15 pF, R_L = 280 Ω	7	10.5	8.5	13

*Courtesy of Texas Instruments, Incorporated

APPENDIX 1-15*

absolute maximum ratings

Supply Voltage	+18V
Power Dissipation	600 mW
Operating Temperature Ranges	
LM555C	$0°C$ to $+70°C$
LM555	$-55°C$ to $+125°C$
Storage Temperature Range	$-65°C$ to $+150°C$
Lead Temperature (Soldering, 10 sec)	$300°C$

electrical characteristics ($T_A = 25°C$, $V_{CC} = +5V$ to $+15V$ unless otherwise specified)

PARAMETER	CONDITIONS	LIMITS						UNITS
		LM555			LM555C			
		MIN	TYP	MAX	MIN	TYP	MAX	
Supply Voltage		4.5		18	4.5		16	V
Supply Current	$V_{CC} = 5V$, $R_L = \infty$		3	5		3	6	mA
	$V_{CC} = 15V$, $R_L = \infty$ (Low State) (Note 1)		10	12		10	15	mA
Timing Error, Monostable								
Initial Accuracy	R_A, $R_B = 1k$ to $100k$,		.5	2		1		%
Drift with Temperature	$C = .1\mu F$, (Note 2)		30	100		50		ppm/°C
Drift with Supply			.005	.02		.01		%/V
Threshold Voltage			.667			.667		$\times V_{CC}$
Trigger Voltage	$V_{CC} = 15V$	4.8	5	5.2		5		V
	$V_{CC} = 5V$	1.45	1.67	1.9		1.67		V
Trigger Current			.5			.5		μA
Reset Voltage		.4	.7	1	4	.7	1	V
Reset Current			.1			.1		mA
Threshold Current	(Note 3)		.1	.25		.1	.25	μA
Control Voltage Level	$V_{CC} = 15V$	9.6	10	10.4	9	10	11	V
	$V_{CC} = 5V$	2.9	3.33	3.8	2.6	3.33	4	V
Output Voltage Drop (Low)	$V_{CC} = 15V$ $I_{SINK} = 10$ mA		.1	.15		.1	.25	V
	$I_{SINK} = 50$ mA		.4	.5		.4	.75	V
	$I_{SINK} = 100$ mA		2	2.2		2	2.5	V
	$I_{SINK} = 200$ mA		2.5			2.5		V
	$V_{CC} = 5V$ $I_{SINK} = 8$ mA		.1	.25				V
	$I_{SINK} = 5$ mA					.25	.35	V
Output Voltage Drop (High)	$I_{SOURCE} = 200$ mA $V_{CC} = 15V$		12.5			12.5		V
	$I_{SOURCE} = 100$ mA $V_{CC} = 15V$	13	13.3		12.75	13.3		V
	$V_{CC} = 5V$	3	3.3		2.75	3.3		V
Rise Time of Output			100			100		ns
Fall Time of Output			100			100		ns

Note 1: Supply current when output high typically 1 mA less at $V_{CC} = 5V$.
Note 2: Tested at $V_{CC} = 5V$ and $V_{CC} = 15V$.
Note 3: This will determine the maximum value of $R_A + R_B$ for 15V operation. The max total $(R_A + R_B) = 20$ MΩ.

*Courtesy of **National Semiconductor Corp.**

SERIES 54/74 FLIP-FLOPS

recommended operating conditions

		SERIES 54/74			'70			'72, '73, '76, '107	
			MIN	NOM	MAX	MIN	NOM	MAX	
Supply voltage, V_{CC}		Series 54	4.5	5	5.5	4.5	5	5.5	
		Series 74	4.75	5	5.25	4.75	5	5.25	
High-level output current, I_{OH}					−400			−400	
Low-level output current, I_{OL}					16			16	
Pulse width, t_w	Clock high			20			20		
	Clock low			30			47		
	Preset or clear low			25			25		
Input setup time, t_{setup}				$20^\uparrow$			$0^\uparrow$		
Input hold time, t_{hold}				$5^\uparrow$			$0^\downarrow$		
Operating free-air temperature, T_A		Series 54	−55		125	−55		125	
		Series 74	0		70	0		70	

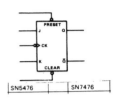

SN5476   SN7476

$^{\uparrow\downarrow}$The arrow indicates the edge of the clock pulse used for reference: $^\uparrow$ for the rising edge, $^\downarrow$ for the falling edge.

electrical characteristics over recommended operating free-air temperature range (unless otherwise noted)

PARAMETER		TEST CONDITIONS†	'70			'72, '73, '76, '107		
			MIN	TYP‡	MAX	MIN	TYP‡	MAX
V_{IH} High-level input voltage			2			2		
V_{IL} Low-level input voltage					0.8			0.8
V_I Input clamp voltage		V_{CC} = MIN, I_I = −12 mA		*	−1.5		*	−1.5
V_{OH} High-level output voltage		V_{CC} = MIN, V_{IH} = 2 V, V_{IL} = 0.8 V, I_{OH} = MAX	2.4	3.4		2.4	3.4	
V_{OL} Low-level output voltage		V_{CC} = MIN, V_{IH} = 2 V, V_{IL} = 0.8 V, I_{OL} = 16 mA		0.2	0.4		0.2	0.4
I_I Input current at maximum input voltage		V_{CC} = MAX, V_I = 5.5 V			1			1
I_{IH} High-level input current	D, J, K, or $\overline{K}$	V_{CC} = MAX, V_I = 2.4 V			40			40
	Clear				80			80
	Preset				80			80
	Clock				40			80
I_{IL} Low-level input current	D, J, K, or $\overline{K}$	V_{CC} = MAX, V_I = 0.4 V			−1.6			−1.6
	Clear				−3.2			−3.2
	Preset				−3.2			−3.2
	Clock				−1.6			−3.2
I_{OS} Short-circuit output current*	Series 54	V_{CC} = MAX	−20		−57	−20		−57
	Series 74		−18		−57	−18		−57
I_{CC} Supply current (Average per flip-flop)		V_{CC} = MAX, See Note 1		13	26		10	20

switching characteristics, V_{CC} = 5 V, T_A = 25°C

PARAMETER¶	FROM (INPUT)	TO (OUTPUT)	TEST CONDITIONS	'70			'72, '73 '76, '107		
				MIN	TYP	MAX	MIN	TYP	MAX
f_{max}				20	35		15	20	
t_{PLH}	Preset	Q	C_L = 15 pF, R_L = 400 Ω, See Note 2			50		16	25
t_{PHL}	(as applicable)	$\overline{Q}$				50		25	40
t_{PLH}	Clear	$\overline{Q}$				50		16	25
t_{PHL}	(as applicable)	Q				50		25	40
t_{PLH}	Clock	Q or $\overline{Q}$		10	27	50	10	16	25
t_{PHL}				10	18	50	10	25	40

POSITIVE-NAND GATES AND INVERTERS WITH TOTEM-POLE OUTPUTS

recommended operating conditions

PARAMETER			SERIES 54 / SERIES 74 ('00,'04,'10,'20,'30) MIN·NOM·MAX	SERIES 54H / SERIES 74H ('H00,'H04,'H10,'H20,'H30) MIN·NOM·MAX	SERIES 54L / SERIES 74L ('L00,'L04,'L10,'L20,'L30) MIN·NOM·MAX	SERIES 54LS / SERIES 74LS ('LS00,'LS04,'LS10,'LS20,'LS30) MIN·NOM·MAX	SERIES 54S / SERIES 74S ('S00,'S04,'S10,'S20,'S30,'S133) MIN·NOM·MAX	UNIT
Supply voltage, V_{CC}	54 Family		4.5 · 5 · 5.5	4.5 · 5 · 5.5	4.5 · 5 · 5.5	4.5 · 5 · 5.5	4.5 · 5 · 5.5	V
	74 Family		4.75 · 5 · 5.25	4.75 · 5 · 5.25	4.75 · 5 · 5.25	4.75 · 5 · 5.25	4.75 · 5 · 5.25	V
High-level output current, I_{OH}	54 Family		−400	−500	−100	−400	−1000	µA
	74 Family		−400	−500	−200	−400	−1000	µA
Low-level output current, I_{OL}	54 Family		16	20	2	4	20	mA
	74 Family		16	20	3.6	8	20	mA
Operating free-air temperature, T_A	54 Family		−55 ··· 125	−55 ··· 125	−55 ··· 125	−55 ··· 125	−55 ··· 125	°C
	74 Family		0 ··· 70	0 ··· 70	0 ··· 70	0 ··· 70	0 ··· 70	°C

electrical characteristics over recommended operating free-air temperature range (unless otherwise noted)

PARAMETER		TEST FIGURE	TEST CONDITIONS[†]	SERIES 54 / SERIES 74 MIN·TYP[‡]·MAX	SERIES 54H / SERIES 74H MIN·TYP[‡]·MAX	SERIES 54L / SERIES 74L MIN·TYP[‡]·MAX	SERIES 54LS / SERIES 74LS MIN·TYP[‡]·MAX	SERIES 54S / SERIES 74S MIN·TYP[‡]·MAX	UNIT
V_{IH}	High-level input voltage	1, 2		2	2	2	2	2	V
V_{IL}	Low-level input voltage	1, 2	54 Family	0.8	0.8	0.7	0.7	0.8	V
			74 Family	0.8	0.8	0.7	0.8	0.8	V
V_I	Input clamp voltage	3	$V_{CC} = $ MIN, $I_I = $ §[§]	−1.5	−1.5		−1.5	−1.2	V
V_{OH}	High-level output voltage	1	$V_{CC} = $ MIN, $V_{IL} = V_{IL}$ max, $I_{OH} = $ MAX — 54 Family	2.4 · 3.4	2.4 · 3.5	2.4 · 3.3	2.5 · 3.4	2.5 · 3.4	V
			74 Family	2.4 · 3.4	2.4 · 3.5	2.4 · 3.2	2.7 · 3.4	2.7 · 3.4	V
V_{OL}	Low-level output voltage	2	$V_{CC} = $ MIN, $V_{IH} = 2$ V, $I_{OL} = $ MAX — 54 Family	0.2 · 0.4	0.2 · 0.4	0.15 · 0.3	0.25 · 0.4	· 0.5	V
			74 Family	0.2 · 0.4	0.2 · 0.4	0.2 · 0.4	0.35 · 0.5	· 0.5	V
I_I	Input current at maximum input voltage	4	$V_{CC} = $ MAX, $V_I = 5.5$ V	1	1	0.1	0.1	1	mA
I_{IH}	High-level input current	4	$V_{CC} = $ MAX, $V_{IH} = 2.4$ V / $V_{IH} = 2.7$ V	40	50	10	20	50	µA
I_{IL}	Low-level input current	5	$V_{CC} = $ MAX, $V_{IL} = 0.3$ V / 0.4 V / 0.5 V ('LS30 / Others)	−1.6	−2	−0.18	−0.4 / −0.36	−2	mA
I_{OS}	Short-circuit output current♦	6	$V_{CC} = $ MAX — 54 Family	−20 ··· −55	−40 ··· −100	−3 ··· −15	−6 ··· −40	−40 ··· −100	mA
			74 Family	−18 ··· −55	−40 ··· −100	−3 ··· −15	−5 ··· −42	−40 ··· −100	mA
I_{CC}	Supply current	7	$V_{CC} = $ MAX					See table on next page	mA

[†] For conditions shown as MIN or MAX, use the appropriate value specified under recommended operating conditions.

[‡] All typical values are at $V_{CC} = 5$ V, $T_A = 25°$C.

[§] $I_I = −12$ mA for SN54/SN74, −8 mA for SN54H'/SN74H', and −18 mA for SN54LS'/SN74LS' and SN54S'/SN74S'; duration of short-circuit should not exceed 1 second.

♦ Not more than one output should be shorted at a time, and for SN54H'/SN74H' and SN54S'/SN74S' parts date-coded 7332 or higher.

* The input clamp voltage specification is effective for Series 54/74 and 54H/74H parts date-coded 7332 or higher.

*Courtesy of Texas Instruments, Incorporated

supply current¶

TYPE	I_{CCH} (mA) Total with outputs high TYP	MAX	I_{CCL} (mA) Total with outputs low TYP	MAX	I_{CC} (mA) Average per gate (50% duty cycle) TYP
'00	4	8	12	22	2
'04	6	12	18	33	2
'10	3	6	9	16.5	2
'20	2	4	6	11	2
'30	1	2	3	6	2
'H00	10	16.8	26	40	4.5
'H04	16	26	40	58	4.5
'H10	7.5	12.6	19.5	30	4.5
'H20	5	8.4	13	20	4.5
'H30	2.5	4.2	6.5	10	4.5
'L00	0.44	0.8	1.16	2.04	0.20
'L04	0.66	1.2	1.74	3.06	0.20
'L10	0.33	0.6	0.87	1.53	0.20
'L20	0.22	0.4	0.58	1.02	0.20
SN54L30	0.11	0.33	0.29	0.51	0.20
SN74L30	0.11	0.2	0.29	0.51	0.20
'LS00	0.8	1.6	2.4	4.4	0.4
'LS04	1.2	2.4	3.6	6.6	0.4
'LS10	0.6	1.2	1.8	3.3	0.4
'LS20	0.4	0.8	1.2	2.2	0.4
'LS30	0.35	0.5	0.6	1.1	0.48
'S00	10	16	20	36	3.75
'S04	15	24	30	54	3.75
'S10	7.5	12	15	27	3.75
'S20	5	8	10	18	3.75
'S30	3	5	5.5	10	4.25
'S133	3	5	5.5	10	4.25

¶ Maximum values of I_{CC} are over the recommended operating ranges of V_{CC} and T_A; typical values are at V_{CC} = 5 V, T_A = 25°C.

switching characteristics at V_{CC} = 5 V, T_A = 25°C

TYPE	TEST CONDITIONS#	t_{PLH} (ns) Propagation delay time, low-to-high-level output MIN	TYP	MAX	t_{PHL} (ns) Propagation delay time, high-to-low-level output MIN	TYP	MAX
'00, '10	C_L = 15 pF, R_L = 400 Ω		11	22		7	15
'04, '20			12	22		8	15
'30			13	22		8	15
'H00	C_L = 25 pF, R_L = 280 Ω		5.9	10		6.2	10
'H04			6	10		6.5	10
'H10			5.9	10		6.3	10
'H20			6	10		7	10
'H30			6.8	10		8.9	12
'L00, 'L04	C_L = 50 pF, R_L = 4 kΩ		35	60		31	60
'L10, L20			35	60		35	60
'L30			9	20		70	100
'LS00, LS04	C_L = 15 pF, R_L = 2 kΩ		9	20		10	20
'LS10, LS20			9	20			20
'LS30			3	4.5		25	35
'S00, 'S04	C_L = 15 pF, R_L = 280 Ω	2	3	4.5	2	3	5
'S10, 'S20	C_L = 50 pF, R_L = 280 Ω	2	4.5		2	5	
'S30, 'S133	C_L = 50 pF, R_L = 280 Ω		5.5	6		6.5	7

Load circuits and voltage waveforms are shown on pages 148 and 149.

typical values are at V_{CC} = 5 V, T_A = 25°C.

schematics (each gate)

CIRCUIT	R1	R2	R3	R4
'00, '04, '10, '20, '30	4 k	1.6 k	130	1 k
'L00, 'L04, 'L10, 'L20, 'L30	40 k	20 k	500	12 k

'00, '04, '10, '20, '30 / 'L00, 'L04, 'L10, 'L20, 'L30 CIRCUITS

Input clamp diodes not on SN54L'/SN74L' circuits.

'H00, 'H04, 'H10, 'H20, 'H30 CIRCUITS

'LS00, 'LS04, 'LS10, 'LS20, 'LS30 CIRCUITS

'S00, 'S04, 'S10, 'S20, 'S30, 'S133 CIRCUITS

Resistor values shown are nominal and in ohms.

444

APPENDIX 1-18*

3-INPUT GATES   MECL MC300 series

MC306 · MC307

Expandable 3-input gates that provide the positive logic "NOR" function and its complement simultaneously. MC307 omits output pull-down resistors, permitting reduction of power dissipation.

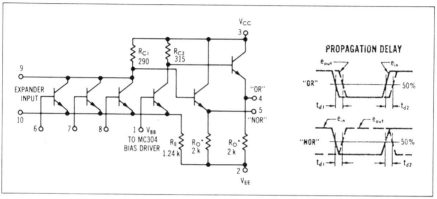

PROPAGATION DELAY

ELECTRICAL CHARACTERISTICS

Test Conditions Vdc ±1%

@ Test Temperature	V_H	$V_{I\,max}$	V_L	V_{EE}	V_{BB}
−55°C	—	−0.945	−1.450	−5.20	−1.25
+25°C	0.690	−0.795	−1.350	−5.20	−1.15
+125°C	—	0.655	−1.300	−5.20	−1.00

Characteristic	V_H Pin No	$V_{I\,max}$ Pin No	V_L Pin No	V_{EE} Pin No	V_{BB} Pin No	dV_{in} Pin No	I_L Pin No	Ground Pin No	Symbol Pin No in ( )	−55°C Min	−55°C Max	+25°C Min	+25°C Max	+125°C Min	+125°C Max	Unit
Power Supply MC306	—	—	—	2,6,7,8	1	—	—	3	I_E(2)	—	8.85	—	8.85	—	8.15	mAdc
Drain Current MC307	—	—	—	2,6,7,8	1	—	—	3	I_E(2)	—	3.6	—	3.6	—	3.3	mAdc
Input Current	6	—	—	2,7,8	1	—	—	3	I_{in}(6)	—	—	—	100	—	—	μAdc
	7	—	—	2,6,8	1	—	—	3	I_{in}(7)	—	—	—	↓	—	—	↓
	8	—	—	2,6,7	1	—	—	3	I_{in}(8)	—	—	—	↓	—	—	↓
"NOR" Logical "1" Output Voltage	—	—	6	2,7,8	1	—	—	3	V_1(5)	−0.825	−0.945	−0.690	−0.795	−0.525	−0.655	Vdc
	—	—	7	2,6,8	1	—	—	3	V_1(5)	↓	↓	↓	↓	↓	↓	↓
	—	—	8	2,6,7	1	—	—	3	V_1(5)	↓	↓	↓	↓	↓	↓	↓
"NOR" Logical "0" Output Voltage	6	—	—	2,7,8	1	—	—	3	V_0(5)	−1.560	−1.850	−1.465	−1.750	−1.340	−1.675	Vdc
	7	—	—	2,6,8	1	—	—	3	V_0(5)	↓	↓	↓	↓	↓	↓	↓
	8	—	—	2,6,7	1	—	—	3	V_0(5)	↓	↓	↓	↓	↓	↓	↓
"OR" Logical "1" Output Voltage	6	—	—	2,7,8	1	—	—	3	V_1(4)	0.825	−0.945	−0.690	−0.795	−0.525	−0.655	Vdc
	7	—	—	2,6,8	1	—	—	3	V_1(4)	↓	↓	↓	↓	↓	↓	↓
	8	—	—	2,6,7	1	—	—	3	V_1(4)	↓	↓	↓	↓	↓	↓	↓
"OR" Logical "0" Output Voltage	—	—	6	2,7,8	1	—	—	3	V_0(4)	−1.560	−1.850	−1.465	−1.750	−1.340	−1.675	Vdc
	—	—	7	2,6,8	1	—	—	3	V_0(4)	↓	↓	↓	↓	↓	↓	↓
	—	—	8	2,6,7	1	—	—	3	V_0(4)	↓	↓	↓	↓	↓	↓	↓
"NOR" Output Voltage Change (No load to full load)	—	—	6	2,7,8	1	—	5③	3	$\triangle V_1$(5)	—	−0.055	—	−0.055	—	−0.060	Volts
"OR" Output Voltage Change (No load to full load)	6	—	—	2,7,8	1	—	4③	3	$\triangle V_1$(4)	—	−0.055	—	−0.055	—	−0.060	Volts
"NOR" Saturation Breakpoint Voltage	—	—	—	2,7,8	1	6①	—	3	V_1(5)	—	−0.40	—	−0.55	—	−0.68	Vdc
	—	—	—	2,6,8	1	7①	—	3	V_1(5)	—	↓	—	↓	—	↓	↓
	—	—	—	2,6,7	1	8①	—	3	V_1(5)	—	↓	—	↓	—	↓	↓

Switching Times	Pulse In	Pulse Out	V_L	V_{EE}	V_{BB}	dV_{in}	I_L	Ground	Symbol	−55°C Typ	−55°C Max	+25°C Typ	+25°C Max	+125°C Typ	+125°C Max	Unit
Propagation Delay Time	6	4	—	2,7,8	1	—	—	3	t_{d1}(4)	7.0	11.0	7.0	11.5	9.5	14.5	ns
	6	5	—	2,7,8	1	—	—	3	t_{d1}(5)	5.5	11.0	5.5	10.5	7.0	12.5	
	6	4	—	2,7,8	1	—	—	3	t_{d2}(4)	5.5	10.0	5.5	11.0	7.0	12.5	
	6	5	—	2,7,8	1	—	—	3	t_{d2}(5)	7.0	10.5	7.0	11.0	9.5	14.5	
Rise Time	6	4	—	2,7,8	1	—	—	3	t_r(4)	6.0	8.5	6.0	10.0	8.0	13.0	
	6	5	—	2,7,8	1	—	—	3	t_r(5)	7.5	11.5	7.5	12.5	9.5	15.0	
Fall Time	6	4	—	2,7,8	1	—	—	3	t_f(4)	6.5	10.5	6.5	12.0	9.0	15.0	
	6	5	—	2,7,8	1	—	—	3	t_f(5)	6.5	12.0	6.5	12.5	9.0	15.0	↓

*Courtesy of Motorola, Inc.

µA710
HIGH-SPEED DIFFERENTIAL COMPARATOR
FAIRCHILD LINEAR INTEGRATED CIRCUITS

ABSOLUTE MAXIMUM RATINGS

Positive Supply Voltage	+14.0 V
Negative Supply Voltage	−7.0 V
Peak Output Current	10 mA
Differential Input Voltage	±5.0 V
Input Voltage	±7.0 V
Internal Power Dissipation	
TO-99 [Note 1]	300 mW
Flat Package [Note 2]	200 mW
Operating Temperature Range	−55°C to +125°C
Storage Temperature Range	−65°C to +150°C
Lead Temperature (Soldering, 60 sec.)	300°C

ELECTRICAL CHARACTERISTICS ($T_A = +25°C$, $V^+ = 12.0V$, $V^- = -6.0V$ unless otherwise specified)

PARAMETER (see definitions)	CONDITIONS (Note 4)	MIN.	TYP.	MAX.	UNITS
Input Offset Voltage	$R_s \leq 200\Omega$		0.6	2.0	mV
Input Offset Current			0.75	3.0	µA
Input Bias Current			13	20	µA
Voltage Gain		1250	1700		
Output Resistance			200		Ω
Output Sink Current	$\Delta V_{in} \geq 5$ mV, $V_{out} = 0$	2.0	2.5		mA
Response Time [Note 3]			40		ns

The following specifications apply for $-55°C \leq T_A \leq +125°C$:

PARAMETER	CONDITIONS	MIN.	TYP.	MAX.	UNITS
Input Offset Voltage	$R_s \leq 200\Omega$			3.0	mV
Average Temperature Coefficient of Input Offset Voltage	$R_s = 50\Omega$, $T_A = 25°C$ to $T_A = +125°C$		3.5	10	µV/°C
	$R_s = 50\Omega$, $T_A = 25°C$ to $T_A = -55°C$		2.7	10	µV/°C
Input Offset Current	$T_A = +125°C$		0.25	3.0	µA
	$T_A = -55°C$		1.8	7.0	µA
Average Temperature Coefficient of Input Offset Current	$T_A = 25°C$ to $T_A = +125°C$		5.0	25	nA/°C
	$T_A = 25°C$ to $T_A = -55°C$		15	75	nA/°C
Input Bias Current	$T_A = -55°C$		27	45	µA
Input Voltage Range	$V^- = -7.0V$	±5.0			V
Common Mode Rejection Ratio	$R_s \leq 200\Omega$	80	100		dB
Differential Input Voltage Range		±5.0			V
Voltage Gain		1000			
Positive Output Level	$\Delta V_{in} \geq 5$ mV, $0 \leq I_{out} \leq 5.0$ mA	2.5	3.2	4.0	V
Negative Output Level	$\Delta V_{in} \geq 5$ mV	−1.0	−0.5	0	V
Output Sink Current	$T_A = +125°C$, $\Delta V_{in} \geq 5$ mV, $V_{out} = 0$	0.5	1.7		mA
	$T_A = -55°C$, $\Delta V_{in} \geq 5$ mV, $V_{out} = 0$	1.0	2.3		mA
Positive Supply Current	$V_{out} \leq 0$		5.2	9.0	mA
Negative Supply Current			4.6	7.0	mA
Power Consumption			90	150	mW

***Courtesy of Fairchild Semiconductors**

absolute maximum ratings

Total Supply Voltage (V_{84})	36V
Output to Negative Supply Voltage (V_{74})	40V
Ground to Negative Supply Voltage (V_{14})	30V
Differential Input Voltage	±30V
Input Voltage (Note 1)	±15V
Power Dissipation (Note 2)	500 mW
Output Short Circuit Duration	10 sec
Operating Temperature Range	$0°C$ to $70°C$
Storage Temperature Range	$-65°C$ to $150°C$
Lead Temperature (soldering, 10 sec)	$300°C$

electrical characteristics (Note 3)

PARAMETER	CONDITIONS	MIN	TYP	MAX	UNITS
Input Offset Voltage (Note 4)	$T_A = 25°C$, $R_S \leq 50K$		2.0	7.5	mV
Input Offset Current (Note 4)	$T_A = 25°C$		6.0	50	nA
Input Bias Current	$T_A = 25°C$		100	250	nA
Voltage Gain	$T_A = 25°C$		200		V/mV
Response Time (Note 5)	$T_A = 25°C$		200		ns
Saturation Voltage	$V_{IN} \leq -10$ mV, $I_{OUT} = 50$ mA $T_A = 25°C$		0.75	1.5	V
Strobe On Current	$T_A = 25°C$		3.0		mA
Output Leakage Current	$V_{IN} \geq 10$ mV, $V_{OUT} = 35V$ $T_A = 25°C$		0.2	50	nA
Input Offset Voltage (Note 4)	$R_S \leq 50K$			10	mV
Input Offset Current (Note 4)				70	nA
Input Bias Current				300	nA
Input Voltage Range			±14		V
Saturation Voltage	$V^+ \geq 4.5V$, $V^- = 0$ $V_{IN} \leq -10$ mV, $I_{SINK} \leq 8$ mA		0.23	0.4	V
Positive Supply Current	$T_A = 25$ C		5.1	7.5	mA
Negative Supply Current	$T_A = 25$ C		4.1	5.0	mA

Note 1: This rating applies for ±15V supplies. The positive input voltage limit is 30V above the negative supply. The negative input voltage limit is equal to the negative supply voltage or 30V below the positive supply, whichever is less.

Note 2: The maximum junction temperature of the LM311 is 85°C. For operating at elevated temperatures, devices in the TO-5 package must be derated based on a thermal resistance of 150°C/W, junction to ambient, or 45°C/W, junction to case. For the flat package, the derating is based on a thermal resistance of 185°C/W when mounted on a 1/16-inch-thick epoxy glass board with ten, 0.03-inch-wide, 2-ounce copper conductors. The thermal resistance of the dual-in-line package is 100°C/W, junction to ambient.

Note 3: These specifications apply for $V_S = \pm15V$ and $0°C < T_A < 70°C$, unless otherwise specified. The offset voltage, offset current and bias current specifications apply for any supply voltage from a single 5V supply up to ±15V supplies.

Note 4: The offset voltages and offset currents given are the maximum values required to drive the output within a volt of either supply with 1 mA load. Thus, these parameters define an error band and take into account the worst case effects of voltage gain and input impedance.

Note 5: The response time specified (see definitions) is for a 100 mV input step with 5 mV overdrive.

*Courtesy of National Semiconductor Corp.

HEWLETT **hp** *PACKARD*

COMPONENTS

0.3" SOLID STATE SEVEN SEGMENT INDICATOR | 5082-7740

TENTATIVE DATA AUGUST 1973

Features

- COMMON CATHODE
- RIGHT HAND DP
- EXCELLENT CHARACTER APPEARANCE
  - Continuous Uniform Segments
  - Wide Viewing Angle
  - High Contrast
- IC COMPATIBLE
  - 1.7V per Segment
- STANDARD 0.3" DIP LEAD CONFIGURATION
  - PC Board or Standard Socket Mountable
- CATEGORIZED FOR LUMINOUS INTENSITY
  - Assures Uniformity of Light Output from Unit to Unit within a Single Category

Description

The HP 5082-7740 is a common cathode LED numeric display with a right hand decimal point. The large 0.3" high character size generates a bright, continuously uniform 7 segment display. Designed for viewing distances of up to 10 feet, this single digit display has been human engineered to provide a high contrast ratio and wide viewing angle.

The 5082-7740 utilizes a standard 0.3" dual-in-line package configuration that allows for quick mounting on PC boards or in standard IC sockets. Requiring a forward voltage of only 1.7V, the display is inherently IC compatible allowing for easy integration into electronic calculators, credit card verifiers, TVs, radios, and digital clocks.

Package Dimensions

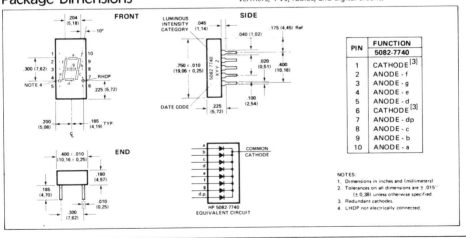

PIN	FUNCTION 5082-7740
1	CATHODE [3]
2	ANODE - f
3	ANODE - g
4	ANODE - e
5	ANODE - d
6	CATHODE [3]
7	ANODE - dp
8	ANODE - c
9	ANODE - b
10	ANODE - a

NOTES:
1. Dimensions in inches and (millimeters)
2. Tolerances on all dimensions are ±.015"
   (± 0,38) unless otherwise specified.
3. Redundant cathodes.
4. LHDP not electrically connected.

*Courtesy of Hewlett Packard, Inc.

Absolute Maximum Ratings

Power Dissipation T_A = 25°C . . . . 400 mW
Operating Temperature Range . . . . −20°C to +85°C
Storage Temperature Range . . . . −20°C to +85°C
Average Forward Current/Segment or Decimal Pt. T_A = 25°C [1] . . . . 25 mA
Peak Forward Current/Segment or Decimal Pt. T_A = 25°C (Pulse Duration ⩽ 500 μs) . . . . 150 mA
Reverse Voltage/Segment or Decimal Pt. 6V
Max. Solder Temperature 1/16″ Below Seating Plane (t ⩽ 5 sec.) [2] . . . . 230°C

NOTES: 1. Derate from 25°C at .25 mA/°C per segment or D.P.   2. Clean only in Freon TF, Isopropanol, or water.

Electrical/Optical Characteristics at T_A=25°C

Description	Symbol	Test Condition	Min.	Typ.	Max.	Units
Luminous Intensity/Segment [1]	$I_{\nu \, AVE}$	I_{PEAK} = 100 mA 10% Duty Cycle	50	150		μcd
		I_F = 20 mA DC		250		
Peak Wavelength	λ_{PEAK}			655		nm
Forward Voltage/Segment or D.P.	V_F	I_F = 100 mA		1.6	2.3	V
Reverse Current/Segment or D.P.	I_R	V_R = 6V			100	μA
Rise and Fall Time [2]	t_r, t_f			10		ns
Temperature Coefficient of Forward Voltage	$\Delta V_F /°C$			−2.0		mV/°C
Temperature Coefficient of Luminous Intensity	$\Delta I_\nu /°C$			−1.0		%/°C

NOTES: 1. The digits are categorized for luminous intensity such that the variation from digit to digit within a category is not discernible to the eye. Intensity categories are designated by a letter located on the right hand side of the package.
2. Time for a 10%-90% change of light intensity for step change in current.

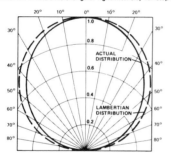

Figure 1. Normalized Angular Distribution of Luminous Intensity.

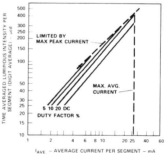

Figure 2. Typical Time Averaged Luminous Intensity per Segment versus Average Current.

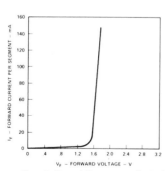

Figure 3. Forward Current versus Forward Voltage.

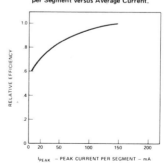

Figure 4. Relative Efficiency (Luminous Intensity per Unit Current) versus Peak Current per Segment.

For more information, call your local HP Sales Office or East (201) 265-5000 · Midwest (312) 677-0400 · South (404) 436-6181 · West (213) 877-1282. Or, write: Hewlett-Packard, 1501 Page Mill Road, Palo Alto, California 94304. In Europe, Post Office Box 85, CH-1217 Meyrin 2, Geneva, Switzerland. In Japan, YHP, 1-59-1, Yoyogi, Shibuya-Ku, Tokyo, 151.
Printed in U.S.A.

SERIES 1603-02
REFLECTIVE

LIQUID CRYSTAL DISPLAYS

- • **Optimum Readability**
- • **Single Plane Viewing**
- • **MOS Compatibility**
- • **No Back Lighting Required**

- • **Low Profile**
- • **Microwatt Power Consumption**
- • **Low Cost**
- • **Ideal for High Ambient Light Environment**

The IEE Series 1603-02 is a 3½ decade Liquid Crystal Display with four floating decimals and an overflow plus or minus one (±). A colon is incorporated in the display for additional application-advantages such as clocks, etc.

The IEE Series 1603-02 reflective Liquid Crystal Display consists of a layer of micron-thin liquid crystal material confined between two sheets of glass, one sheet having a clear conductive electrode and the other a reflective coating etched in a segmented pattern to create a digital display. The organic material requires extremely low power in order to be an effective display. Upon generation of an electric field, the liquid layer becomes turbulent and scatters light. This scattering effect (caused by a continuous change in the index of refraction) appears as an optically dense area; by selectively energizing appropriate segments, a digital format is obtained. When the applied field is removed, the liquid crystal material returns to its original quiescent, transparent condition.

Reflective displays eliminate the need for backlighting, which makes them excellent displays for use in portable equipment or in equipment where low power consumption is a definite consideration.

For optimum life, liquid crystals are operated on A.C. (40-100 Hz) which, coupled with a 15-30 volt range, make them directly compatible with MOS logic. Numerous companies are engaged in the manufacture of standard MOS circuits for use with Liquid Crystal Displays and this list may be obtained from Industrial Electronic Engineers, Inc. (IEE) upon request.

ELECTRICAL SPECIFICATIONS

Operating Voltage: A.C. 40-100 Hz. Typical 24 Volt (Peak to Peak). Maximum 40 Volt. Minimum 10 Volt.

Power Consumption: 20 microwatts per segment (maximum).

Rise Time: 50 milliseconds.

Decay Time: 150 milliseconds.

Contrast Ratio: 15:1 minimum.

Life: 10,000 hrs. minimum at 24 V.A.C.

Operating Temperature: 5°C to 55°C.

Storage Temperature: −10°C to 70°C.

Relative Humidity: 0 to 100%.

MECHANICAL SPECIFICATIONS

Package Size: See diagram.
Character Size: .433″ (11 MM).
Character Width: .260″ (7.6 MM).
Segment Width: .055″ (1.4 MM).
Decimal Point Width: .06″ (1.5 MM).
Decimal Point Height: .08″ (2 MM).

Contacts:

The conductive electrodes on the Series 1603-02 Liquid Crystal Displays are terminated in an edge board configuration having .050″ (1.3 MM) spacing, which allows the use of an edge connector or a spring contact right angle connector to be used in conjunction with printed circuit board patterns.

ORDERING INFORMATION

	Part No.
Liquid Crystal	1603-02
Right Angle Connector	22076-01
Mounting Kit (PC Board with right angle connector attached)	22077-01

AVAILABILITY:

Series 1603-02 displays and optional hardware (connector, PC boards) are available for customer evaluation from shelf stock. For large quantity requirements and/or special designs, consult IEE for information.

*Courtesy of Industrial Electronic Engineers, Inc.

TTL
MSI

TYPES SN5490A, SN5492A, SN5493A, SN54L90, SN54L93, SN7490A, SN7492A, SN7493A, SN74L90, SN74L93
DECADE, DIVIDE-BY-TWELVE, AND BINARY COUNTERS

BULLETIN NO. DL-S 7211807, DECEMBER 1972

'90A, 'L90 . . . DECADE COUNTERS

'92A . . . DIVIDE-BY-TWELVE
 COUNTER

'93A, 'L93 . . . 4-BIT BINARY
 COUNTERS

description

Each of these monolithic counters contains four master-slave flip-flops and additional gating to provide a divide-by-two counter and a three-stage binary counter for which the count cycle length is divide-by-five for the '90A and 'L90, divide-by-six for the '92A, and divide-by-eight for the '93A and 'L93.

All of these counters have a gated zero reset and the '90A and 'L90 also have gated set-to-nine inputs for use in BCD nine's complement applications.

To use their maximum count length (decade, divide-by-twelve, or four-bit binary) of these counters, the B input is connected to the Q_A output. The input count pulses are applied to input A and the outputs are as described in the appropriate function table. A symmetrical divide-by-ten count can be obtained from the '90A or 'L90 counters by connecting the Q_D output to the A input and applying the input count to the B input which gives a divide-by-ten square wave at output Q_A.

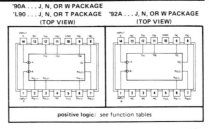

'90A . . . J, N, OR W PACKAGE
'L90 . . . J, N, OR T PACKAGE (TOP VIEW) '92A . . . J, N, OR W PACKAGE (TOP VIEW)

positive logic: see function tables

'93A . . . J, N, OR W PACKAGE (TOP VIEW) 'L93 . . . J, N, OR T PACKAGE (TOP VIEW)

positive logic: see function tables

NC—No internal connection

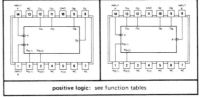

TYPES	TYPICAL POWER DISSIPATION
'90A	145 mW
'L90	20 mW
'92A, '93A	130 mW
'L93	16 mW

functional block diagrams

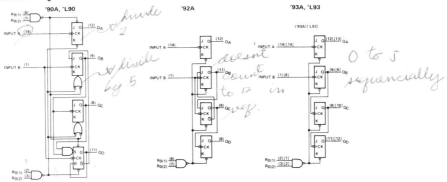

'90A, 'L90   '92A   '93A, 'L93

⟜ . . . dynamic input activated by transition from a high level to a low level.

The J and K inputs shown without connection are for reference only and are functionally at a high level.

TYPES SN5490A, SN5492A, SN5493A, SN54L90, SN54L93, SN7490A, SN7492A, SN7493A, SN74L90, SN74L93
DECADE, DIVIDE-BY-TWELVE, AND BINARY COUNTERS

'90A, 'L90 BCD COUNT SEQUENCE (See Note A)

COUNT	OUTPUT			
	Q_D	Q_C	Q_B	Q_A
0	L	L	L	L
1	L	L	L	H
2	L	L	H	L
3	L	L	H	H
4	L	H	L	L
5	L	H	L	H
6	L	H	H	L
7	L	H	H	H
8	H	L	L	L
9	H	L	L	H

'90A, 'L90 BI-QUINARY (5-2) (See Note B)

COUNT	OUTPUT			
	Q_A	Q_D	Q_C	Q_B
0	L	L	L	L
1	L	L	L	H
2	L	L	H	L
3	L	L	H	H
4	L	H	L	L
5	H	L	L	L
6	H	L	L	H
7	H	L	H	L
8	H	L	H	H
9	H	H	L	L

'90A, 'L90 RESET/COUNT FUNCTION TABLE

RESET INPUTS				OUTPUT			
$R_{0(1)}$	$R_{0(2)}$	$R_{9(1)}$	$R_{9(2)}$	Q_D	Q_C	Q_B	Q_A
H	H	L	X	L	L	L	L
H	H	X	L	L	L	L	L
X	X	H	H	H	L	L	H
X	L	X	L	COUNT			
L	X	L	X	COUNT			
L	X	X	L	COUNT			
X	L	L	X	COUNT			

'92A COUNT SEQUENCE (See Note C)

COUNT	OUTPUT			
	Q_D	Q_C	Q_B	Q_A
0	L	L	L	L
1	L	L	L	H
2	L	L	H	L
3	L	L	H	H
4	L	H	L	L
5	L	H	L	H
6	H	L	L	L
7	H	L	L	H
8	H	L	H	L
9	H	L	H	H
10	H	H	L	L
11	H	H	L	H

'93A, 'L93 COUNT SEQUENCE (See Note C)

COUNT	OUTPUT			
	Q_D	Q_C	Q_B	Q_A
0	L	L	L	L
1	L	L	L	H
2	L	L	H	L
3	L	L	H	H
4	L	H	L	L
5	L	H	L	H
6	L	H	H	L
7	L	H	H	H
8	H	L	L	L
9	H	L	L	H
10	H	L	H	L
11	H	L	H	H
12	H	H	L	L
13	H	H	L	H
14	H	H	H	L
15	H	H	H	H

'92A, '93A, 'L93 RESET/COUNT FUNCTION TABLE

RESET INPUTS		OUTPUT			
$R_{0(1)}$	$R_{0(2)}$	Q_D	Q_C	Q_B	Q_A
H	H	L	L	L	L
L	X	COUNT			
X	L	COUNT			

NOTES: A. Output Q_A is connected to input B for BCD count.
B. Output Q_D is connected to input A for bi-quinary count.
C. Output Q_A is connected to input B.
D. H = high level, L = low level, X = irrelevant

schematics of inputs and outputs

'90A, '92A, '93A EQUIVALENT OF EACH INPUT

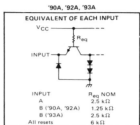

INPUT	R_{eq} NOM
A	2.5 kΩ
B ('90A, '92A)	1.25 kΩ
B ('93A)	2.5 kΩ
All resets	6 kΩ

'L90, 'L93 EQUIVALENT OF EACH INPUT EXCEPT A AND B OF 'L93

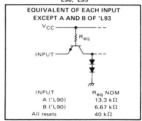

INPUT	R_{eq} NOM
A ('L90)	13.3 kΩ
B ('L90)	6.67 kΩ
All resets	40 kΩ

'L93 EQUIVALENT OF A OR B INPUT

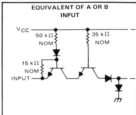

'90A, '92A, '93A, 'L90, 'L93 TYPICAL OF ALL OUTPUTS

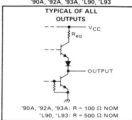

'90A, '92A, '93A: R = 100 Ω NOM
'L90, 'L93: R = 500 Ω NOM

Resistor and Capacitor Values

APPENDIX 2-1
TYPICAL STANDARD RESISTOR VALUES

Ω	Ω	Ω	$k\Omega$	$k\Omega$	$k\Omega$	$M\Omega$	$M\Omega$
—	10	100	1	10	100	1	10
—	12	120	1.2	12	120	1.2	—
—	15	150	1.5	15	150	1.5	15
—	18	180	1.8	18	180	1.8	—
—	22	220	2.2	22	220	2.2	22
2.7	27	270	2.7	27	270	2.7	—
3.3	33	330	3.3	33	330	3.3	—
3.9	39	390	3.9	39	390	3.9	—
4.7	47	470	4.7	47	470	4.7	—
5.6	56	560	5.6	56	560	5.6	—
6.8	68	680	6.8	68	680	6.8	—
—	82	820	8.2	82	820	—	—

APPENDIX 2-2
TYPICAL STANDARD CAPACITOR VALUES

pF	pF	pF	pF	μF	μF	μF	μF	μF	μF	μF
5	50	500	5000		0.05	0.5	5	50	500	5000
—	51	510	5100		—	—	—	—	—	—
—	56	560	5600		0.056	0.56	5.6	56	—	5600
—	—	—	6000		0.06	—	6	—	—	6000
—	62	620	6200		—	—	—	—	—	—
—	68	680	6800		0.068	0.68	6.8	—	—	—
—	75	750	7500		—	—	—	75	—	—
—	—	—	8000		—	—	8	80	—	—
—	82	820	8200		0.082	0.82	8.2	82	—	—
—	91	910	9100		—	—	—	—	—	—
10	100	1000		0.01	0.1	1	10	100	1000	10000
—	110	1100		—	—	—	—	—	—	
12	120	1200		0.012	0.12	1.2	—	—	—	
—	130	1300		—	—	—	—	—	—	
15	150	1500		0.015	0.15	1.5	15	150	1500	
—	160	1600		—	—	—	—	—		
18	180	1800		0.018	0.18	1.8	18	180	—	
20	200	2000		0.02	0.2	2	20	200	2000	
22	220	2200		—	0.22	2.2	22	—	—	
24	240	2400		—	—	—	—	240	—	
—	250	2500		—	0.25	—	25	250	2500	
27	270	2700		0.027	0.27	2.7	27	270	—	
30	300	3000		0.03	0.3	3	30	300	3000	
33	330	3300		0.033	0.33	3.3	33	330	3300	
36	360	3600		—	—	—	—	—	—	
39	390	3900		0.039	0.39	3.9	39	—	—	
—	—	4000		0.04	—	4	—	400	—	
43	430	4300		—	—	—	—	—	—	
47	470	4700		0.047	0.47	4.7	47	—	—	

Answers to Problems

1-4	0.3 V, 2 ms, 40%, 0.67
1-5(b)	40%
1-6	5.25 V, 9.52%, 5 μs, 10 μs, 54.3%
1-7	0.33 V, 2.59 V
1-8	100, 6.6
1-9	1 μs, 2.4 MHz
1-11	1 MHz, 15.9 Hz
1-12	0.7 μs, 0.63%
1-13	433 Hz, 70 kHz
2-3	0.754 V
2-4	0.493 mA
2-5	263.9 ms
2-6	14.52 μs, 6.6 μs, 33 μs
2-7	6.54 V
2-8	9.71 V, 5.29 V
2-9	3.5 mA, 1.04 mA
2-10	5.64 μs, 0.1 V
2-11	7.8 V, 14.52 μs
2-12	3.12 V, 2.18 μA
2-16	0.185 V, 0.405 V
2-20	0.18 V
2-21	1.8 V, 0.9 V
3-7	20 kΩ
3-8	50 kΩ

3-12	560 Ω, 1N746
3-15	8 kΩ, 1.79 mA
3-17	6.8 kΩ
3-19	1N751, 1 kΩ
3-23	0.1 μF, 100 kΩ
3-26	1 μF, 47 kΩ
3-27	1N757, 1 μF, 120 kΩ
4-4	30.3, 182 μA
4-5	16.2 μA, 114 μA
4-6	0.228 mW, 1.25 μW, 2.28 mW, 1.25 μW
4-8	14.999 V, 13.5 V, 15 V, 2.026 μs
4-10	80.5 pF, 80.5 kHz
4-12	20 V, 106 mV
5-3	1 kΩ, 21.5 kΩ
5-4	60.4 kHz
5-5	22 kΩ, 560 kΩ, 17.28 V
5-8	12 kΩ, 680 kΩ, 7500 pF
5-9	12 kΩ, 10 kΩ, 0.18 μF
5-10	820 Ω, 10 kΩ, 0.47 μF
5-12	270 kΩ, 330 kΩ, 0.015 μF, 56 μs
5-13	150 kΩ, 180 kΩ, 1μF
5-15	3.9 kΩ, 510 pF, 100 Ω
5-16	− 19.89 V, 6.63 V
6-3	$R_E = 4.7$ kΩ, $R_L = 8.2$ kΩ, $R_1 = 47$ kΩ, $R_2 = 47$ kΩ
6-4	$R_{L1} = 3.3$ kΩ, $R_1 = 56$ kΩ
6-6	20 pF
6-8	[3.9 kΩ, 180 kΩ], [39 kΩ, 220 kΩ]
6-10	2.2 kΩ, 500 Ω, 1.8 kΩ
6-12	[39 kΩ, 100 kΩ], [39 kΩ, 150 kΩ]
6-13	56 kΩ, 270 kΩ, 180 kΩ
6-16	2.2 kΩ, 1.5 kΩ
7-2	18 kΩ, 1 μF, 120 kΩ, 0.15 μF
7-4	3.9 kΩ, 3.9 kΩ, 2.7 kΩ
7-5	5 kΩ, 1.8 kΩ
7-7	1.6 mA, 0.7 mA
7-9	47 kΩ, 0.008 μF, 16.45 V
7-11	9 V, 1.145 kHz
7-12	12 kΩ, 27 kΩ, 390 pF
7-14	1.2 μF, 5.6 kΩ, 30 μF, 15 kΩ, 0.68 μF
7-16	0.68 μF, 100 kΩ, 1.8 μF, 56 kΩ, 0.18 μF

7-18   [56 kΩ, 150 kΩ], [0.05 μF, 22 kΩ, 4 μF], [100 kΩ]
7-20   15 kΩ, 150 kΩ, 0.1 μF, 106 Hz
7-23   [47 kΩ, 220 kΩ], [22 kΩ, 0.1 μF]
7-24   [20 kΩ, 39 kΩ], [10 kΩ, 18 kΩ]
7-25   [390 Ω, 1 kΩ, 3.3 kΩ], [39 kΩ, 39 kΩ], [39 kΩ, 0.02 μF]

8-2   3.9 kΩ, 3.9 kΩ, 39 kΩ, 22 kΩ
8-3   0.0025 μF
8-4   309 pF
8-5   10 kΩ, 1800 pF
8-8   5.6 kΩ, 150 kΩ, 5.6 kΩ, 1500 pF
8-9   0.004 μF, 680 Ω
8-11   1.8 kΩ, 116.2 kΩ, 1207 pF
8-15   68 kΩ, 68 kΩ, 68 kΩ, 0.015 μF
8-18   22 kΩ, 0.02 μF
8-22   0.01 μF, 6.8 kΩ, 15 kΩ
8-23   151 μs, 76.2%
8-25   33 kΩ, 0.05 μF, 1.2 kΩ, 10 kΩ
8-26   0.05 μF, 820 Ω, 30 kΩ

9-4   5.6 kΩ, 56 kΩ, 33 kΩ
9-5   4.7 kΩ, 47 kΩ, 27 kΩ
9-8   56 pF
9-9   120 pF, 174 kHz
9-12   160 pF
9-13   120 pF, 174 kHz

10-2   8.2 kΩ, 0.9 V, 9 V
10-4   1.5 kΩ
10-10 7

11-4   20 kΩ, 15 kΩ, 6.7%, 0.03%
11-6   2 kΩ, 3.3 kΩ, 10%, 0.005%
11-8   0.035%, 0.00027%
11-10 0.047%, 0.0001%
11-15 10 kΩ, 47 kΩ, 8.2 kΩ, 0.11%
11-17 0.039 μF, 10.73 μs, 0.08%

12-12 560 Ω, 39 kΩ

13-4   15, 7
13-9   100 μA

14-14 6, 50

Index